Seashore Plankton

Planktonic epitokous stages of the polychaete *Myrianida prolifera*. Stock worms, which live on seashore hydroids, grow stolons, like miniature worms, on the rear of their body (see Figure 14.17). These detach from the main body as non-feeding sexual forms and swim up into the plankton, where they reproduce and disperse. At the top of the photo is a male stolon (called a polybostrichus), with distinct forked antennae. At the bottom is a female (the sacconereis) brooding her eggs; this is a month-long process, after which she dies. This mode of reproduction is called epitoky.

Seashore Plankton

A Practical Introduction to Biology and Identification

Julian Cremona

PELAGIC PUBLISHING

First published in 2026 by
Pelagic Publishing
20–22 Wenlock Road
London N1 7GU, UK

www.pelagicpublishing.com

Seashore Plankton: A Practical Introduction to Biology and Identification

https://doi.org/10.53061/KGFB3717

A CIP record for this book is available from the British Library

ISBN 978-1-78427-619-5 Pbk
ISBN 978-1-78427-624-9 ePub
ISBN 978-1-78427-625-6 PDF

EU Authorised Representative: Easy Access System Europe – Mustamäe tee 50, 10621 Tallinn, Estonia, gpsr.requests@easproject.com

Cover photograph: a hydromedusa in the genus *Amphinema*

Typeset in Myriad by BBR Design, UK

5 4 3 2 1

Printed and bound in India by Replika Press Pvt. Ltd.

FSC www.fsc.org MIX Paper | Supporting responsible forestry FSC™ C016779

To Brenda

The Mauve Stinger *Pelagia noctiluca* is a jellyfish usually found off-shore but as the seas warm it has become more common washed up on western shores in autumn. The numerous raised dots on the surface are sting cells, making it an unwelcome visitor. The orange structures are sensory organs, the rhopalia, which can detect light and gravity. The large, flat oral arms help trap zooplankton for food. This is a 1 cm diameter juvenile just developed from the ephyra larva.

Contents

FIGURE P.1 A west-coast UK rocky shore at low tide. The blue water is a medium full of hidden life, including both plankton arriving from the open sea as well as the larvae of seashore organisms.

Preface

What incalculable numbers of these microscopical animals!

Charles Darwin, *Voyage of a Naturalist*

I watched as the Blue Tits picked their way across the shingle to the water's edge before they slowly immersed themselves in the cold December sea. Most days the ladies swim out towards the rising sun. They eyed me, as they often do, as I paced back and forth along the nearby jetty, dragging a net through the seawater of Milford Haven. On occasion this group, and others, ask if I have caught anything, spying the slightly murky water in my jars, the results of my regular sampling. As the swimmers enter the sea I reflect on how much of the sort of material contained in my sample jars they must have imbibed. Looking at the sun glinting on the calm, flat water, it can be difficult to imagine the astonishingly biodiversity ecosystem that lies beneath the surface.

Nature-lovers and naturalists who walk through forest and heath can observe that the environment around them has structure and organisation. The harvesters of the light – trees, shrubs, herbaceous plants – clearly delineate the communities living there. Even if we do not understand the woodland ecosystem, for example, we are at least familiar with the basics; after all, we can see the trees and many of the animals within. Even on a rocky shore the seaweeds create a distinct habitat within which different animals live alongside each other and form vertical zones of communities running down the beach.

By contrast, the seawater abutting our coastline may not be recognised as an ecosystem in its own right. Structure appears to be lacking and the species, if you can see them, are in an apparent state of chaos. Twice daily the incoming tide covers the shore in a gauzy veil that trickles through the fingers, and it will require more effort on the part of the naturalist to observe the life within. If you make that small step, however, the rewards are incredible, as an entirely new, previously hidden world of life opens up to you.

This book is about the organisms that dwell in these waters. Many invertebrate species live along the shore and most have a larval stage that develops and disperses in this obscure medium, forming components of the plankton – that is, creatures that drift at the whim of the current. Whether the coast is formed of sediment or a rocky shore, for much of the year the animal life living there liberates vast numbers of young stages into the water column. Some are much more spectacular to look at than the adult of their species, with most, indeed, bearing little resemblance to their parents. Drifting through the water, these larvae can live for hours, days or many weeks, spreading out along the coast, feeding and growing, metamorphosing through different phases as they mature, before settling back on the shore. A study of this wealth of microscopic life allows us to complete the story of the seashore and of marine ecology. This is a hidden world few get the chance to see.

The purpose of this volume is to demonstrate how straightforward it is to view this amazing biodiversity. It guides the reader through observing, identifying and understanding planktonic life and, to this end, it is divided into three sections. The first deals with practical issues, beginning with a definition of plankton and a brief summary of the background history of plankton studies. This is followed by the practicalities of methodology: collection of samples, methods of observation and the often tricky identification of the material. You will be taken through types of sampling net and their use, and ways to magnify the collected samples, from uncomplicated digital USB microscopes plugged into a computer to inexpensive optical microscopes. Ways of recording organisms photographically, whether with a smartphone or mirrorless camera, are also briefly dealt with. Microscopy can seem a real barrier for many amateur naturalists – especially where cost is concerned – but with the right assistance it need not be. Hopefully, any worries readers may have regarding these issues will be dispelled in this section, along with the information about extra resources given at the end of the book. The second section, the bulk of the chapters, systematically reviews the planktonic life that occurs on the shore, along with their biology. The short final section concludes the book with a holistic view of plankton ecology and the influence of changing climate as well as other human effects.

This book is for the amateur, naturalists as well as ecologists, as well as professional coastal biologists seeking information on plankton. You might imagine that such professionals would know all about plankton already but, traditionally, the subject has been dealt with mostly under the umbrella of oceanography rather than coastal biology. Students studying for a typical biology degree may, if they are fortunate, receive a few hours of practical plankton study in the entire course. Oceanographers, as the name suggests, travel away from the coast to study plankton in more detail within the oceans and at great depths. I first read *The Open Sea: The World of Plankton*, written in 1956 by the great oceanographer Professor Alister Hardy, when I was 16, and it was a total inspiration. At the time I understood barely half of its 335 pages, but it is testament to its author that at that age I could at least understand that much. Hardy's very readable book explained the study of plankton as it exists within oceanography. The study of shore plankton is subtly different. For one thing, it is more accessible and, for another, more diverse. If you want to understand the seashore you need to study plankton.

I hope that the contents of *this* book will dismiss any trepidation about observing coastal plankton. For example, a common misconception is that you need a boat. Oceanographers use expensive and complex methods, including a wide variety of boats, in their work, as they typically work offshore. Fifty years ago, convinced that I had to have a boat, I purchased an old fibreglass rowing boat that I could carry around on top of an equally old Land Rover. Travelling around the British Isles, I attempted to seek out coastal plankton. It was not without its dangers and, on one occasion, I needed saving off the coast of Dingle in western Ireland: fortunately, as my tiny boat was being swept out towards America, a fisherman in a curragh saw me and rowed out to rescue me.

When I worked as director at Dale Fort Field Centre I had access to boats to collect plankton along the coast, but in more recent years I have developed simple techniques to collect samples from dry land. I have also learnt to appreciate the plankton biodiversity found in tidal water at the shoreline, which is far greater than that of the deeper water beyond. Seawater lapping the shore has several sources of organisms: oceanic currents and wind bring material to the coast; then there are larvae released by rocky and sedimentary seashore creatures; and rivers and estuaries bring a surprising amount of material to the shore throughout the year, particularly during periods of heavy rain. All this adds to the incredible diversity that can be collected easily in a net in a matter of minutes. All the species with photographs discussed in this book have been collected using simple nets from the shore. Enjoy the exploration of a new world.

Acknowledgements

Over the years much of my work with plankton has been with a multitude of students, whom I thank for their assistance, patience and enquiring minds. I have also met many biologists and ecologists who have helped me in various ways: John Archer-Thomson, Steve Morrell and Phil Wensley stand out. Both John and Steve read the manuscript of this book and provided many useful comments and additions. I am indebted to them both for their help, although any mistakes are entirely my own. I have known Mark Burton for many years and thank him for his help and that of the Skomer Marine Reserve Team for regularly providing me with monthly samples from off Skomer Island, which form a valuable extension to my shore plankton work. Field Studies Council Dale Fort Field Centre have continued to allow me access to their site so that I can regularly sample plankton. I miss my great friend, biologist and sounding board, the late Dr Robin Crump. Finlay Pratt and Clare Cremona have helped with some general photography and collections. There is a wide plankton and microscopist community that enjoys exchanging information, especially about identifications. My thanks to them. Elizabeth Beston has been most helpful, particularly with diatoms, and has kindly allowed me to use four of her photographs. Thanks to a number of specialists at the Marine Biological Association who have helped with some difficult specimen identification, and Gill Mapstone of the Natural History Museum in London, an expert in siphophores. I am grateful to the equipment staff and others at NHBS who have provided equipment. My friends at the Quekett Microscopical Club have for several decades now been of huge assistance with my microscopy – in particular, Phil Greaves, Carel Sartory, Ray Sloss and Spike Walker. My local friend Mike Crutchley has always been available with technical microscope advice. I would like to thank Nigel Massen and the staff at Pelagic, particularly the editing done by Sarah Harrison and David Hawkins. To my wife, Brenda, my biggest supporter, guide, critique-giver and source of inspiration: thank you.

Photographic credits

All photographs by Julian Cremona except the following:

1.1 NASA Earth Observatory; 1.2A Nikolay Kovalenko – Unsplash; 1.13 Steve Wells (CC0); 2.1 Port Macquarie Museum, New South Wales; 2.3 National Portrait Gallery of Australia; 2.4 NASA Earth Observatory; 3.2 NHBS.com; 4.3 NHBS.com; 7.17a and 7.17b Elizabeth Beston; 7.18a Elizabeth Beston; 8.13 Hushaan fromtinyisles – Unsplash; 10.8 Elizabeth Beston; 11.1 Isabella Tuiran – Unsplash; 11.3 Mariusz Smenzyk – Unsplash; 11.4 Chris Hardy – Unsplash; 11.8 Jose Manuel Santos Alfonso – Unsplash; 11.9 Atanas Zaro – Unsplash; 11.24 Blair Morris – Unsplash; 19.5 NASA Earth Observatory; 19.6 Cameron Witney – Unsplash; 20.1 Michal Mikulec – Unsplash; 20.8 Kenny van Roosbroek – Unsplash

A sample of plankton at ×100 magnification showing different golden brown diatoms around a unicellular protist, *Acanthometra pellucida*. The latter has spines of strontium that act as a supporting structure for the cell. Diatoms are producers and form the base of the planktonic food web. Width of photograph 500 µm.

PART I

What is plankton?

FIGURE 1.1 Satellite image from 7 April 2025, when the UK was unusually clear of cloud cover. With warmth and suitable nutrients the phytoplankton blooms, creating the green smudges visible along the coast and in the North Sea. The pale, almost white, bloom at the entrance to the English Channel around Cornwall consists of offshore phytoplankton, primarily coccolithophores, a strange group of algae in the phylum Haptophyta. It is these organisms that formed the chalk laid down back in the Cretaceous period. Large coastal estuaries such as the Bristol Channel are brown with silt and nutrients. (NASA)

CHAPTER 1

A drifting life

> Taken directly from the Greek πλαγκτός [planktos] it should not be translated as just meaning wandering or drifting for it has a much more passive sense, meaning those which are made to wander; it is applied to all those organisms which are powerless to prevent themselves being carried at the mercy of the moving waters.
>
> Sir Alister Hardy, writing about plankton as Chief Zoologist to the Discovery Oceanographic Expedition of 1925–27

The sea never stops moving. Wind-induced waves lap the shore, the tide makes its vertical movement twice a day, slackening briefly between oscillations to the horizontal, and local currents sweep along the coast. Even on a calm day the sea never ceases its motion. When the coastguard is called upon to search for a person floating at sea it can be half an hour before the helicopter or lifeboat reaches the last known point of location and probably the person will have gone. In all likelihood they will have been moved to a completely different area. This is why coastguards produce complex search plans based on known water movements. For organisms that are large and powerful enough to swim relatively independent of the current, like mackerel, these water movements will offer minimal resistance. We call these creatures nekton. By contrast, a jellyfish, even though it can swim to a degree, lacks the strength to overcome the movement of the waters and this is why we might find them stranded on a beach. Jellyfish, therefore, are an example of plankton: drifting organisms that are unable to determine their long-term direction and are at the whims of the current. Even large jellyfish like the Barrel Jellyfish *Rhizostoma pulmo* are drifters and are termed megaplankton (see Table 1.1). Plankton is usually used as a collective term for organisms that live in this way, while the singular name is plankter; terms used in this book. Elsewhere plankton is also occasionally used in the singular.

FIGURE 1.2 The Barrel Jellyfish *Rhizostoma pulmo*. (A. Nikolay Kovalenko)

In addition to the local coastal currents we've already met, major ocean currents exist across all the world's oceans, making up a huge conveyor system. These are particularly important as, unlike coastal currents, they may run for thousands of miles. The Gulf Stream is such a current. Starting near the Caribbean, it moves northwards along the North American coast before going east across the Atlantic to arrive in northern Europe. It brings us not only warm water but also plankton, carried on the current to arrive at our shores. Even one of the heaviest bony fish in the world, the Sunfish *Mola mola*, follows the Gulf Stream to Ireland and Britain, where they can be found sunning themselves along the west coast in summer; as a result, the species has sometimes been called a megaplankton. However, although they are not fast swimmers they can leave the current, and probably follow it, in fact, to feast on the drifting jellyfish. They are, therefore, not true plankton, unlike the jellyfish and other plankton that arrive on European shores. So, for example, siphonophores, like the beautiful blue By-the-wind-sailors *Velella velella* (Chapter 11) may be born in the mid-Atlantic, but currents can deposit them in mass strandings on the European seashore in early summer.

The open ocean itself is known as the pelagic zone (and pelagic creatures are those that live in open water), and is subdivided by oceanographers according to depth and proximity to the shore. The sea surrounding most of northern Europe is relatively shallow. This is the neritic zone, above the continental shelf, where the waters are less than 200 m in depth. It is within the neritic zone that our seashore or near-shore plankton live.

Plankton from seashores and estuaries

A collection of plankton from a shore context could contain material that has travelled great distances from the open sea offshore and plankters moved by local currents along the coast. Sheltered marine waterbodies such as lagoons and estuaries may have plankton that drift onto nearby coastal shores. Rivers and estuaries, meanwhile, will contain material originating from dry land. Sampling in or near estuaries thus yields a markedly different array of organisms to a rocky shore on the open coast. The size of the estuary will determine the level of biodiversity and the extent of its effect, such as creating low salinity, along the coast. The larger the estuary, the greater the range of habitats, such as saltmarsh and mudflats, which will function to increase species diversity.

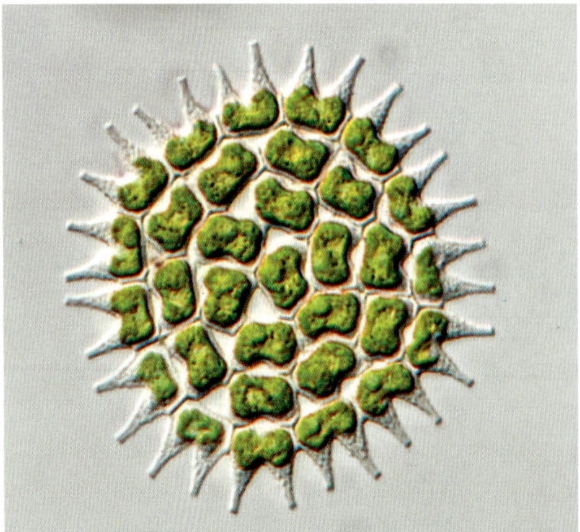

FIGURE 1.3 *Pediastrum* is a green alga found in freshwater plankton that enters estuaries, especially during flooding. It is a coenobial colony made up of a number of individual cells and can survive for weeks in marine plankton.

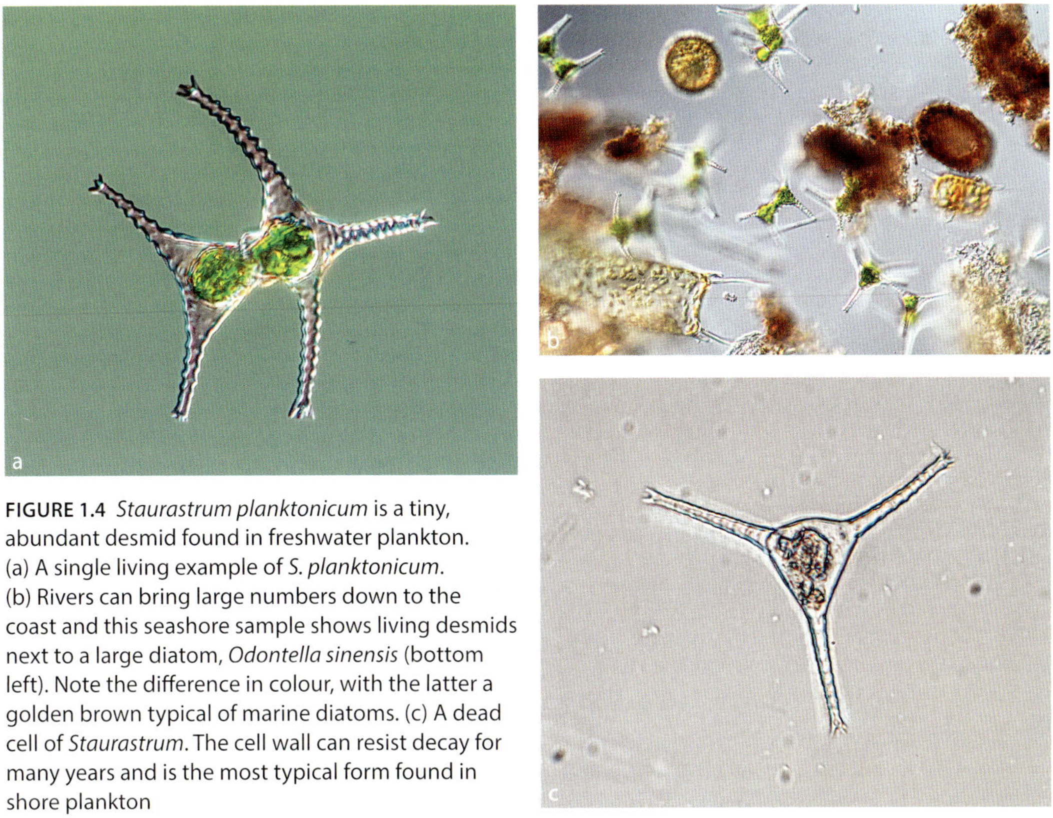

FIGURE 1.4 *Staurastrum planktonicum* is a tiny, abundant desmid found in freshwater plankton. (a) A single living example of *S. planktonicum*. (b) Rivers can bring large numbers down to the coast and this seashore sample shows living desmids next to a large diatom, *Odontella sinensis* (bottom left). Note the difference in colour, with the latter a golden brown typical of marine diatoms. (c) A dead cell of *Staurastrum*. The cell wall can resist decay for many years and is the most typical form found in shore plankton

Heavy rain and flooding inland will increase the chances of finding freshwater material in coastal samples. Similarly, tidal action will flush individuals of many species that typically inhabit saltmarsh and mud habitats into the plankton. Most harpacticoid copepods are sediment dwellers, but will occasionally be common in the shore plankton, along with several diatom species such as those in *Entomoneis* and *Surirella*. The volume of water from a spring tide, for example, can wash large numbers of organisms into the freely drifting plankton before they gradually settle back again into the sediment.

Creatures that arrive in these coastal contexts from areas of fresh water may be killed by the change in salinity, but their bodies can be surprisingly resistant to decay in salt water. For example, desmids, a group of microscopic green freshwater algae (not to be confused with diatoms) may be so robust that their cell walls can remain intact for many years (see Figure 1.4). This means the structure will still be identifiable even though there are no living parts inside.

In fact, on all seashores there is a continual exchange of organisms between the benthos (the bottom substrate) and the water column. Crustaceans such as *Idotea* spp. and members of Cumacea (see Figure 13.2) occasionally leave the seabed and swim up into the water, dwelling there for a short time as they disperse and perhaps try to find a mate, before settling back down to the benthos again. These creatures we call tychoplankton, temporary visitors. In contrast, the likes of water fleas (cladocerans) are holoplankton, permanent members of the plankton, as all stages of their development live as part of this community. A third type of organism found among the plankton, and among the most prolific forms along the shore, is the larval forms produced by adult animals living in the benthos. These are known as meroplankton. For example, the

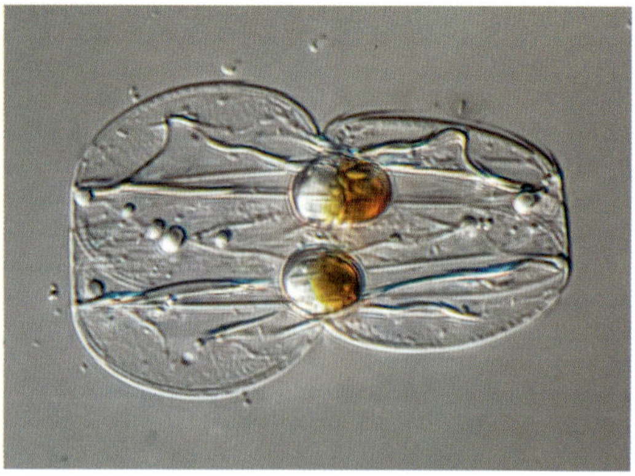

FIGURE 1.5 The diatom *Entomoneis* is common in the surface of estuarine mud and is often washed into the plankton.

FIGURE 1.6 *Idotea baltica*, an isopod crustacean that occasionally swims up into the plankton from seaweed and from offshore.

Edible Periwinkle *Littorina littorea*, common on a variety of shores, has a larva that lives in the plankton for several weeks, feeding and dispersing, before settling to become an adult. In fact, the majority of seashore invertebrates have a planktonic larva, whether on rocky or depositing shores. The Dogwhelk *Nucella lapillus* is one of the few exceptions, instead laying egg capsules on rocks with abundant food inside for the developing embryos. The number of eggs the dogwhelk produces is low compared with most other shore species, as the chances of the offspring's survival are so much greater than when living in the plankton. The juvenile snail emerges from the egg as a 'crawlaway', a tiny mollusc that wanders off across the rocks in search of diminutive prey such as young barnacles.

Eggs and larvae

Most invertebrates undergo significant change in body structure over the course of the lifecycle. Traditionally, invertebrate lifecycles are taught in schools as if metamorphosis is exclusive to insects, but many other groups, including crustaceans, marine worms (polychaetes) and echinoderms, undergo far more complex and dramatic changes in form during their development. There may be a pre-larval form followed by a series of larvae, often undergoing a number of stages and even quite catastrophic changes in structure.

For invertebrates a life as plankters in the water column is ideal for dispersal to new environments, although it has the disadvantage that drifters, unable to choose where they will end up, have a higher chance of dying on the way. It is therefore essential for adults to create large numbers of offspring – not just hundreds but often millions of progeny – to increase the chance of survival. An individual Common Starfish *Asterias rubens* can produce almost 3 million eggs per year. While remaining in the plankton for a long time decreases the individual's chances of survival, for the species this strategy will mean better gene flow. Dropping out of the plankton soon after being released by the parents risks the mixing of similar genes, which reduces variation. Reaching a distant population, therefore, has an advantage for the species as a whole. However, life as a holoplankter also presents problems when it comes to reproduction, with limited visibility and mobility in the water column making it difficult to find the opposite sex; many holoplanktonic species are thus hermaphrodites, with an option to self-fertilise, or they may use pheromones for attraction.

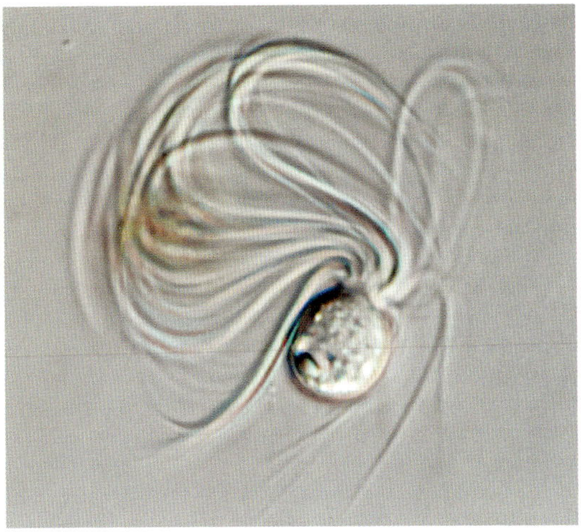

FIGURE 1.7 Flagellates, which measure between 2 and 20 μm, form a significant proportion of the nanoplankton. Cell diameter 10 μm.

FIGURE 1.8 A lecithotrophic egg of the Small Periwinkle *Melaraphe neritoides* with the developing larva inside. Diameter 200 μm.

The period an organism spends living in the plankton is therefore important and will depend on the type of egg and, consequently, the larva that is produced. The starfish mentioned above exemplifies broadcast spawning, where extremely small eggs containing a minimal food supply are released into the water for external fertilisation. Most fertilised eggs die in some way, but those that hatch will produce a larva that now needs to utilise other plankters as a food source. These eggs and larvae are referred to as planktotrophic: relying on the plankton for food and growth. They may live in the plankton for many weeks, a behaviour that increases the chance of their being consumed by a predator.

By contrast, the Small Periwinkle *Melarhaphe neritoides* releases an egg, one at a time into the plankton, that is considerably larger than those of the starfish, although still invisible to the naked

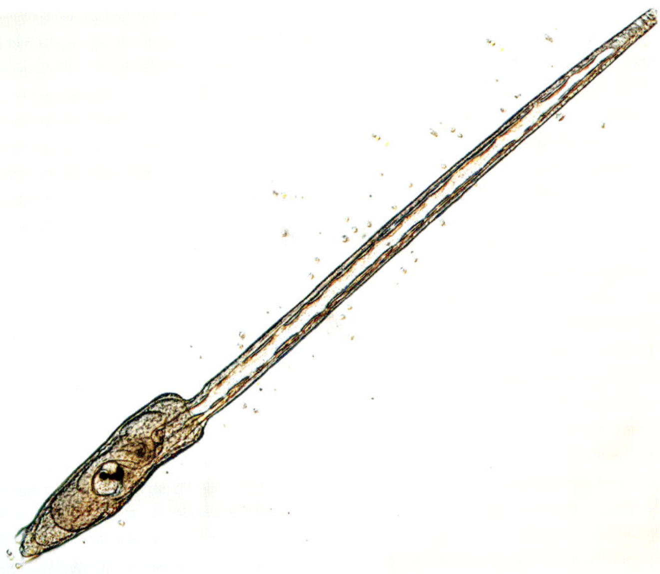

FIGURE 1.9 The lecithotrophic 'tadpole' larva of a sea squirt *Ciona* (a tunicate). Note the body still has yolk cells attached and, accordingly, during its short time in the plankton the animal does not feed. Length 400 μm.

Classifying plankton

We have seen how it is possible to categorise plankton based on the degree of time they spend in the water column, from brief incidental appearances (tychoplankton) to partial lifecycle (meroplankton) to living permanently in the plankton (holoplankton). However, a more useful organisation of plankton is by size (Table 1.1). Viruses are especially common in the sea but are beyond the scope of this book, as they require highly specialised methods of collection and observation. Likewise, the picoplankton is barely visible with a compound microscope and in most cases will appear as small moving dots in the background when viewing samples. It is a pity that observing the organisms in this category is so difficult, because research in recent years has revealed an astonishing biodiversity. Our practical quest begins with the third category, nanoplankton, while the final group, the megaplankton, will make only very occasional appearances in the net, although their larvae may often be found.

Micro-, meso- and macroplankton are the primary members covered in this book and relatively easy to see under the microscope. Within these groups we can recognise several trophic or feeding levels. The diatoms and most dinoflagellates, which constitute the phytoplankton, trap sunlight to create organic molecules; these are the producers. The consumers are more typically the zooplankton, such as the copepods and arrow worms. Figure 1.10 shows a plankton sample, within which we see both producers and consumers, along with detritivores, which feed on general detritus and particulate matter. This is a typical plankton community.

In any sample, mixed with these organisms will be a considerable amount of inorganic and organic debris. Some of the organic matter derives from the organisms living within

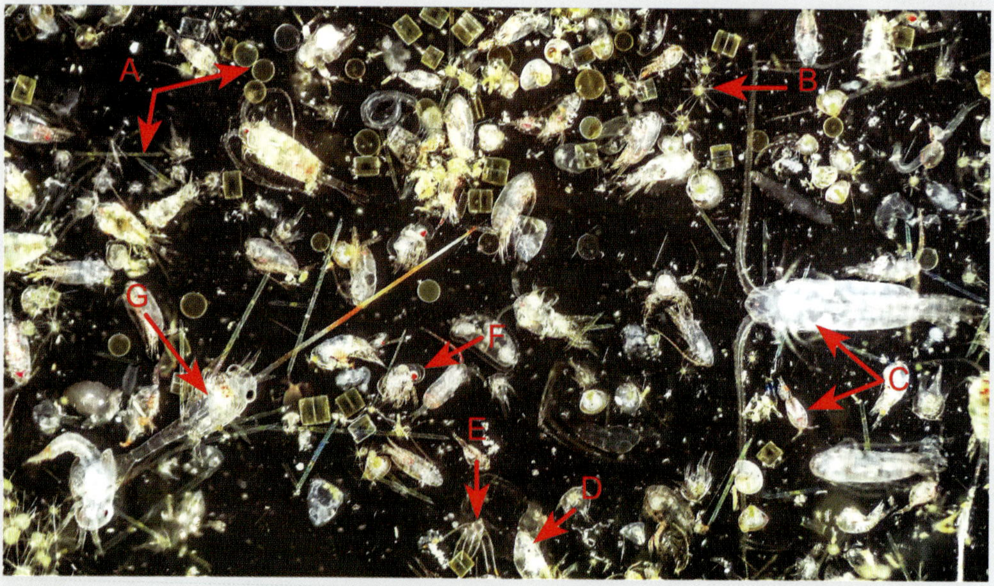

FIGURE 1.10 A late summer sample of plankton. Key: A. diatoms of the phytoplankton; B. radiolarian, unicellular bacterial feeders; C. a pair of different copepods; D. cyprid larva of a barnacle; E. pluteus larva of an echinoderm; F. cladoceran, or water flea; G. larva of a porcelain crab. Width of photo approximately 8 mm.

TABLE 1.1 Plankton classification based on size.

TYPE	SIZE RANGE	EXAMPLES
Femtoplankton	0.02–0.2 µm	Viruses
Picoplankton	0.2–2.0 µm	Bacteria and cyanophytes
Nanoplankton	2.0–20 µm	Flagellates
Microplankton	20–200 µm	Tintinnids, diatoms and dinoflagellates
Mesoplankton	0.2–20 mm	Copepods, hydromedusae and many larvae
Macroplankton	20–200 mm	Arrow worms
Megaplankton	0.2–2.0 m	Large jellyfish

the plankton, such as faecal matter from copepods, while other material will originate from plants and animals along the shorelines or from terrestrial sources. In the latter case, pollen and even the scales of lepidopterans occur commonly. We will look at these non-plankton oddities later. Much of the organic matter will be unrecognisable; it is in the process of being broken down into smaller and smaller pieces that eventually form an important food source referred to as Particulate Organic Matter (POM). In addition, a considerable amount of dissolved organic matter (DOM) is produced in the process of decomposition. This material is added to by messy feeders: much of the finer material generated during the process of consumption will dissolve and disperse in the water. This is still useful food for those organisms that rely on diffusion – a process much like living in a bath of food, whereby nutrients pass into the animal through the cell membrane. Microorganisms such as bacteria, cyanobacteria and fungi are vital in the breakdown and recycling of much of this organic matter. Their small size makes them difficult to study and so their discussion within this book is limited to a brief examination of their role in recycling within the plankton in the final section.

FIGURE 1.11 A sample dominated by the faecal material of copepods.

eye and requiring a microscope to be seen. This second type of planktonic egg, which contains considerable yolk, is called a lecithotrophic egg. Some tunicate larvae, similarly, are notable for the number of yolk cells that remain attached to their body for some hours after hatching. With sufficient energy stores available, the ensuing embryo and larva will be able to develop rapidly without feeding on other plankters, ensuring that the length of time the larva spends in the plankton will be short – sometimes just an hour or two. In the case of the Small Periwinkle this period may be around 24 hours before metamorphosis occurs. The larva then sinks to the bottom, where a juvenile forms and the lifecycle is completed.

There are plenty of examples of lifecycles that mix and match the reproductive strategies described above. A female crab, for example, may retain 1,000 fertilised eggs beneath her abdomen, where she protects and nurtures them, having provided plenty of yoke. When the larvae hatch after several months they are well developed as they enter the plankton. Now life will mean feeding as a planktotroph on other plankton, passing through a variable number of instar stages and trying to survive weeks or even months as a pelagic organism. During that time the juvenile crab may leave the coastal zone and be taken by the currents way offshore into deep water, with decreasing chances of survival.

Plankton and changing factors

The ocean is an unusual ecosystem in that it has no clearly delineated parameters or boundaries like those that can be seen in other habitats. It is dynamic and in a constant state of flux, with the organisms within constantly being juggled and resorted in their seawater medium. Although there is no structural organisation or zoning in the seashore space there are instead temporal zones. For instance, phytoplankton will move towards the light during the day, while zooplankton are known to descend in the water during the latter part of the night to lessen their predation.

Week by week, changes in the constituent species of plankton gradually occur and, although there may be several peaks in biodiversity and biomass during the year, plankton is always present in neritic environments. As in any habitat, a series of abiotic factors influences the organisms. Plankters need to minimise the tendency to sink and so the viscosity of the water is particularly important, as it changes with temperature over the year. The salinity of the water along the shoreline will also be influenced by proximity to the land, with water running on to the shore and estuaries bringing river water to the coast. This will have a strong effect on the constituents of both phytoplankton and zooplankton. The pH of seawater is relatively stable because of the volume of water involved and the presence of different minerals but this too can change, especially with nutrient runoff from the land. These issues will be discussed in more detail in the last section of the book.

Marine neuston, pleuston and plankton

The term neuston is given to the life living in, on or just under the surface of the seawater. These organisms may have come from land or from the benthos of the coast and many are incidentals. Samples taken above the rocky shore may include the occasional Marine Springtail *Anurida maritima*, a blue-grey insect 2–3 mm long. The shore in late spring can be covered in them, crawling over the substrate, and on a high tide they may enter the surface tension as a means of dispersing along the coast. The small snail *Hydrobia* does the same, suspended on a mucus raft. As the springtail is above the surface it can be said to be epineustonic, while the *Hydrobia*, which hangs beneath, is hyponeustonic. In the surface tension itself can be huge numbers of bacteria

FIGURE 1.12 Marine Springtail *Anurida maritima* living in the surface tension. Note that they have been shedding their exoskeletons – the white exuviae below them. Average length 3 mm.

and protists. Much larger organisms that use the surface tension for buoyancy, such as the floating *Velella velella* (Figure 11.20), are often referred to as pleuston; these organisms are affected more by factors such as wind than those in the water. Many of the organisms described here may have a flexible relationship with the terms used to define them, and some may fit several descriptors. Most, however, feed on the plankton beneath and their larval stages will often disperse as part of the plankton.

Marine pond skater *Halobates*

An insect that has fascinated me for years is *Halobates*, a relative of the pond skater, which occasionally makes landfall. These pale and wingless surface dwellers ride the surface tension of ocean waters, living their entire lifecycle here using floating debris (including plastics) on which to lay eggs. Like plankton, these insect are therefore drifters, dependent on surface currents and wind that determine where they go across the ocean.

FIGURE 1.13 Hawaiian Sea Skater *Halobates hawaiiensis*. (Steve Wells CC0)

FIGURE 2.1 Portrait of John Vaughan Thompson (1779–1847). The original is held in the Port Macquarie Museum, New South Wales.

CHAPTER 2

A brief history of discovery

> … a small muslin towing net on crossing the Ferry at Passage, such a capture of minute animals was made as furnished a treat which few can ever expect to meet, and could hardly be excelled for the variety, rarity, and interesting nature of the animals taken.
>
> John Vaughan Thompson, a naval surgeon based in Cork Harbour, credited with being the scientist to develop and use a plankton net, 1828 (www.dib.ie)

Humans have probably observed plankton for as long as they have been sea-faring. In antiquity the Greeks and Phoenicians navigated their way around the Mediterranean and out into the Atlantic, where they recorded strange creatures and, crucially, bioluminescence, a sign that they were finding plankton. Over the centuries mariners have regularly reported seeing strange lights in the water. It was the work of the seventeenth-century Dutch microscopist Antonie van Leeuwenhoek on microscopic life in fresh water, however, that really began the documentation of plankton. Similar work on microbes continues today in Amsterdam, where the ARTIS-Micropia, the world's only museum dedicated to microbes and microscopic life, can be found. Originally founded in 1838, the recently redesigned museum is an incredible homage to plankton and other microscopic protists. Possibly the first science report identifying phytoplankton was made by Joseph Banks, the famous naturalist on Cook's *Endeavour* in 1768; the ship's sailors referred to 'sea sawdust', which is now thought to be colonies of the cyanobacteria *Trichodesmium*.

A notable figure in the history of early plankton studies is John Vaughan Thompson, an amateur planktonologist and army surgeon posted to Ireland in 1816, where he made a series of discoveries in what would later be recognised as plankton. At the time many of the invertebrate larval forms encountered in the plankton were being recorded as distinct species, as their connection to the adults found on the shore was not recognised. Thompson, whose knowledge of crustacean for the time appears astonishing, put forward the idea of metamorphosis occurring in crustaceans and the presence of larvae, while in 1830 he determined that barnacles, usually considered molluscs, were crustaceans. His first real publication, in September 1828, was a series of memoirs collectively called *Natural History of Nondescript or Imperfectly Known Animals* – perhaps not the best title to bring him scientific favour. And, despite the scope of his discoveries, scientists of the time rejected much of his work, leading to his eventual disillusionment. After his death in 1847 it would be another century before his work was fully recognised and appreciated.

Other plankton pioneers include the French naturalists François Péron and Charles Lesueur, who used nets to capture phytoplankton during an expedition to Australia in the early 1800s and in 1809 described the exceptional zooplankton diversity found in the bays of Villefranche and Cap de Nice in the Mediterranean.

Several years after their work, Charles Darwin made diary entries while on HMS *Beagle* concerning the small creatures he found at sea. Darwin took an interest in all form of nature, but one of the groups that especially fascinated him was the barnacle. Between 1846 and 1854 he published a series of detailed monographs unravelling the natural histories of barnacles and their planktonic larval stages, a story told in Rebecca Stott's fascinating book *Darwin and the Barnacle* (2003). His friend Joseph Hooker, meanwhile, was on his own voyage to Antarctica (1839–1843), where he

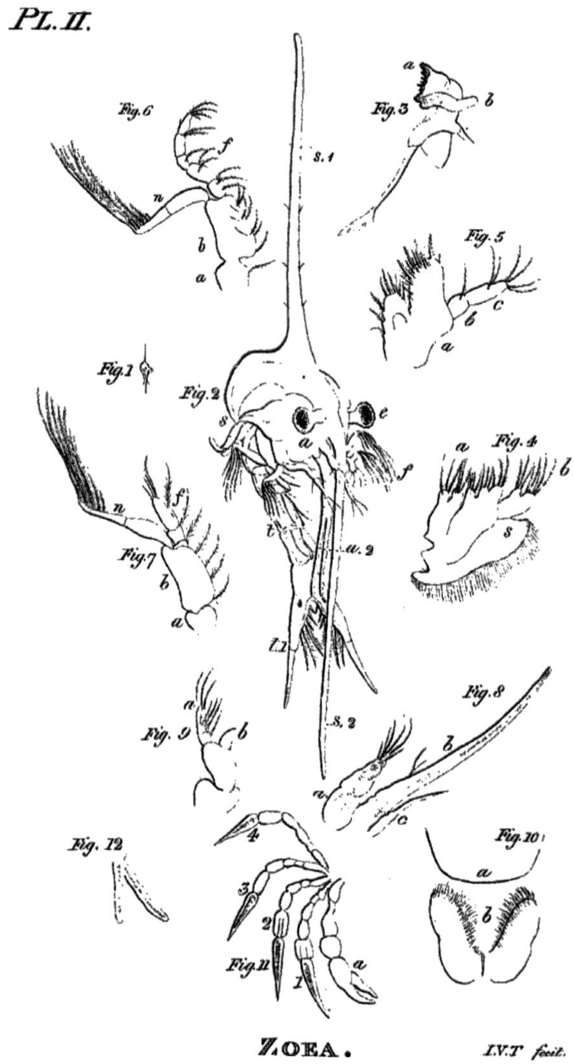

FIGURE 2.2 One of many exquisite drawings from John Vaughan Thompson's work on crustacea, made in 1828. This shows a planktonic crab larva, or zoea.

trawled diatoms with nets. His study of these microscopic plants discussed their huge abundance in the sea and their place in the food webs of the polar regions, as well as the importance of silica in the structure of diatoms' cell walls.

At around the same time as these scientists were working, the German scientist Johannes Müller initiated regular plankton sampling in the small Helgoland Archipelago in the North Sea. Visiting between 1845 and 1854, he used a fine mesh net to collect a bewildering number of what he referred to as *Auftrieb*, or 'floating stuff' – planktonic larvae and adults – that were new to science. Müller, the first to use such fine nets, continued to develop techniques to sample and observe microscopic marine life. His work led to the establishment of a permanent biological station on Helgoland. Referred to by many scientists as the 'cradle of plankton research', this island, more than 150 years later, continues his important work.

In 1854 Müller inspired his countryman Ernst Haeckel – a man who, it would later be said, merged science with art – with the words 'as soon as you have entered into this pelagic wonderland you will see that you cannot leave it' (from *Planktonic Studies* written by Haekel, 1893). Haeckel's

FIGURE 2.3 François Péron. Together with Charles Lesueur, Péron studied plankton in the Mediterranean. Conrad Westermayr (engraver) after Jean Henri Cless, *François Péron* c. 1805. (National Portrait Gallery of Australia)

subsequent plankton work showed a particular fascination for single-celled organisms such as radiolarians and included beautiful and complex drawings and paintings. His book *Art Forms of Nature*, published in 1904 and still in print today, shows diatoms and other plankton in exquisite detail.

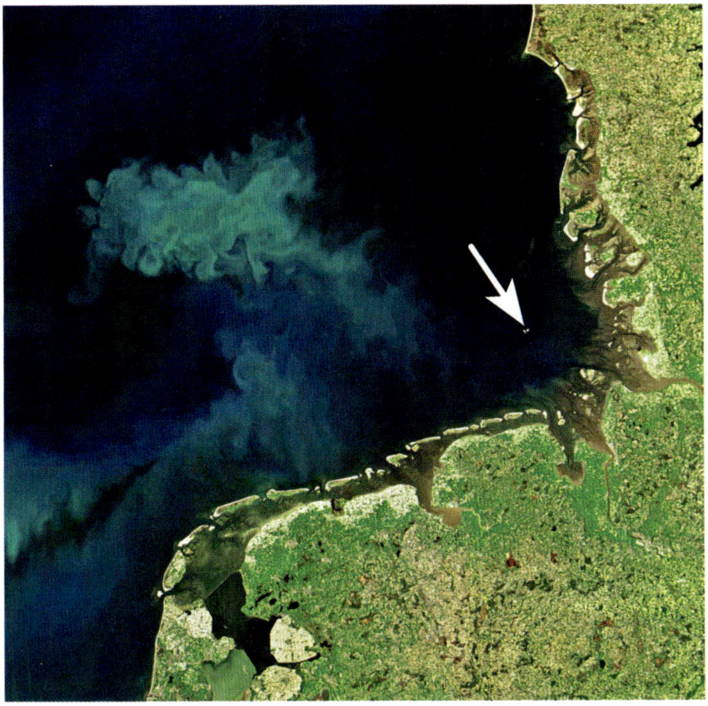

FIGURE 2.4 The Helgoland archipelago, marked with an arrow. Here Müller researched plankton in the 1840s; a biological station was established in the late nineteenth century. Note the plankton bloom to the west. (NASA)

FIGURE 2.5 A radiolarian, the inspiration for Ernst Haeckel's incredible drawings.

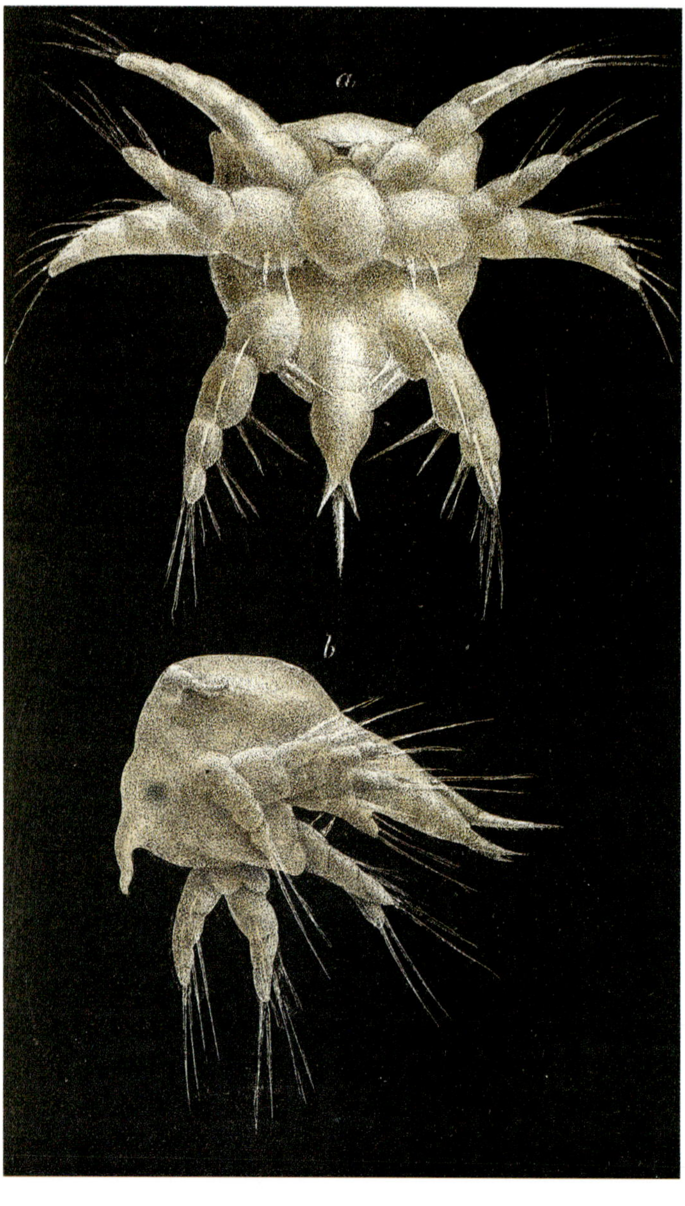

FIGURE 2.6 An engraving of barnacle nauplii larvae, as drawn by Philip Henry Gosse (from *Tenby*, 1856).

Early oceanographic voyages

British scientists did not take plankton particularly seriously until 1872, when the HMS *Challenger* expedition, led by Sir John Murray, explored the world's oceans over a period of four years. This was the first expedition dedicated solely to marine science. Charles Wyville Thomson, the expedition's chief scientist, organised the trawling of the ocean depths to collect sediment samples, which included microfossils. (Samples that are now housed in the Natural History Museum in London.) It was realised early on that many free-drifting organisms existed in the open sea, although the principal aim of the research was to establish the basis of oceanography. The importance of plankton to fisheries was realised by several scientists including Victor Hensen, who also coined the term plankton, describing it as the 'blood of the sea'. Funded by Kaiser Wilhelm II and the German government, he led the first oceanic study dedicated specifically to plankton, which took place in the North Atlantic in 1889. These studies laid the foundations of our modern understanding of the ocean, while Ernst Haeckel used much of the material from them for his drawings.

Plankton study in the nineteenth century had begun with naturalists sporadically investigating their local seashores and by the close of the century it had become organised, professional oceanography, the study of oceanic ecosystems. Our particular interest is with the shore plankton, however, and it is the work of Philip Henry Gosse that is of most relevance to this book. He was one of the greatest naturalists of the nineteenth century, who worked on all aspects of nature but especially microscopic seashore life, and exchanged letters regularly with Charles Darwin discussing their work (see Figure 2.6). Gosse was intrigued by metamorphosis and spent considerable time observing barnacle larvae, drawing them and confirming some of the observations made by Charles Darwin. He noted very early on, for example, that if a sample was placed in a glass jar by a window both zooplankton and barnacle larvae congregated on the brightly lit side.

All his books are fascinating to read and he was a prolific writer about his nature travels. In the 1840s he was living in London on limited means, but he realised that he could write and the public enjoyed his works on natural history. During a period of low mental health he took his wife and son to live in Devon for a year; the resulting book, *A Naturalist's Rambles on the Devonshire Coast*, which contains some amazing illustrations, was published in 1853. In 1856 he produced a similar account of a summer spent in Tenby, on the Pembrokeshire coast. In both cases he not only described the life that he found on the coast but also produced incredibly detailed drawings and engravings, to which he laboriously added colour by hand. Figure 2.7 shows a full-page engraving including a hydromedusa that he collected from Tunnel Rocks, near Ilfracombe, on the north coast of Devon. Like so many organisms he and others found, only later would it be recognised as having previously been named: Gosse gave it the scientific name *Saphenia titania*, but it had already been discovered and named *Amphinema*, in 1810, by François Péron and Charles Lesueur.

This highlights one of the many conundrums of the time. Communicating new discoveries and namings across the scientific community was difficult and in the mid-nineteenth century many species were being discovered with little chance of classification in the system we use today. For example, the arrow worms, organisms living permanently in the plankton, are today placed in their own distinct phylum, the Chaetognatha; within this, the genus *Sagitta* (now *Parasagitta*) had

FIGURE 2.7 A page from Gosse's *A Naturalist's Rambles on the Devonshire Coast* showing, at the bottom, a hydromedusa drawn and coloured by Philip Henry Gosse, 1853. Compare this with the photograph of *Amphinema* in Figure 11.15b.

been discovered as early as 1781. As this classification was unknown to Gosse, in his two-volume *Manual of Marine Zoology for the British Isles*, published in 1855/56, he includes it with the true worms, Polychaeta. Subsequently, in his book on Tenby, no fewer than seven pages are given over to regular finds of *Sagitta* spp.; in this volume he included it as a mollusc. Similarly, Thomas Huxley was convinced that it was a form of arachnid.

As a result of the understanding of the importance of plankton to fish stocks the Plymouth Marine Laboratory was established in 1886 to research the management of the fishing industry. In addition, female scientists, hitherto largely excluded from this field, began to make their mark from the 1920s; several outstanding scientists included, at the Port Erin marine laboratories on the Isle of Man, Catherine Herdman, and Sheina Marshall, based at Millport in Scotland. We will meet Sheina again in Chapter 13, when we look at her research on copepods. At the Plymouth laboratory Marie Lebour researched both phytoplankton and other marine plankton and in 1930 wrote an authoritative tome *The Planktonic Diatoms of Northern Seas*. Published by the Ray Society, this comprehensive book remains one of the best guides to diatoms, particularly as the diagrams can be used to assist identification.

In the first quarter of the twentieth century, oceanography was revolutionised by Sir Alister Hardy. Collecting plankton at sea from a large vessel had always been difficult and, when he joined the RRS *Discovery* on the two-year exploration of the Antarctic, he designed the Continuous Plankton Recorder. Prior to Hardy's invention, the entire contents of the net were collected en masse in a container. Hardy's device made use of a continuously moving band of silk which trapped the plankton and stored it in a preservative for later examination; the idea borrows the principle of a net, with the silk acting as a filter and the action of pulling it through the sea pushing large volumes of water through it. He later went on to become an expert on whales and wrote the definitive books on plankton, notably *The Open Sea: The world of plankton*. As the twentieth century progressed studies became concerned less with the constituents of plankton and more with understanding the ecosystem. Advancements were made in researching plankton distribution and vertical migration, primary production and the microbial loop, also known as the microbe food web. Despite progress, much remains unknown, and exploration of the intricate world of plankton continues.

FIGURE 3.1 A view down Milford Haven at dawn and high tide. The long pontoon is ideal for collecting plankton and is above both rocky and sandy shores. Additionally, to the left is a slipway that could be used in a more limited way. At the end of the peninsula, right, is a small jetty that provides sampling access.

CHAPTER 3

Collecting from the shore

The basic aim of plankton collection is to filter as large a volume of seawater as possible. For instance, buckets of seawater could be poured through a funnel containing a filter suitable for capturing the plankton. Various methods have been employed on ships, including the use of a pump to draw water out of the sea and through filters that collect the plankton, after which the water is returned to the sea. A net, however, is the obvious piece of sampling equipment. There are many caveats, nevertheless. There is a difference between collecting nearshore and offshore plankton. The latter may require moving many miles out into the open sea in a suitably large, ocean-going ship that makes the use of a small net impractical. Instead, professional oceanographers use large *bongo* nets pulled in pairs behind vessels. In addition, as seen in Chapter 2, Alister Hardy designed the Continuous Plankton Recorder to be towed behind ocean-going commercial boats. The majority of plankton nets are designed for towing behind a small boat just offshore, but in this book we will be dealing with those nets operated from the land, which are straightforward to use but come with some drawbacks.

Plankton nets

If your aim is simply to take a quick look at plankton, you can make your own net. A pair of tights is required. Cut one of the legs off at the top and stretch and fold this thicker area over a strong wire frame in the region of 15 cm in diameter, before stitching in place. Enough wire is necessary to create a suitable handle. Next, push a small jar down into the foot to collect the sample. Wrap a thin wire or elastic band tightly around the outside of the foot to secure the jar in place and prevent the water and plankton going down the side of the jar. Sweep the whole contraption slowly back and forth through the water for some minutes and it is amazing what can be captured in this way, even though repeated use is usually limited by the fragility of the netting, the mesh size, the wire handle and the degree of stretching that takes place. Figure 3.4 shows commercial net mesh sizes as well as an example of tights fabric. The latter, it can be seen, is not far from a zooplankton mesh.

While much enjoyment can be had from creating a homemade net, keen amateurs will find that a commercially produced plankton net gives easier, better and more consistent results. Most commercial plankton nets, as already noted, are designed to be towed by a boat, but can also be used from the shore; suppliers usually have a wide range of alternative sizes and filters. Typically the net fits around a round stainless-steel frame to which is attached a harness of three cords that are connected to a single small ring. A long cord or rope is attached to the ring for pulling the net through the water. There are also small conical nets available that will fit standard commercial pond nets with a long wooden handle enabling it to be swept back and forth through the water. In the case of both the homemade net and this one, the handle, even if long, places a limitation on reach. These nets are suitable for large rock pools or places where you can get close to the water. The problems come when working from platforms such as pontoons and jetties, where the water surface will usually be some metres away. An alternative method which suits these handled nets is to hold the net in place in a strong tidal current (see below), letting the water flow through for five minutes or so.

FIGURE 3.2 A plankton net fitted to a standard pond net frame and handle. (NHBS)

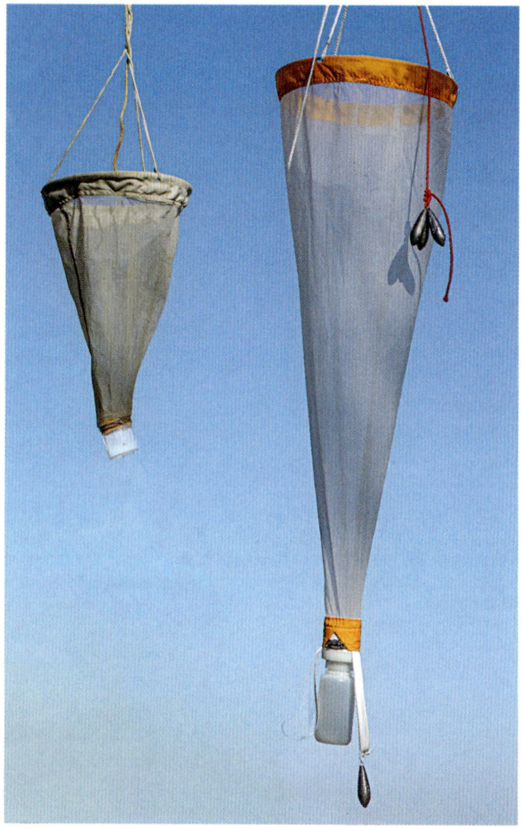

FIGURE 3.3 Two phytoplankton (53 µm mesh) nets side by side. The longer one has a 30 cm diameter, tapering to the adapter to which a bottle can be attached. This is ideal for calm conditions and long pontoons, but is less manoeuvrable than the smaller one. The latter has a 20 cm diameter and is shorter, so works well in confined areas, windy conditions and rough weather.

For flexibility and consistency it is best to use a tow or pull net, and so we will concentrate on these for the rest of this chapter. There are several further items necessary to achieve a serviceable net. Firstly, some form of weight is required to hold the net at a depth of about half a metre or so below the surface. Any suitable heavy metal object that can be easily tied to the small towing ring will do. In the past, I have used a pair of old stainless-steel spanners. Probably the most convenient solution is to buy steel fishing weights (as shown in Figure 3.5). If no weight is attached, the net will tend to lift and breach the surface of the water when it is pulled, picking up any debris floating there. The length of cord between the weight and the ring should be enough for the former to hang just below the net, so as not to obstruct the opening. As a rough guide, a small net requires around 300–400 g, while a larger one would need 600–700 g.

The second item that can be very useful is some form of lightweight pole that can be used in a way similar to a fishing rod. If you simply hold the net's cord in your hand and walk along a platform it is likely that the net will snag on the edge or you will risk falling in as you attempt to avoid snagging the net. The cord should be wrapped around the pole's end so that the net can be held well clear of where you are walking. A trekking pole is ideal.

Do not skimp on the length of cordage used for pulling the net: 7 or 8 m can be necessary for a net 30 cm in diameter, and the height of some piers may necessitate substantially more.

Choosing a pull net

Pulling a net designed for a boat along the edge of a platform can be surprisingly tricky. A good deal of effort is required, with a chance of falling into the water while avoiding trip hazards. It is tempting to choose a large net as it will collect more material in the same time as a small one; frame diameters can vary from 20 to as much as 40 cm, and the larger the net the greater the volume of water will be filtered in a given time period. But there are other considerations (Table 3.1). A wider frame opening means a longer net and, while a large net is fine in a boat, when pulling from land many variables, such as wind, water currents and swell, come into play. The bigger the net the more unmanageable it can become. So, working from an unobstructed pontoon in calm conditions is perfect for any size of net. Increasing the working height above the water to several metres or more – for example, a pier – is fine unless there is a strong wind, which will make lowering a large net difficult. Figure 3.8 shows collection from a jetty in strong wind and Atlantic swell – not ideal with a large net. Think of the locations and conditions in which you are most likely to use the net.

It is important also to choose net-mesh and filter size carefully. The usual options, in microns, are 20, 35, 53, 100, 150 or 230. The filter will be the same mesh size as the net. Choose the mesh that best suits your requirements in terms of the size of the organisms you wish to collect. In general, phytoplankton is typically collected with a mesh size of 53 microns, while for zooplankton, which are larger, a mesh of 230 microns is more suitable. If your interest is in a diverse range of plankton the phytoplankton net will be the better option, as this will collect a much wider range of plankton sizes, including zooplankton. It is surprising how even 53-micron mesh will retain some 20–30 micron-sized organisms as they get trapped amongst the larger material; similarly, zooplankton nets may retain animals down to 100–150 microns and some of the large diatoms, but anything smaller will pass through the net or filter. I am interested in the general diversity of plankton and, although I have several nets, I tend to use only the 53-micron mesh net. It collects a very wide range of organisms, from single-celled ciliates and small diatoms up to all crustacean larvae and medusae, although the smallest creatures will not be in large numbers.

It is important to bear in mind that, regardless of mesh size, it is impossible to get a true estimate of abundance with nets or even other samplers. This particularly applies to what is termed 'soft microplankton' (easily damaged organisms) like naked protists. Abundance can be established with tintinnids (a ciliate protist), as it is more robust due to the shell casing or lorica that they secrete around themselves. They vary in size from around 25 to a

TABLE 3.1 A comparison of the two nets shown in Figure 3.3.

	20 CM DIAMETER	30 CM DIAMETER
Plankton volume	Low: requires 10 minutes pull	High: requires 5 minutes pull
Pulling effort	Easy and flexible	Less flexible; strength required
Manouvreability	Good	Poor
Ideal locations	Any: pier, jetties, pontoons	Long pontoons, minimal turns
Weather conditions	Any: calm, windy, turbulent	Calm conditions

maximum of 500 micron and so a very fine mesh of 20 micron would be needed to obtain an accurate estimate of abundance. The 53-micron mesh would collect some tintinnids, but would seriously under-sample the population. A 20-micron mesh, when compared with 53 microns, also trebles the number of rotifers collected. Quantitative sampling can become very complicated when considering copepods, where studies show that the mesh needs to be less than 75% of the copepod carapace width to catch 95% of individuals. If you intend to make surveys of specific species careful research on species and mesh sizes will be needed; otherwise, keep to a general net like the 53-micron phytoplankton one.

Using a net

The net tapers away from the mouth to a small aperture (the cod end). Some nets, usually older ones, may have a fixed filter, but ideally the cod end has an adapter for attaching different collecting options. Typically a filter is attached to the adapter; plankton collects on the filter, which is unscrewed and flushed with seawater to wash the sample carefully into a small container. A wash-bottle is best, but a large pipette or syringe will work fine. Personally, I prefer to attach a screw-on bottle or small container at the cod end. After collection the plankton sample is in the bottle, which can be easily unscrewed so that the contents can be tipped into a second container for transport.

The collection technique is essentially the same for both bottle and filter. As you pull the net along any plankton or material in the water will catch on the side of the net as the water passes through. As you remove the net from the water gently jerk the cord repeatedly as the net breaks the surface to dislodge the plankton from the sides of the net. Work slowly and continue with this jerking action even when the entire net is clear of the water surface, to allow time for the water drain through the netting and the material within the net to collect in the bottle or filter – the small mesh size means that water does take time to filter through, so be patient. A final flick should dislodge any plankton around the rim of the bottle or filter. If you are using a bottle the water within will look slightly dirty – this is your collected material. Unscrew the bottle or filter, taking care to hold the junction with the net over a suitable container as you do so, so that the water immediately above is caught as well. The bottle contents can be decanted into the container and sealed with a lid. The filter can be turned over and the contents flushed with seawater into the container and sealed.

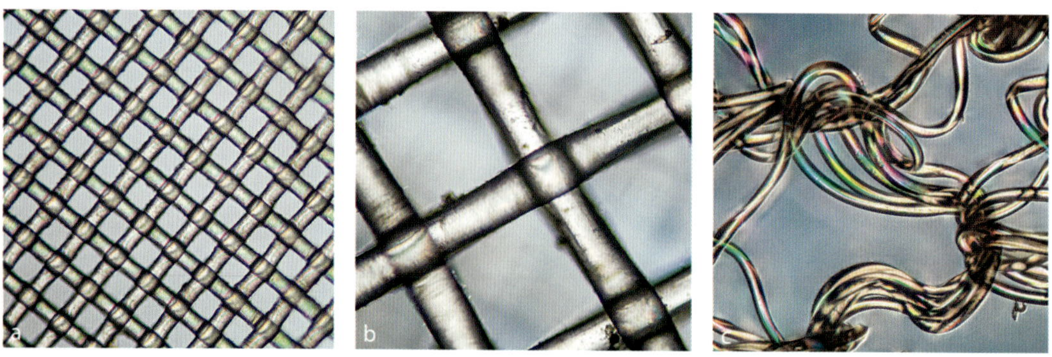

FIGURE 3.4 Magnified mesh of plankton nets, all at ×100. (a) 50 μm (b) 230 μm and (c) material from a pair of tights.

FIGURE 3.5 A plankton net set-up. Attached to the harness ring is a red cord to suspend the weights, in this case 3 × 5 oz/142 g. The filter has been removed from the adapter and is in the middle of the photograph, and a bottle is ready to screw on instead. The two white loops of fabric allow the attachment of more weight (in this case 2 × 5 oz/142 g) to keep the bottle low in the water, allowing air bubbles to escape.

FIGURE 3.6 A weighted net at a suitable depth for pulling, avoiding the detritus floating at the surface.

Net maintenance

Having purchased a good plankton net, make sure that, after use, you wash it thoroughly with clean, fresh water, paying particular attention to cleaning the mesh well, where organic material will have become caught. A friend of mine, Phil Greaves, washes the net once a year in a biological detergent to completely remove organic material. Consider biosecurity (washing in a disinfectant) if you sample with the same net in different locations so that species, particularly invasive ones, are not distributed. Seawater will soon degrade the stainless steel, which becomes pitted, so run plenty of fresh water over it. Hang up to air dry. Following this procedure should give your net many years of use.

Be sure to check any securing knots on the cords and harness regularly and particularly before use. More than once I have had to return to a site at low water to search for my lost net, which had escaped during sampling as a result of poor knot management! While transporting your net home to wash make sure it is not touching any non-stainless-steel iron work, as it will soon become contaminated with rust.

The collection process

Unfortunately, many people buy a net, pull it through the water for several minutes and say they found very little. Taking a trawl through water near pontoons and jetties takes very little time, but it does require thought and care to achieve good results. The process is more complex than towing from a boat and for this reason this section is quite detailed.

Choosing a good collection site is important. The main options are outlined in Figures 3.7, 3.8, 3.9 and 3.10, but each site will have its own advantages and drawbacks. Take a pontoon as an example: you need some depth to fully submerge the net and from a pontoon there is a clear drop from the edge into deep water. However, pontoons also include plenty of trip hazards, such

FIGURE 3.7. Collection site 1. Pontoon or pier: usually found in sheltered areas and will enable the collection of a diverse range of shore plankton, depending on the types of shore nearby. These form one of the easier sites.

FIGURE 3.8. Collection site 2. Jetty: although not very long these may be closer to the open coast, as can be seen here from the turbulent water. These are often harder to use as multiple turns are required and manoeuvring in the waves can be tricky, but yield good results with open-coast species and larvae.

FIGURE 3.9. Collection site 3. Causeway or bridge: these are generally found near estuaries and consequently need careful timing. In the image here the tide is coming in and so the net is placed upriver of the bridge, where it is held against the bottom with a particularly heavy weight. The tidal water, coming from the right, filters through the net. Sample diversity will be limited by the low salinity, but many estuarine diatoms, nematodes, ciliates, flatworms, rotifers, harpacticoid copepods and ostracods, can be expected, as well as debris and sediment. Reversing the direction of the net as the tide goes out will result in a shift towards freshwater species.

FIGURE 3.10. Collection site 4. Promenade or sea wall: these are commonplace and often the only option available. The site shown here is an old boat dock that has been repurposed as a public walkway and is sheltered from the wind, allowing a large net to be used. Depending on the site, parts may slope down to the shore rather than being vertical, and it is worth checking at low tide to make sure that there is sufficient depth at high tide. Strong currents at high tide may exist on exposed corners, allowing the net to be suspended at a single point and reducing the problem of hitting the structure. Results can be pleasingly good, but also variable, depending on the nature of nearby shores.

as stanchions, metal rings and other devices for mooring boats. If there is swell coming in from the sea pontoons can also move with the water, so you might consider a buoyancy aid if you feel a little unsteady, although if you use a pole to hold the net at a distance from your body you do not need to be close to the edge. Alternatives include places with a protective rail, such as piers and promenades. Harbours can be ideal; in fact, anywhere along the coast with a defined edge where you can drop your net in the water can work.

Local invertebrate populations around collection sites will be releasing eggs and larvae that can dominate your samples at certain times of the year. Consequently, it is useful to have several places to collect from, preferably with different types of shore nearby. Using a barnacle-studded harbour wall may produce inordinate volumes of barnacle larvae in samples, while a promenade above a sandy beach will have bivalve larvae. A jetty I use in winter (Figure 3.8) has large numbers of the Small Periwinkle *Melarhaphe neritoides* attached to the surface and between November and February they release large numbers of eggs into the plankton.

The state of the tide presents practical issues and so having a tide timetable is essential. Figure 3.11 shows a tide graph illustrating the position of the water's edge throughout the day. An ideal time for collection is at maximum water height, because there should be sufficient depth to use the net even from the edge of a promenade. A high spring tide occurring around dawn is a good time for zooplankton that exhibit vertical migration over a 24-hour period, rising at night and descending during the day. By sampling early in the day, they will be nearer the surface. Time of day is not too crucial for shore plankton, but a high tide is necessary for water depth and bringing fresh material across the shore to your sampling location, maximising the range of species and larvae present. Additionally, sampling early gives you the rest of the day to look at fresh plankton, as it can quickly deteriorate in storage.

As well as taking your net and pole to the site, you need suitable containers with tight lids for transporting your sample, which should be kept in a suitable carrying box with spare cordage and a small knife or scissors. The collecting process does not take long. Find a suitable stretch of walkway, unwind your cord and slowly drop the net into the water. The phytoplankton net, because of its small mesh size, will not allow water to easily pass through, and it will float initially. With the help of the attached weight it should start to sink but if you have a bottle at the end it is likely to be upturned and full of air, keeping it close to the surface. Any air, especially at the cod end, should therefore be allowed to escape if possible. If you are close to the water surface, then hold the bottle beneath the surface and allow it to fill. Dexterous net handling is helpful here, otherwise: when the net sinks in deep water a quick vertical pull can flip it the right way around to release air.

Once you are sure any large air bubbles have been removed pay out more cordage and begin to walk slowly along beside the water so that the net moves just beneath the surface and remains stable as you walk. If the net begins to rise to the surface you are probably walking too fast; a slow, consistent speed is needed. Ideally, you should pull the net in this way for approximately five to ten minutes. Unless you are on a long pontoon, at some point you will need to turn around. Again, this is where the pole is useful. As you stop to make your turn the net will slow, beginning to nose-dive, and to prevent this you must quickly reverse your walking direction. If you raise the pole as you turn around the net's momentum will raise the opening and keep it going in an arc back the way you came. It is important to keep the net's speed and position consistent as you turn to reduce the chance of any of your catch, particularly large zooplankton such as arrow worms, escaping. With practise this will become much easier. I will usually do two five-minute samples with a 30 cm diameter net, which will allow for any inadvertent release of plankton from the bottle as a result of poor technique.

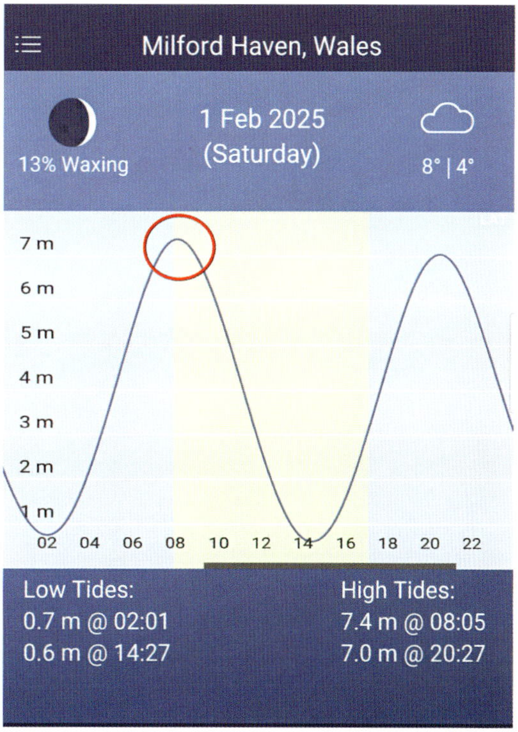

FIGURE 3.11 Graph of tide times on an ideal sampling day, with the red circle indicating the best time. This is a screenshot from the free Android app Tides Charts Near Me. Many others are available.

A static-net tip

An alternative to pulling the net is to place it into a current of water such as exists in an estuary or at the corner of sea walls when the tide is going out on a spring tide. If there is no current and walking is not an option consider letting the net sink as far into the depths as possible before pulling it vertically up to collect the sample. Repeat a number of times. Obviously this needs practice and deep water, but I know of several people for whom this is the only viable method. Essentially, by whatever means, simply filter as much seawater as possible.

Collections are very straightforward if the sea is calm, but this is quite a rarity. There is a chance that swell and waves below the pontoon or jetty will pull at the net. As you walk you may feel a slackening of the cord, meaning a wave is carrying the net. At this point you need to speed up to stop the bottle inverting. Alternatively, you may feel a wave pulling at the net, in which case you need to slow down. If the water is forcing against the mouth of the net a 'bow wave' of pressure builds and prevents water entering the net, sweeping down the outside instead. It all depends on the state of the water and often the time of year. There will always be some plankton even in mid-winter, but at these times combating wave conditions can be difficult.

At all times be aware of your safety. Being close to the water's edge presents a clearly visible risk, but less predictable occurrences such as freak waves and their reflections off nearby stone walls may also create dangers. If the conditions are rough pontoons will ride the waves, snaking up and down, and be very unstable. This is why in winter pontoons are usually removed but, even if they are present, I try to find stable jetties instead. These are often significantly shorter and therefore successful sampling requires many more turns of the net, but with practise the collection of good samples are still possible.

A final point about choosing good locations is to check ahead of sampling that they are available for public use, as some jetties and pontoons are private. The latter tend to be locked with a gate but look around and there may be contact details. Marina pontoons are private but prior contact will often allow access.

One last point about your plankton sample. Mostly it will appear to be dirty water, but have a quick look to see if there are any larger organisms within. Large predatory animals such as comb jellies (ctenophore) or polychaetes (bristle worms) may, by the time you have sat down to observe your sample at home, have consumed most of the life in the sample. If you spot something of this sort it is prudent to transfer it to a separate container as soon as possible.

Dead or alive?

Keep your samples cool. Fortunately, I live about 20 minutes from my primary sampling points, and I place the samples in a domestic fridge as soon as I arrive home. When I am ready to look at the samples they can be removed, one at a time, from the fridge and taken to the microscope area. At room temperature plankton samples will deteriorate; single-celled animals such as tintinnids die very quickly. After 24 hours, even if kept in a fridge, little will still be alive other than copepods and a few diatoms lying at the bottom. To keep the sample as healthy as possible, particularly in summer, use a cool bag with freezer blocks near (but not too close) to the plankton samples during transit. Thermos flasks were commonly used to transport samples and they can still be useful if cold blocks are not available.

If you must keep samples for any length of time – if you have a long travel time, for example – the only real option is to add a preservative. For long-term storage formalin is the best method, but it does have health implications and must be used carefully. A good alternative is 3% Lugol's Iodine Solution, which is sold as a medication, usually in a bottle with a dropper, and is easily purchased online. The addition of a few drops not only works as a fixative but also helps improve detail in dead diatoms, although it does cause some cell distortion. Formalin fixes the specimens but quickly causes colour to be lost and some finer detail to disappear, making identification difficult, especially with protists. If fixatives are needed, then the use of both chemicals, in separate samples, may be the best solution.

Buying all the gear for plankton trawling can be expensive, especially if you suspect that this is something that you will not do very often. If, on the other hand, you just want a brief look to see what larger plankton species might be around in the water, go down to a rocky shore at low tide and find a moderately small rock pool. With a simple pipette draw some water from the edge of the pool and among the weed, placing the sample into a jar with a suitable lid. In addition, take a small amount of weed fringing the pool, such as the coral weed *Corallina* spp. Pick this carefully, so that anything contained within does not get shaken out into the pool. Keep the sample in a jar with a small amount of seawater to cover it. It is also worth spending some time looking closely at the water found in rock pools high up on the shore, watching out for small darting specks of light. With a pipette you may be able to catch some of these inhabitants, or try a small tea strainer with a mesh as fine as you can get. Samples from either method can be placed in a small tube of sea water.

When you return home let these jars stand for five minutes to allow some of the debris and plankton to settle at the bottom. Samples of fringe weed should be tipped into a dish and allowed to settle. Although this will not produce high volumes of plankton it may be sufficient to get you started. Magnifying the surface of the weed should show attached benthic diatoms and an amazing variety of small animals such as sea spiders (see Figure 13.4) and bryozoans. Rock-pool life is very rewarding to study but a different subject from that considered here. More information, therefore, can be found in the book *Rock pools* by Peter Haywood, also published by Pelagic, as well as in other seashore books listed in the bibliography.

FIGURE 4.1 A photograph of a plankton sample, magnified approximately ×20, showing an array of diatoms and debris. This image was taken using an inexpensive USB microscope.

CHAPTER 4

How to observe marine plankton

> The contribution to visible life is very small indeed; in fact, more than 99.9% of life on earth is invisible.
>
> Sean Nee, 'More than meets the eye', *Nature* 429 (2004)

Your collected plankton samples are in containers stored in the fridge ready for viewing. Before removing them from their cool place, the first thing to do is to make sure everything else is ready: microscopes, lights and clean glassware. In addition, various sizes of pipettes will be needed and almost certainly there will be some spillages, so keep some old tea towels at hand, which will also be useful when cleaning the glassware. A small amount of absorbent toilet paper could be helpful when putting drops of sample on to microscope slides and removing excess water.

Notice the plural above: one microscope will not cover all eventualities. A 53-micron mesh net may pick up specimens from several centimetres in length down to less than 20 microns. The latter will require significant magnification of ×400 or even ×1,000. In contrast, using those magnifications to observe a crab larva will yield only close-ups of eyes and hairs. A view of the entire larva will require ×10 to ×40 magnification, depending on the species.

Microscopy is a huge topic and this section will barely scratch the surface, but hopefully it will give sufficient pointers that will help to get you started in your exploration of the world of marine plankton (and see Resources, at the end of the book). We will also look here at the basics of taking photographs of magnified material.

USB and HDMI microscopes

The latest digital microscopes are both inexpensive and able to take photographs. They have the ability to stream images to a screen either via USB on a computer or through HDMI to a monitor or television. That is a significant advantage when working with children, people with eyesight issues, those needing spectacles or even simply when spending a lot of time looking at plankton. I use screens for all my microscopes, saving fatigue and potential discomfort that can result from extended periods peering down eyepieces. Software is needed to put an image on the computer screen and allow you to capture a photo or video. You may need to download such software from the manufacturer, although some operating systems, such as Windows, provide a small basic app called Camera as part of the package.

Digital microscopes are often surrounded by advertising hype suggesting that they can achieve very high magnification, but this should be largely ignored. To produce these high values requires the use of digital zooming: that is, an optically produced image is enlarged with software. Usually the result is exceptionally poor and therefore it is best to avoid any form of digital magnification. This is particularly so with those even less expensive handheld digital microscopes that profess to magnify by thousands of times. Above ×10 to ×20 magnification it will be almost impossible to hold the microscope sufficiently still to get an image. And yet many USB microscopes have adverts showing astonishing photographs on the screen. Beware that it is all too easy to digitally add photos to these and may not in fact have been taken with the device in question. One particular handheld version available online shows an incredible photo on the screen that was clearly taken

FIGURE 4.2 A close-up of the controls on an HDMI inspection camera cube.

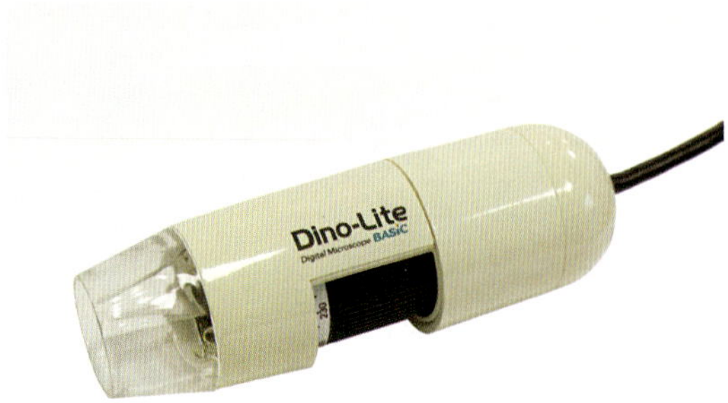

FIGURE 4.3 An example of a quality USB camera for plugging into a computer port to view and take photographs. This can be used by hand or attached to a stand. (NHBS)

by a scanning electron microscope! An online search for USB microscopes will bring up a huge variety of names and possibilities, so some due diligence is required to sort the wheat from the chaff and find something that suits your requirements, remembering that quality magnification rarely goes above ×20 to ×30.

HDMI microscope set-up

Figure 4.4 shows an HDMI microscope set-up. The main sample is in the dish at bottom left. The microscope consists of a black zoom lens with a grey camera cube and controller at the top. From there a HDMI feed (the thick black cable; the thin cable is power) goes to a monitor set behind, where the livestream can be seen. This screen could be a television or a computer monitor of any size. The lens is attached to a rack that moves vertically to focus on the subject, while the lens itself can be rotated to optically zoom, changing the magnification.

To the side of the stand are two wooden blocks topped with a dense white foam material. These are homemade supports for the operator's wrists, enabling careful and controlled manipulation of the sample. Small amounts of the sample have been transferred by pipette to the small dishes (tea-light holders) on the homemade stage, which has a central hole through which light passes from beneath. The dish and specimens are therefore back-lit, producing a clearer image. To achieve the best results from an HDMI microscope, lighting is very important. Often, as noted, an LED ring will come with the microscope, which could be placed beneath the stage, but if not they are easily sourced separately. The lighting shown in Figure 4.4 is homemade, constructed from the lid of an old coffee container that is painted black inside, creating the dark background. Around the inside of the lid's rim a strip of LEDs has been stuck, and its two electric wires are soldered to a fine cable that passes out to an LED controller and dimmer switch at the back below the screen. Although this is not as bright as a purchased LED ring it creates a perfect light underneath the dish. The homemade illumination was originally designed and produced by Carel Sartory, from the Quekett

My preferred microscope of this type has a primary HDMI connection but does have a USB as well. The USB may sound perfect, as it captures images directly to the computer, but usually the resolution is quite low. HDMI streams directly to any size monitor, filling the screen. The HDMI microscope will be in two parts, a zoom lens and a cube-shaped camera above. The latter attaches by a screw fitting to the lens and on its upper surface are sockets for power and connection, plus a series of buttons to control the image and to take photographs. Video and stills will be recorded at full resolution on to a micro-SD card inserted into a slot at the top. In many cases a remote control will be supplied – a very useful addition that enables photograph to be taken without touching the microscope, which would cause vibration and consequent blurred images.

A variety of stands and bases are available. Parts can be purchased separately or all together. Often lighting is better added afterwards, although many come with an LED light ring arranged around the lens to help with illumination, which can be is useful. The camera cube can also be purchased separately and added to other optical devices, including compound microscopes. Although the camera sensor is small, the resolution can be anywhere between 12 to 40 megapixels – sufficient quality to be used for some of the photographs in this book.

Finding a suitable HDMI microscope online can be difficult, as they rarely have a brand name and most parts are generic. Consider using eBay as a starting point, searching with terms such as *HDMI USB type C industry microscope camera inspection*. All parts should be relatively inexpensive, although price largely depends upon the complexity of the camera resolution and stand.

Microscopical Club (see Resources, at the back of this book).

The ideal needles to sort through specimens are placed in the foreground of Figure 4.4: these are 3D printer needles (used for unclogging nozzles), which can be easily bought online (and see below). Once a subject has been isolated using the needles it can be transferred with a pipette to a microscope slide or another dish. The remote control, below the right wrist support, is used for taking photographs or video without having to touch the top controls on the camera. Top light is provided by two LEDs clipped to a bar at the back. These, along with tea-light holders and glass dishes, are available from IKEA or other online retailers. The tube of green liquid on the right is ordinary window cleaner for cleaning glassware.

FIGURE 4.4 A Hayear HDMI 4K inspection microscope set-up ready for sorting through plankton.

Optical microscopes

Optical microscopes produce a higher-quality image, can have better lighting and require cameras to be secondarily attached to be able to stream video or take photographs. There are two types of optical microscope: stereoscopic and compound.

Stereoscopic microscopes

Stereoscopic microscopes are used for low magnifications around ×10 to ×30, depending on the microscope, and so would be used in the first instance for looking at a sample in a dish, similarly to the HDMI microscope. It has a binocular head, with each eyepiece passing through special optics to a pair of objectives, creating a three-dimensional view. There will usually be one or two pairs of objectives, although modern examples may have a zoom for increasing magnification. As the stereoscopic microscope is usually used to begin the process of sorting through your sample the magnification may need to be changed, which, unless you have a more sophisticated model, can be a rather tedious process; in contrast, the HDMI microscope can easily be zoomed with a steady increase in magnification. Generally, for further detailed examination, from ×40 up to ×1,000, the compound microscope is preferred.

Compound microscopes

Unlike the stereoscopic microscope, a compound microscope is essentially one tube with an eyepiece at the top – which may, for the user's comfort, come in the form of a binocular head with two eyepieces – and a rotating turret at the bottom, offering a number of objective lenses with different magnifications. Typically, this microscope will offer magnifications of between ×40 and ×400, calculated by multiplying the magnification of the eyepiece, say ×10, with that of the lens, such as ×20. The combined magnification in this case would be ×200.

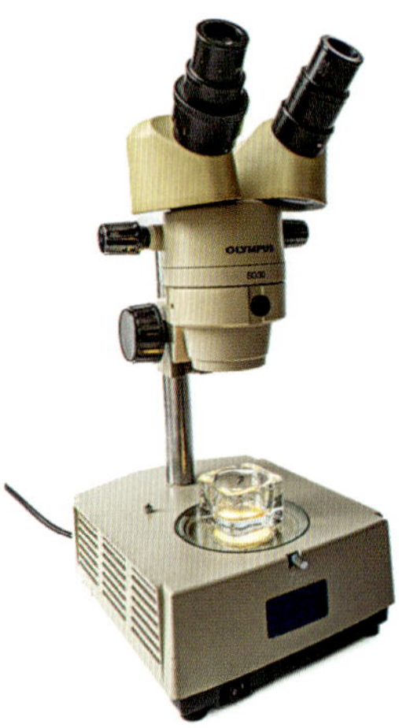

FIGURE 4.5 A stereoscopic microscope, 1990s. The small knob switches between two magnifications.

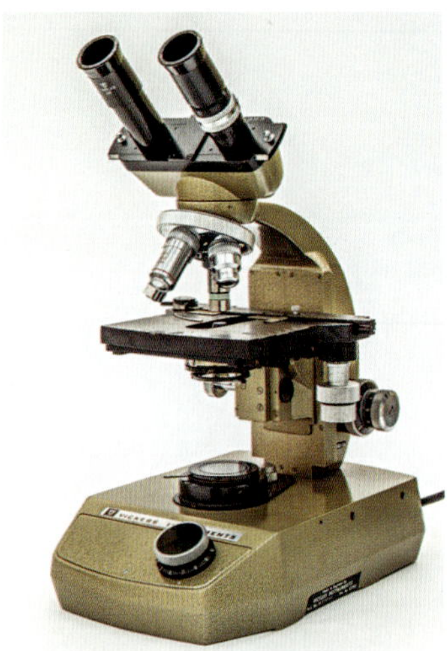

FIGURE 4.6 A Vickers compound microscope, *c.*1970.

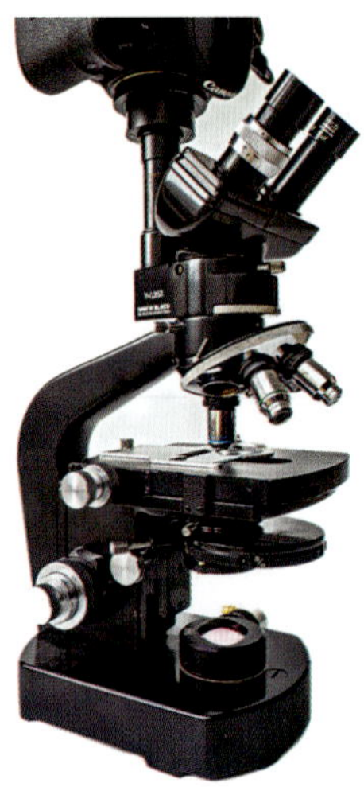

FIGURE 4.7 A Wild M20 trinocular compound microscope with phase contrast, 1960s. A camera is attached to the third tube by a universal adapter.

To use the compound microscope, you would drop your plankton on to a glass microscope slide placed on a stage above an aperture through which light can come from below. Beneath the stage the light may be directed upwards, either by a mirror or using a light such as an LED fitted into the microscope base. Between the light and the stage may be a basic device for reducing or increasing the brightness – or, better still, what is called a sub-stage condenser.

The most basic type of compound microscope available, of a type suitable for students – probably a monocular with no sub-stage condenser – will be relatively inexpensive. More sophisticated research microscopes are generally binocular, with a greater range of objective lenses and a sub-stage condenser able to give better control of the light. A further addition is a mechanical stage, which holds the glass slide firmly with a spring-loaded clip and has two knobs that, when rotated, move the specimen in a controlled way in all directions across the stage. Another useful option in a compound microscope is the addition of a third tube, located behind the two eyepieces. Instead of an eyepiece, this is designed to attach a camera and is called a trinocular.

Good illumination is critical for a clear image and different forms of lighting exist, some of which are beyond the scope of this volume. The basic form of lighting described so far, where a light source is directly under the stage, is called bright field illumination, and provides illumination evenly across the entire field, although it is possible to over-illuminate the specimen, which obliterates the detail. Student microscopes may have a rotating disc with holes of different sizes or a slider/dial to vary the brightness.

Often, however, for very translucent specimens, a dark field is better. Here a stop filter is placed within the sub-stage condenser, restricting the amount of light in the centre or stopping it completely. This creates a cylinder of light with a dark centre. When this rim of light hits the specimen it illuminates the detail more from the side, against a black central background. Many of the organisms in fresh plankton are translucent and so, especially with increasing magnification, the specimen loses detail as contrast diminishes. Dark field is helpful in this situation.

Some compound microscopes offer a specialist technique known as 'phase contrast' that was developed in the 1960s to further assist in overcoming the problem of detail loss. This requires special objectives which intensify the contrast to produce a clearer monochrome image. Several decades later this was improved further by utilising polarised light and prisms to create not only enhanced contrast but also intensified colour. This technique, called differential interference contrast (DIC), is extremely effective and many of the photographs taken for this book used this type of lighting.

If you are new to microscopy perhaps your first thought would be to buy a new microscope. Talking with a supplier may help you understand what you need, but it is worth joining a microscope group such as the Quekett Microscopical Club (details in Resources). With regular meetings both face to face and online, they can provide extensive help and support. In addition, clubs provide access to quality second-hand microscopes. Personally, I have never purchased a new microscope. The second-hand market provides opportunities to have lighting such as phase contrast and even DIC at a significantly reduced cost compared with buying new. Figure 4.7 shows a 1960s Wild M20 microscope which not only has a third tube to attach a camera but also phase contrast. It is relatively common to find similar examples of high-quality optics as well as any accessories that may be needed later on, such as alternative objectives and sub-stage condensers. Research thoroughly and look out for second-hand Zeiss, Reichert, Nikon, Olympus, Leica, Leitz, Wild and Vickers equipment. Ensure that the optics are clean (fungus can grow well inside optics) and free of scratches and that all knobs move smoothly.

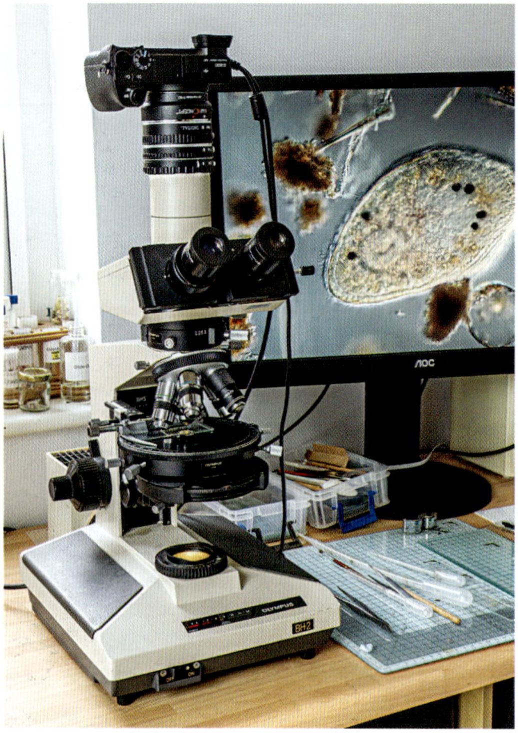

FIGURE 4.8 Olympus BH2 research microscope fitted with DIC lighting, *c.*1985. The camera is attached to the trinocular tube and an HDMI cable streams a live view to a large monitor at the side. The black rings below the camera are inexpensive extension rings to vary the magnification and help fit the specimen to the field of view. Photos are taken with a remote control to avoid vibration. On the screen is a flatworm.

When using a compound microscope start with a low magnification, such as a ×10 objective lens. When starting out in microscopy it is worth practising, if possible, with a prepared slide where the specimen is preserved and permanently mounted. Place this on the stage with the illumination coming through the centre, on to your specimen. There are two sets of focusing knobs on the side of the microscope. At first, use the larger coarse-focus knob to move the specimen close to the end of the objective, watching carefully from the side so that you do not touch the slide. Depending on the microscope, focusing can move either the stage or the lens. Looking through the eyepiece at the specimen, which is likely to be completely out of focus, slowly use the coarse-focus knob again to separate the objective away from the subject. When the specimen comes into focus switch to using the fine adjustment, which can be tweaked to produce a sharp image. If you have multiple objective lenses on a turret be careful when rotating a new lens into place, as it may hit the slide. You may need to move the specimen away from the new objective.

Inverted microscopes

One other microscope type that should be mentioned here is the inverted microscope, a variation on the compound form. These are set up so that both the light-controlling condenser and the light source are above the stage where the specimen is located. The objective lenses are beneath. Inverted microscopes are ideal for looking at organisms in water, such as plankton, as it is possible to place thin glass dishes such as Petri dishes directly on to the stage. Because the lens is beneath the stage, looking up through the bottom of the dish, there is much more space to move the specimen around.

Sorting and viewing plankton samples

Plankton enthusiasts, professional and amateur alike, all have their own method of dealing with plankton which will partly be governed by what they want to get out of their sample. For example, I am interested in viewing and studying the diversity of plankton life. Some planktonologists recommend that the sample should be diluted down if you are doing systematic and quantitative sampling, but this probably will not be necessary if you are simply looking for interesting organisms.

A good selection of different-sized glass dishes will be needed and although petri dishes are ideal most people do not have these in their home. Traditionally biologists have used watch glasses to place small amounts of material under a stereoscopic microscope. While it is still possible to buy these, small glass dessert dishes, ideally with flat bottoms, or tea-light holders, as already mentioned, are inexpensive and will suffice. Most of my samples have a volume of around 200 ml or more, which is too large to view. Therefore, they need to be divided up into smaller units. To do this, after leaving the main sample for a minute or so to settle, I decant the top half (primarily seawater) into one dish and the remainder (most of the visible plankton that has sunk to the bottom), into a second. Then I begin to separate material out by pipetting small amounts into a series of smaller dishes to look at under low magnification. Choose pipettes that are as fine tipped as possible: 5 ml and 10 ml sizes are particularly useful, and these can be easily obtained online. The zooplankton in spring is often so dense with crustacean larvae that it may be useful to reduce the frenzy of their movement early on by placing a small light close to the side of the main sample dish, as immediately they will flock to it. They can easily be transferred then to another dish with a pipette.

At this stage I use an HDMI inspection microscope to look at the sample at around ×2 to ×4 magnification, observing the organisms, selecting individuals or groups with a pipette and adding these to another small dish. I try to remove unwanted material from around the main subject by adding small amounts of seawater before placing it under the microscope. Return any material you

FIGURE 4.9 A plankton sample under low magnification, showing the tip of a 0.15 mm diameter 3D printer nozzle-cleaning needle being used to clear debris from around a periwinkle egg.

do not intend to examine immediately to the fridge. If samples are going under the compound microscope I add several drops to a microscope slide before putting a cover slip carefully on top to prevent the water evaporating. Under a stereoscopic microscope cover slips are not usually required. To avoid trapping air bubbles between slide and cover slip, place one edge of the cover slip on the slide first before slowly lowering it down across the sample. Toilet paper can then be used to remove surplus water at the edges of the cover slip. It is very important that all glassware, especially the slide and cover slip, is as clean as possible. Before starting, clean a few of each by adding a drop of window-cleaning fluid and rubbing it around both sides before carefully wiping dry with a clean tea towel – old, thin, well-used tea towels are best for this job. Any cheap window cleaner bought from a supermarket works well and I keep a small amount in a dropper bottle, while boxes of microscope slides and coverslips are best purchased online.

When sorting through material, either in a dish or on a slide, it is often necessary to clear debris away from the main subject or to separate aggregations of detritus. Typical dissecting kit needles, especially mounted ones, are too large for the job. In the nineteenth century microscopists used mounted human hair or eyelashes, but there are good alternatives easily obtainable today. Acupuncture needles are relatively inexpensive and available in several thicknesses. The downside is that they are sold in large batches and are a sterile product, all of which adds to the cost. With the development of 3D printers, a new needle has become available. These are made to unclog printer nozzles but are also ideal for microscopy, and are my preferred option (see Figure 4.4). A basic 3D printer nozzle-cleaning kit contains a number of plastic tubes each containing ten needles of a particular thickness, ranging from 0.15 mm to 0.4 mm. They are similar to acupuncture needles but are not sterile and come in smaller packs, so obtaining a sensible number of needles in a range of sizes is much more affordable.

To help in the work of isolating and separating material, on either side of my microscopes I have variously sized blocks of wood with a strong foam glued to the top surface. These allow me to rest my wrists, allowing more precise control when using needles to separate debris and specimens when magnified (see Figure 4.4).

A trip on the wild slide

I tend to use basic glass microscope slides and only occasionally a cavity slide, which has a small dip in the glass. The space afforded by the cavity may mean that small, active specimens move too much, and these slides may be useful only for organisms of around millimetre or so in size. The combination of basic slide and cover slip is more likely to hold specimens in place and prevent movement. However, if it seems likely that organisms may be crushed, rather than just held in place, add a tiny amount of Vaseline to the four corners of the cover slip before lowering it on to the slide. This provides support for the cover slip, which, if necessary, may be gradually pushed down a little to trap the material. Used carefully, this technique is good at controlling active material while preventing damage, although glassware must be cleaned thoroughly afterwards. A homemade alternative to the cavity slide involves attaching simple round washers of variable thickness to basic slides to create a 'well' into which samples can be placed. Clean the washers well and superglue them to the glass, leaving for a day or two to harden.

Several other techniques can be employed to slow down the movement of organisms on the slide when viewing under a compound microscope. One is the use of Epsom salt. Make up a concentrated solution in seawater and place a drop on the slide near the edge of the cover slip so the solution just runs on to it. On the other side of the cover slip gently place a small piece of toilet paper to absorb some of the water; as it does so, the Epsom salt solution will be drawn across underneath the cover slip. This slowly narcotises the organism, although the time it takes depends on the solution concentration and the specimen. Too concentrated a solution left for too long will eventually kill the animal. To prevent this, place a drop of seawater on the cover slip edge and draw that across with the absorbent paper to replace the salt solution.

As well as searching under a stereoscopic or HDMI microscope for specific organisms, pipette a few drops of the material at the bottom of the sample dish and place them on a slide with a cover slip. Under the compound microscope begin viewing under a ×10 objective and, starting in one corner, slowly move the slide systematically up and down, working your way across the cover slip until you have looked at the complete sample. This is the best way to really analyse what your plankton sample contains. By working methodically, you are more likely to find unusual things that you can then view under higher magnification by changing the objective lens. There will be plenty of things that you cannot identify, which is why making a photographic record of the session is so useful: it will allow you to come back at a later stage to review the subjects. The organisms themselves have a limited survival time and photography removes some of the pressure to make an immediate and identification, unless you decide to fix the specimen with a preservative.

A photographic record

HDMI microscopes have a camera built in, which makes recording video and taking stills straightforward. When using a compound microscope, a smartphone can be a good way of recording images because the lens on the microscope eyepiece is quite small and requires a similarly small camera lens to fit over it. Because the eyepiece is round but the camera sensor is square you will need to zoom and pinch the phone's screen to avoid as much of the dark areas around the image as possible. If the exposure is set to automatic the image may be overexposed (too bright), although sometimes the camera can be encouraged to adjust the exposure by tapping the screen over the part of the image that you wish to see. Modern smartphones have very good control of exposure and normally have settings relating to this factor. Video clips are can also be made and, particularly with moving material, these can be useful for identification purposes. If you intend to do much photography it is worth finding an adapter to connect the smartphone to the microscope. A small compact camera (as long as the lens is small) can also be used in this way.

DSLR or mirrorless camera bodies, with the lens removed, can also be attached to microscopes, usually via an adapter that replaces the eyepiece. The image in the camera's rear screen can guide you in focusing with the microscope's controls. Similarly, some of the cube cameras fitted to HDMI microscopes have the same thread as that found on some trinocular microscopes, meaning that they can be used across various microscopes for photography. In fact, these cameras can be purchased separately specifically for this purpose.

Stereo microscopes are generally not suited for photography, even though it is possible to buy trinocular stereoscopics. At these low magnifications the HDMI microscope really comes into its own, especially if you are using a 4K version with high-resolution images.

Improving depth of field and the field of view

When taking a photograph the camera focuses on a particular point and, moving away from that point, sharpness and detail gradually reduce with distance. This is referred to as depth of field: the depth, from front to back, within the picture that is in sharp focus. If you take a close-up photograph of a butterfly you see that not only does the background appear very soft but even

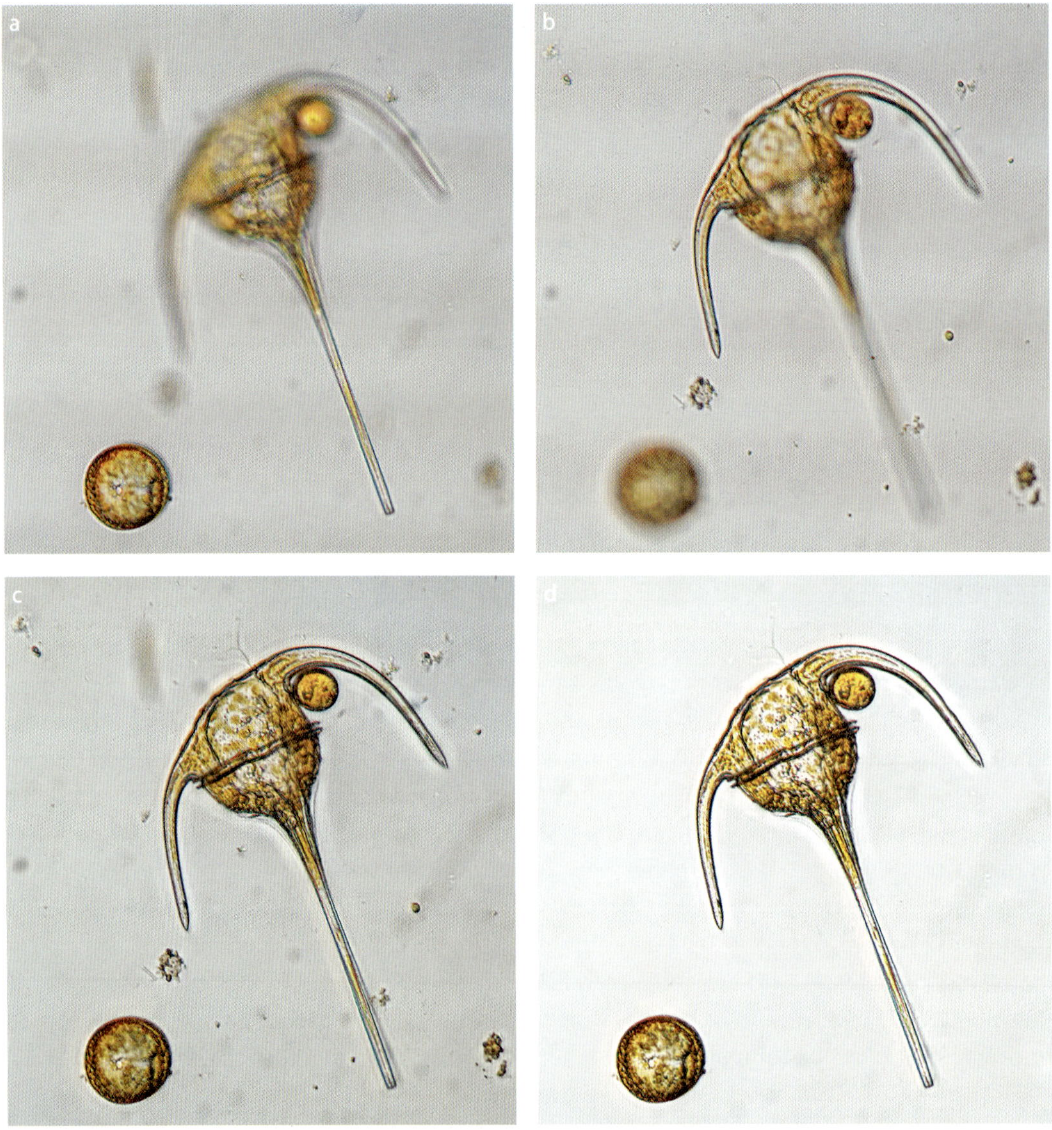

FIGURE 4.10 Focus stacking photos of the dinoflagellate *Tripos* taken at ×200. A series of 14 images was taken where (a) is the first and (b) is the last. Between each photograph the focus was moved very slightly downwards through the specimen. Finally, all 14 photographs were combined to produce a composite image (c). (a) and (b) have just small parts in focus, whereas the last one is sharp throughout: it has good depth of field. As well as photographing every aspect of the *Tripos* the stack has picked up all the detritus around. The fourth photo (d) is the final version of the composite after the debris has been removed from the photo using software cloning. The image has been brightened a little. Note that the flagellum at the top seems to split in two. The problem with stacking a live animal is that the flagellum was moving and has appeared on several of the image stacks. If necessary, cloning can be used to remove these extra details.

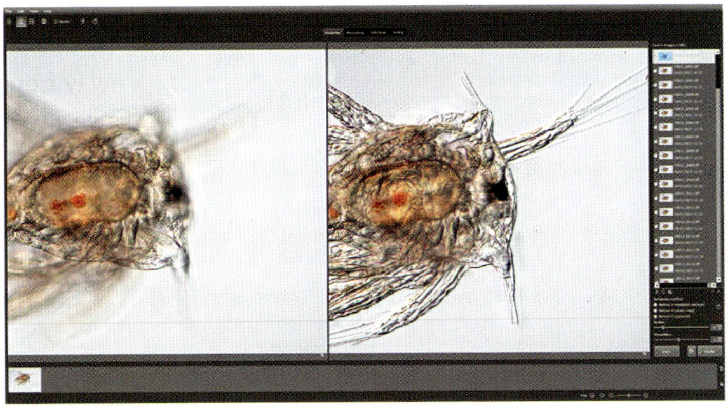

FIGURE 4.11 A screenshot of Helicon Focus software stacking a four-second video clip of a living nauplius barnacle larva, ×200 objective lens. The nauplius was very active and needed a quick stack before it started to move again. The extracted frames (TIFF files) are shown on the right and the split screen shows the first frame on the left and the complete composite on the right.

FIGURE 4.12 A single frame saved from a video clip of a moving dinoflagellate *Dinophysis tripos*. Length 85 μm.

parts of the insect lose sharpness, all due to a limited depth of field. The most important point is that the greater the magnification the weaker the focus becomes, losing definition and detail.

To improve the depth of field a process called focus stacking can be employed. This requires taking a number of photographs at different points of focus through the subject and then combining them into a single composite image on the computer. To do this, begin by focusing just above the specimen, so that barely anything is sharp, and take your first photograph. Then move the fine focus adjustment very slightly so that a fresh part of the subject comes into focus. Take another photograph. Repeat this process until the subject goes completely out of focus. The number of images, your stack, may vary from several to perhaps 20 or more. Suitable software on a computer is then needed to stack them. Several dedicated programmes, such as Helicon Focus or Zerene Stacker, are very good. Photo editors such as Photoshop and Affinity will allow you to create a good composite, but have their limitations. Essentially, these programmes take all the sharp pixels and combine them into one image. Many of the photographs in this book have been created this way.

Video stacking is a similar process. I find it is ideal for microscopy, especially if you are short of time or the subject is likely to move. Wriggling larvae may stop for a few seconds, long enough to

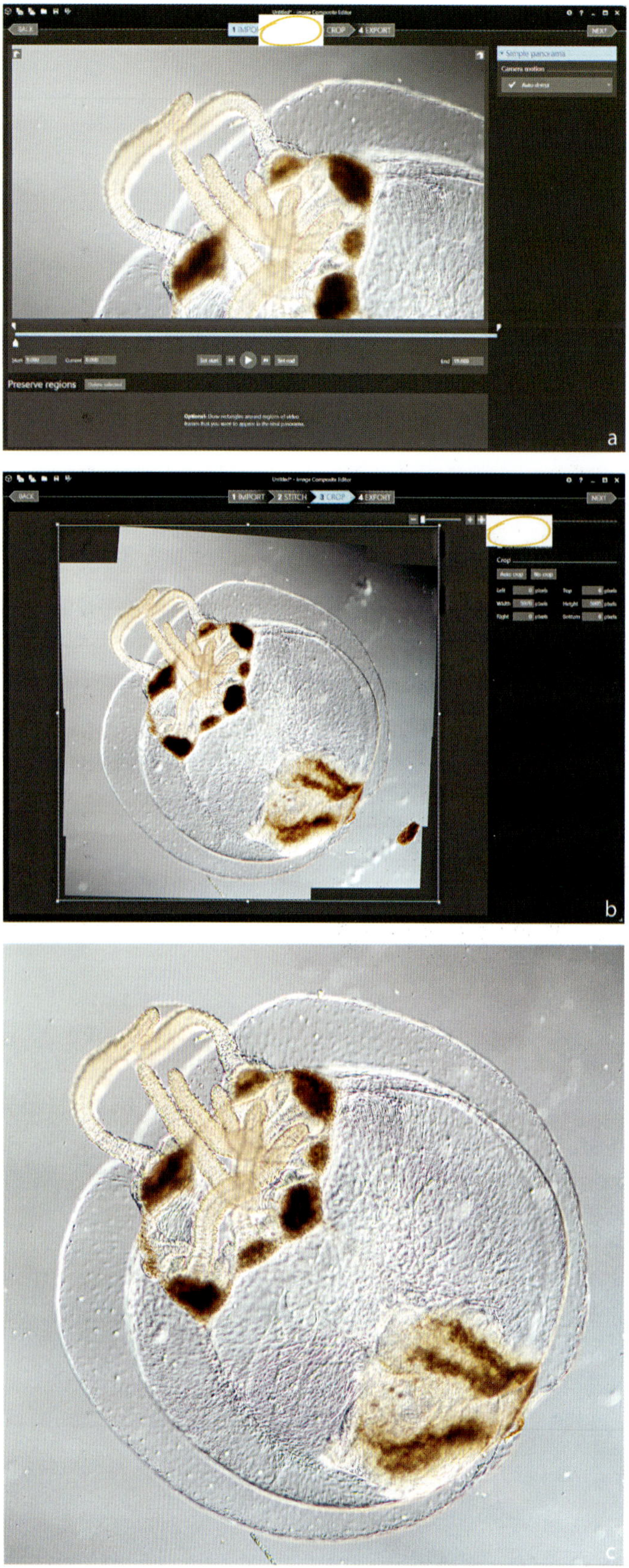

FIGURE 4.13 Video stitching a hydromedusa with Image Composite Editor. The video had zigzagged back and forth over the specimen. (a) The screen after the video is first dropped into the software. Note the limited field of view. (b) After clicking Stitch and going to the third screen, we see gaps where the video was missing. Click Auto Complete to fill these in. (c) The final image is 40 MP and has been cropped and cleaned.

record a short video. Have the camera ready to start recording and watch the animal's behaviour so that the second it stops moving you can start recording. 4K video or higher resolution is best. As with still shooting, begin with the specimen just out of focus and then rotate the fine focus, passing through the specimen until it is out of focus again. Aim for a clip 6–8 seconds long at a frame rate of around 30 frames per second, giving a total of approximately 200 frames. The larger number of frames when compared with a stack of stills will, in theory, provide much more detail. However, each frame is only 8 megapixels, where many cameras will produce frames of 20 megapixels or more. These frames can then be stacked in Helicon Focus software.

Focus stacking with video can have drawbacks if a lot of pieces of debris or moving bacteria are present. With 200 frames in a stack and a tiny protist or bacterium whizzing about, the chances are that there could be as many as 200 protists or bacteria in the eventual composite. However, it is relatively straightforward to clean the picture of extraneous material using clone tools and other removal tools in photo-editing software. Another useful feature of Helicon Focus – or for that matter any programme that can extract frames from a video clip – is that, because each frame is in essence a still photograph, reasonable images of moving organisms that are difficult to photograph can often be obtained by recording a short video and saving the best frame as a still image.

One drawback of compound microscopes is that they do not have a zoom between magnifications and if you cannot fit the entire structure you are interested in within the field of view you may have to take a number of photos, moving the subject for each one, and then stitch them together. Good software to perform the stitching is therefore invaluable and programs do vary. You might think that just changing the objective to a less powerful one will work, but to go from ×10 to ×4 is quite a jump and if the ×10 is a phase contrast or DIC lens the ×4 will not be. The process is just like making a panorama of a landscape where you turn the camera slightly for each shot, but here you move the subject instead of the camera. There needs to be plenty of overlap, ideally of about a third, between each one. Most photo-editing software has a stitching option, although there is one dedicated program that is particularly good and free to download: Image Composite Editor (ICE) by Microsoft. It is one of the most forgiving pieces of software available (see Figure 4.13), with both very good stitching facilities for stills and the capacity to handle video panoramas.

To make a video-stitched image, first look around the organism, checking where the edges are located and what needs to be covered. Then position the field of view at one corner of the specimen, set the video running and slowly move the specimen until the relevant area has been covered. Drop this clip, which may be as large as 250 megabytes, into ICE, and it should stitch all the frames to create the entire image. If there are gaps missing along the edge the programme has an option to auto complete, filling in the 'holes'. Adding so much data together can create high resolutions of more than 100 MP, which may need compressing.

Hopefully this chapter has helped you to get the most out of looking at plankton. At the end of the book there are details of resources and organisations that can offer further help and support.

FIGURE 5.1 Plankton sample under low magnification, showing the initial chaos before sorting. The larger zooplankton, such as the porcelain crab larva crossing diagonally over an *Obelia* hydromedusa and, to the left, a bryozoan larva, all stand out. Of the phytoplankton, only the bigger diatoms are obvious and among the debris are several protist species. A stacked image taken in April.

CHAPTER 5

Organising chaos

The first time you look at a sample it can seem rather chaotic, as there can be so much material to sort through. This is one of the reasons why you might need to dilute the collection down with clean seawater. Begin by removing the large zooplankton and placing them in separate small dishes with additional seawater. At certain times of year there may be a blizzard of crustacean nauplii larvae, both copepods and barnacles, that jerkily zigzag randomly around the dish. These can be quite disruptive, getting in the way and knocking other material that you are trying to study. Placing a small light at the side of the dish should soon draw them, along with other large zooplankton, away from the rest of the sample. Hundreds of creatures may congregate near the light, so do occasionally check, as crab larva and polychaetes (marine worms) will be there too.

What to expect

As you focus on what has settled at the bottom of your sample you will notice varying amounts of organic debris. The actual quantity changes from sample to sample, month to month, and may well be down to the weather conditions. Strong wind in previous days will create turbid waters that may lift a great deal of material from the benthos. Heavy rain inland may have caused flooding that will flush terrestrial and freshwater organisms and debris down rivers to spill out along the coast. A tremendous amount of this organic material is coalesced particulate matter that is difficult to identify, but other clearly defined organic structures can be made out and you will soon learn to identify these. Pollen is commonly seen. Airborne pollen from grass and trees will be caught in the surface tension of the sea water. Butterfly and moth scales occur as well. Bristles, mainly from polychaetes, are common, with some quite large and distinct chaetae, like those of Sabellid worms. When copepods are thriving in the plankton much of the organic material will be faecal pellets, which can cover the bottom of the sample dish (see Figure 1.11). Arthropods regularly shed their external skeletons and these exuvia are a particular feature of all plankton samples, either intact or broken into small pieces, especially limbs with setae. The most

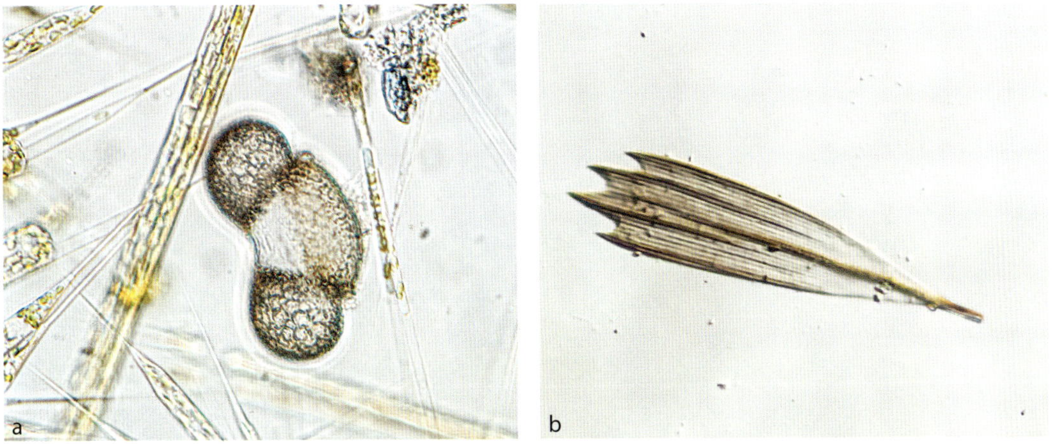

FIGURE 5.2 Organic material from the land. (a) pollen grain (b) lepidopteran scale.

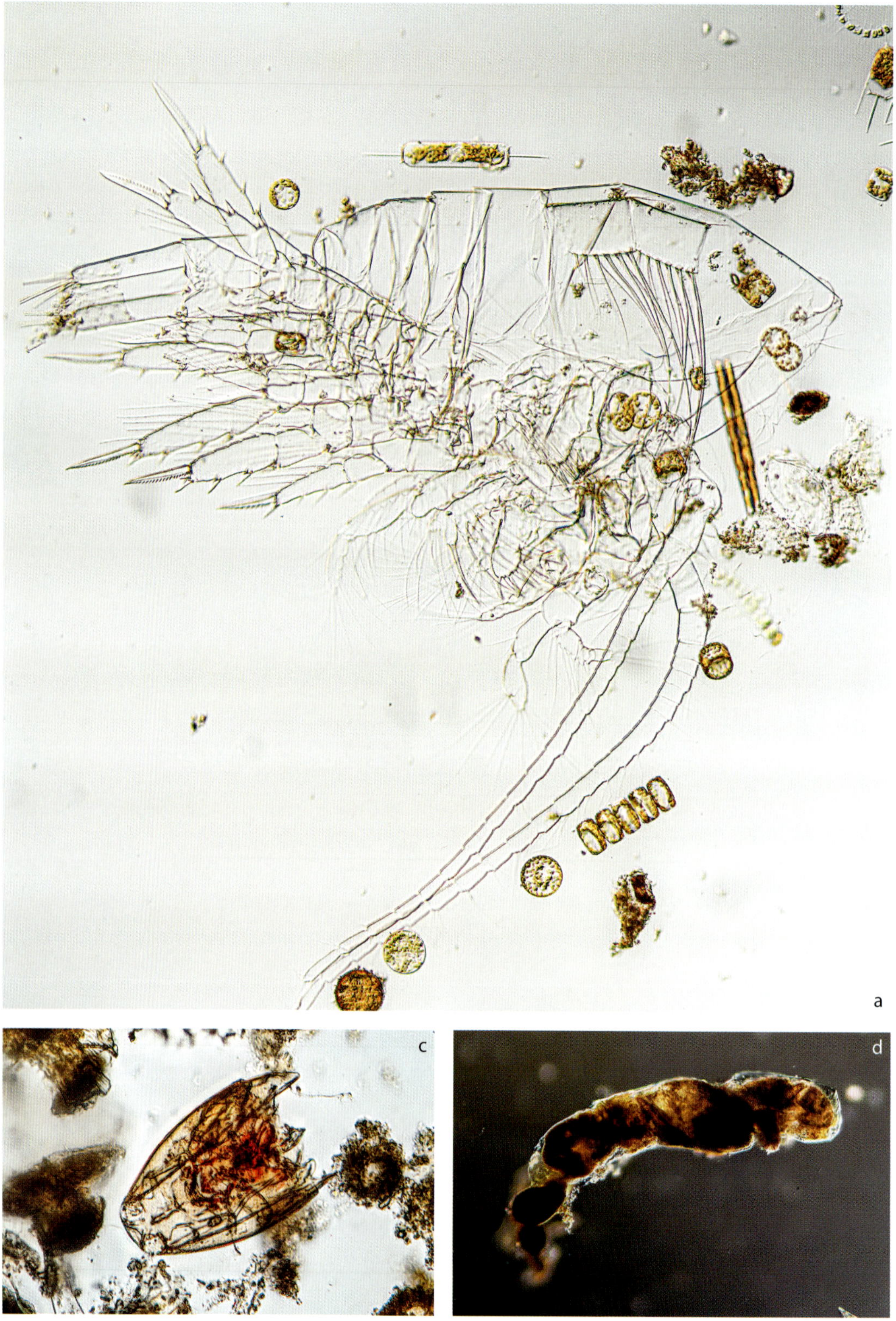

FIGURE 5.3 Organic material from marine arthropods. (a) complete exuvia of a copepod and (b) barnacle limb exuvia, both with diatoms and detritus caught up in the setae. (c) head capsule of a midge larva (d) large copepod faecal pellet, which could be dissected to examine contents.

b

obvious and large exuvia are the limbs of barnacles, half a centimetre or more across, like a big transparent net that catches all sorts of other debris and living organisms. Pieces of crustacean appendage as well as dead bodies can fill with moving nanoplankton feeding on any dead matter inside. On rocky shores there are a surprising number of midge larvae *Clunio marinus* living in rock pools and their exuvia can appear in the shore plankton. The toughest part is the head capsule, which resists decay and is the body element most likely to appear. Material that is resistant to decomposition can occur again and again in samples; for example, the cell wall of the freshwater desmid *Staurastrum* (see Figure 1.4) can often persist for many years after it has been washed into the sea.

As you sift through the debris the larger diatoms among the phytoplankton will stand out, such as *Coscinodiscus* and *Odontella sinensis*, being up to half a millimetre in size. Smaller diatoms will

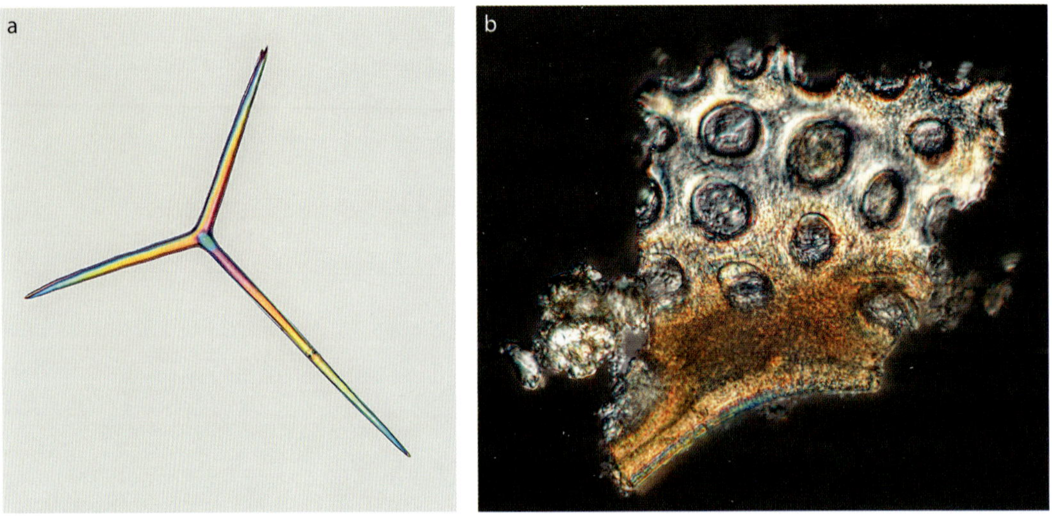

FIGURE 5.4 Organic skeletal material. (a) sponge spicule (b) probable ossicle from holothurian echinoderm.

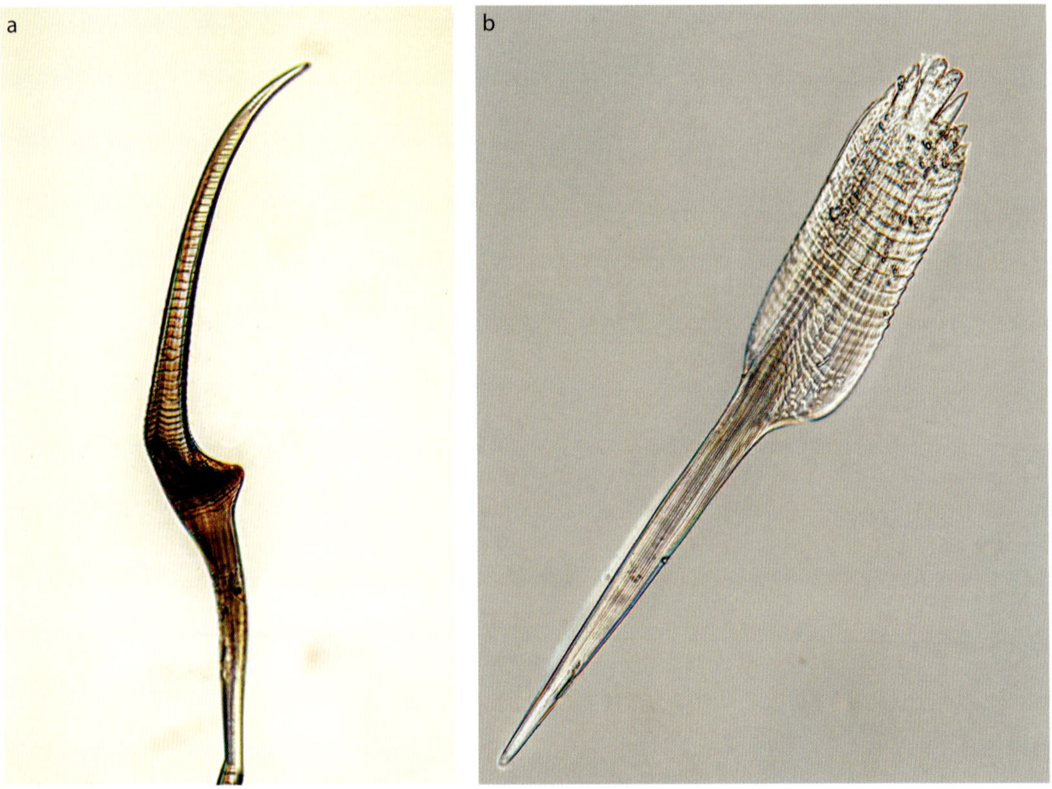

FIGURE 5.5 Specialised chaetae from the Sabellid polychaete *Sabellaria spinulosa*: (a) a bill-hook chaeta; (b) a damaged chaeta lost from the crown.

be difficult to make out and only by taking a few pipette drops of the material at the bottom and looking at them under high magnification will you be able to see these cells, which are 20–60 microns long. Ultimately, the only method for finding the smaller material is to take relatively random pipette samples from the bottom of the dish and view under the compound microscope.

FIGURE 5.6 One of many plates in McIntosh's *The British Marine Annelids* (1915). The red arrows indicate the chaetae shown in Figure 5.5. This ten-volume work is the best source for the identification of polychaete bristles.

Sorting out the organisms

Marine plankton covers probably the widest range of taxonomic groups of any ecosystem. This makes looking at a few drops under the compound microscope quite a daunting prospect if you are hoping to identify what you see. The sample is likely to contain bacteria, single-celled protists, a myriad of arthropods and some chordates. With such a diversity of species, initial identification will be an uphill slog. Keep a record of the size of organisms to hand, as this will help with identification. A problem particular to identification via microscopy is understanding the orientation of a specimen. Once the cover slip comes down over the sample, most organisms will be trapped in a fixed plane. Lifting the cover slip to reset it will inevitably damage or reorientate what you were looking at. Be aware, then, that you may be trying to identify something from a different viewpoint or plane from those shown in identification books. Diatoms, for example, can look quite different from either a girdle or valve view (Figure 7.5). Figure 14.9 shows a view of an organism that has appeared in my samples every April and always looks the same because the orientation has always been the same – from the top. It was some time before a suggestion led to a positive identification of the strange larva of a worm called *Polygordius*. A side view would have been easier. This is why photography is so useful: it gives you a record of your samples going back over previous months and years.

Under high magnification you will see small moving dots, some with beating flagella, that will be extremely difficult to identify. These microbes are likely to be bacteria and possibly fungi, as well as members of a taxon called choanoflagellates, which are picoplankton (discussed briefly in the next chapter). Nanoplankton and the slightly larger microplankton are almost exclusively single-celled organisms classified as protists, most of which are small and easily missed under the microscope unless they are active. So important and extensive are these that they are covered here in three separate chapters. The first group, diatoms, covers a very wide size range and even include some organisms in the mesoplankton. Dinoflagellates are a group of protists, most of which are also classified as phytoplankton. The remaining protists, the heterotrophs, which do not photosynthesise, have their own chapter, which covers ciliates and tintinnids, among other organisms. The subsequent eight chapters cover the increasing complexity of invertebrates until we reach the tunicates and similar organisms in Chapter 18. Each chapter follows a different group, covering some of their ecology, life histories and development.

Taxonomy and terminology

Species classifications are constantly being rethought and revised to meet new criteria or take account of new evidence. This is a problem for a book like this, as, shortly after publication, the taxonomy can become outdated. The chapters below use major and accepted group names. Crustacea was for years a class alongside insects, but now insects are considered to have evolved from the crustacea. Increasingly, crustacea is seen as a phylum or subphylum. At the species level there is a great deal of change and confusion, particularly with diatoms. Where there has been some recent changes I have given both names with the most accepted name first: for example, *Rhizosolenia* (*Guinardia*) *striata*. If in doubt regarding a marine species' current name and classification check the World Register of Marine Species (WoRMS). In discussions I have for many years used the genus name *Ceratium* for a large group of dinoflagellates. In 2013 a large proportion of the species within that genus was changed to *Tripos*, although it was some years before I realised this. Taxonomic chaos!

The topic of systematics looks at classification from the viewpoint of diversity and evolutionary relationships. Diatoms are the single-celled 'plants' of the sea and comprise 80% of the phytoplankton, but where do they fit in with a somewhat convenient group known as algae? Over my lifetime they have constantly moved and currently they reside with the brown seaweeds, like kelp. In 2014, Fabian Burki wrote a fascinating research paper on phylogenetics, the study of evolutionary relationships among various groups of organisms. This research field takes in evolutionary history, characteristic features of the organism and molecular sequencing data (DNA). A quick online search will produce an array of complex phylogenetic trees, what we might loosely call trees of life, showing the different ancestry of specific groups. Although complex, they help place organisms in some sort of order. It is especially helpful to show where much of the huge variety of marine plankton sits in relation to other forms of life on earth. When viewed and compared like this, the large animals that we are so familiar with become just a tiny part of a massively greater whole.

Invertebrates

Over the last few decades the excellent book *Invertebrates*, by Gary and Richard Brusca, has passed through four editions with input from additional authors. It is the one book that covers these metazoans very well plus a large section on protists, unicellular animals. Inside the front cover is a summary of the phyla and class structure that changes through each edition, along with the addition of hundreds of pages (currently standing at 1,104). The second edition, from 2003, is still very useful for the general biologist and used copies are very affordable.

For more titles and help with identification check the Resources and Bibliography at the end of this book.

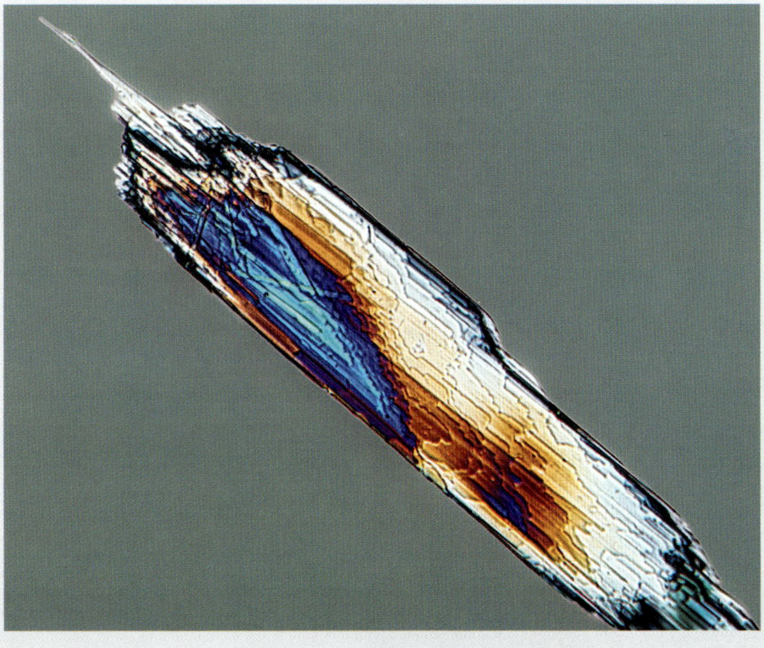

FIGURE 5.7 Inorganic particles can be varied and beautiful within the plankton, especially after water turbulence.

A nectosoma larva of a *Poecilochaetus* polychaete. This spionid larva can be quite common in the summer plankton, staying there for some months and growing more segments before settling. It is transparent except for the chromatophores at the base of the parapods. Length: 4.5 mm.

PART II

Organisms within the plankton

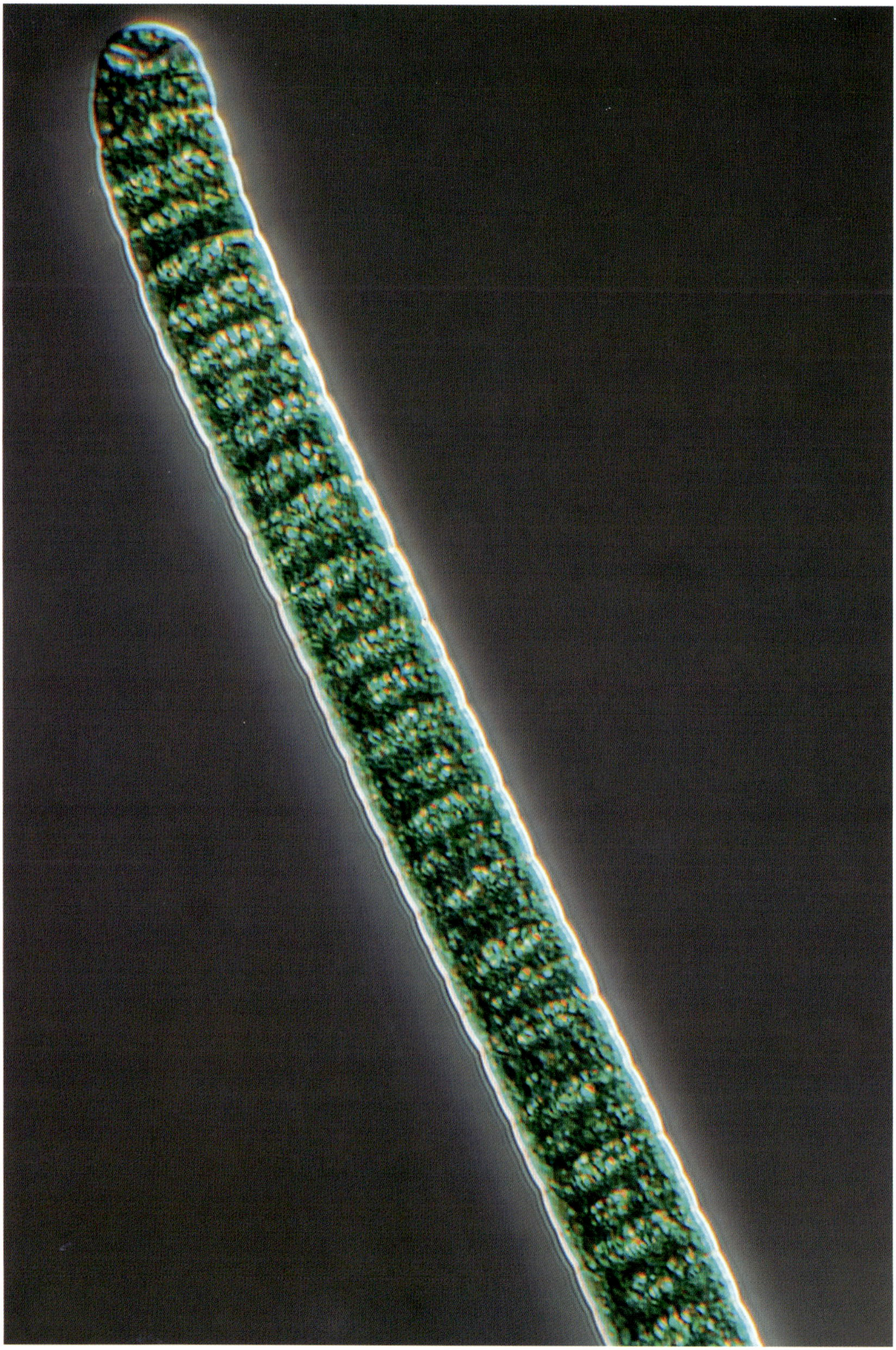

FIGURE 6.1 *Oscillatoria* sp., a filamentous cyanobacterium, common in plankton near estuarine shores. Width approximately 15 μm.

CHAPTER 6

The microbes: fungi, bacteria and viruses

Microbes are the most abundant and important group of organisms in the plankton and yet are difficult to see and study because of their minute size. They are grouped collectively within the femtoplankton and picoplankton.

Fungi

Most of our fungal knowledge has been derived from terrestrial sources. Even here, it is estimated that only 1–10% of all fungal species have been discovered. In the marine environment our knowledge and understanding of this kingdom is even poorer. Marine fungi have largely been ignored. Only since the 1970s have biologists really begun to research marine fungi. At the start of this century around 500 species of marine fungi had been identified. As of July 2024 that figure stood at 2,118 (figures from www.marinefungi.org) and it continues to rise.

Marine fungi are highly specialised and very small, making them difficult to study. Furthermore, they are tricky to define. Many so-called marine species live in terrestrial habitats as well. Trillions of spores are released from terrestrial fungi, a high proportion of which end up in the sea. Originally it was thought that marine samples containing fungal spores were just contaminated from terrestrial sources, but it has now been shown that they are 'amphibious', regularly transitioning between sea and land and back again. Fungi are recognised as amazingly resilient organisms. They can survive high levels of salt in the environment and have the ability to change their membrane composition to deal with high hydrostatic pressure. They are astonishingly good at adapting to their environment, creating variations for environments of every description. For instance, the red yeast *Rhodotorula mucilaginosa* is common in the air, soil and fresh water, with a strain now discovered in oceans that is currently being investigated for its anti-cancer properties.

Almost certainly fungi of various sorts are associated with every group in the plankton. No doubt fungi enter the plankton and are dispersed around the coast and across the oceans, although very little is understood about this process. They are known to be pathogenic on a variety of marine organisms, such as sponges and seaweeds found on rocky shores. *Aspergillus sydowii* is a well-known species on land but only in the last few decades has it been realised that it is abundant in plankton blooms in neritic waters, which can have a negative impact on phytoplankton. It is thought to bloom at sea and then, along with *Penicillium* and *Cladosporium*, infect brown seaweeds and sea fans. As phytoplankton bloom, so the fungi growing on them will also increase, although this lags about a month behind the phytoplankton. The highest concentrations of fungi are found near the surface during a bloom, and disappear as the phytoplankton decline. In 2009 *A. sydowii* created a surface bloom off the east coast of Australia more than twelve times the area of England.

The primary fungal group found along the coast and especially above seashores is the Chytridiomycota, or chytrids. They are also abundant in the sediment of depositing shores. Chytrids are the main group associated with phytoplankton, where they can be parasitic and symbiotic. The former Chytrid infections are lethal and are probably one of the biggest biotic controls on phytoplankton density, by substantially altering the cells' internal constituents and creating lipids in the process. This upgrades the nutritional quality, making the phytoplankton more edible for the zooplankton. In the same way, parasitic forms of fungi make cyanobacteria more susceptible to grazing. Chytrids also modify the amount and composition of dissolved organic carbon (DOC) that their hosts release, which, in turn, affects bacterial community structures. This is part of the microbial loop, in which DOC is moved through planktonic trophic levels and upwards, via micro- and mesoplankton, to ultimately be consumed by nekton such as fish.

Less known are the aphelids, a fungal group parasitic on algae. These are internal parasites on common phytoplankton species that destroy the cell from within, eventually killing the host. Because the host dies, aphelids are classified as being a parasitoid fungal infection rather than as a true parasite.

A strange group called the Cryptomycota continues to be the subject of debate. Mycologists do not currently agree about this group's status as true fungi, as they lack the characteristic chitinous cell walls. However, they are thought to reproduce, like chytrids, with flagellated motile spores that disperse to find a new host. They develop inside the host cell to become amoeboid and engulf the host's organelles, eventually inducing the host to create a cell wall to surround it.

Within marine ecosystems fungi play crucial roles in the marine carbon cycle, as intimated above, and as saprotrophs, parasites, pathogens, commensals and symbionts, as well as passively forming part of the diet of many zooplankton, which consume fungi when they ingest fungi-infected plankton; the gut contents of copepods, for example, often incorporates fungal material. They can also degrade hydrocarbons, such as oil and plastics, and considerable research is being done on the topic.

Bacteria

Bacterial biomass is estimated to represent between 50 and 70% of the living carbon in the neritic zone where light is present, and, as such, are the single most important group of oceanic heterotrophs. Some parasitise fish and there are plenty of invertebrate hosts as well.

By contrast the cyanobacteria are autotrophs, an important photosynthesising group, which include two especially common genera that occur worldwide: *Synechococcus* and *Prochlorococcus*.

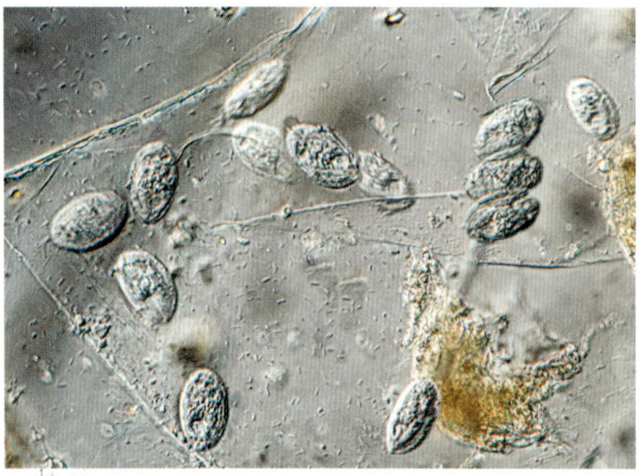

FIGURE 6.2 The ovoid objects are rapidly moving ciliates within a shed exoskeleton. They are feeding on bacteria (the elongated dots), which are present in large numbers. Ciliate size 20 µm.

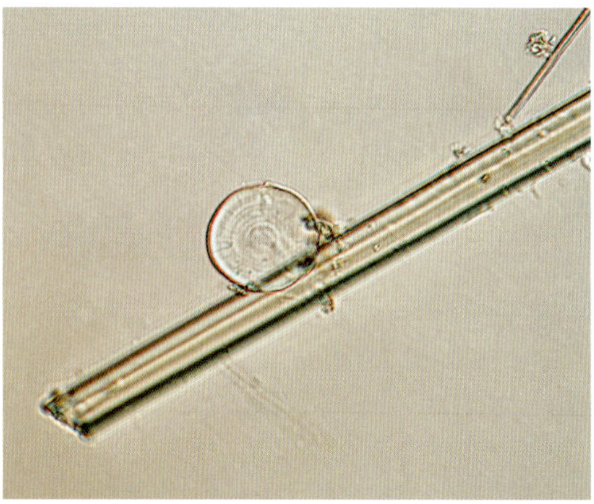

FIGURE 6.3 The round structure is a fungal Chytrid that attaches to diatoms. 40 µm diameter.

The former is one of the most abundant autotrophic members of the picoplankton in the marine ecosystem, contributing nearly 20% of oceanic primary production. It is around 1 micron in length, ovoid in shape, bright green and exists where there is plenty of light, which it harvests using the pigment chlorophyll A and the accessory pigment phycobiliprotein. It is common close to the shore and in estuaries, where eutrophication of the water with nitrate and phosphate from human land-based activities occurs, increasing the chances of a bloom. *Prochlorococcus* is similarly shaped but half the size and has an unusual form of chlorophyll. These cyanobacteria occur together, although *Prochlorococcus* is less abundant in temperate waters, while the reverse is true in the tropics. Their abundance does fluctuate for reasons that are not clearly understood, but grazing and temperature probably play a large part. Viral infections, too, are common, and a likely cause of population density decline.

Despite comprising a major part of the picoplankton, even at magnifications approaching ×1,000 these genera appear only as tiny green dots. In contrast, towards the end of the sampling day, even under a ×40 objective, a myriad of tiny dots can often be seen moving around on the slide. These are likely to be the most abundant group of bacteria, saprophytic forms, which multiply in warmer condition. These help to coalesce particulate organic matter together to form ragged balls of detritus; this microbial loop is discussed later.

Viruses

Viruses are found in every environment on earth, but it is thought that their importance is most evident in the oceans, where they are by far the most abundant life form and a major cause of mortality via infectious disease in crustaceans, whales and many more hosts. On a daily basis, they kill approximately 20% of the microbial biomass, creating a knock-on effect within nutrient and energy cycles. Their main influence, therefore, is upon the composition of marine communities and their associated biogeochemical cycles.

The most abundant species of virus in the ocean, although it is not particularly common along the coast, is *Nitrosopumilus maritimus*, which has a spindle shape and is 65 nanometres long. It infects Archaea-type bacteria, which oxidise ammonia and produce oxygen in the absence of light. With Archaea having a major role in carbon and nitrogen recycling, viral infections will have a major impact on ecosystems.

An unusual virus, called *Emiliania huxleyi* virus 86, in the family Phycodnaviridae, is a giant among viruses and enters coccolithophores, a member of the phytoplankton not common on the seashore. What makes it an intriguing species is that, probably because of its unusual size, it becomes caught in the gut of both copepods and the tunicate *Oikopleura*. It builds up in the gut and is eventually dispersed by being released with the animals' faeces. Viruses are sensitive and may be destroyed by a variety of environmental stresses, such as low salinity.

Most free virus particles in marine systems appear to be pathogens of bacteria and therefore are an essential part of the microbial loop, described above in this chapter. Viruses specific to a particular bacterium kill individual organisms, releasing dissolved organic carbon (DOC). Phytoplankton also create DOC when their cells are grazed and broken open, releasing DOC into the water along with mucilage from the faeces. The effectiveness of the loop therefore changes with bacterial population density. Chapter 19 has more about the role of fungi, bacteria and viruses in the marine carbon cycle.

FIGURE 7.1 A colony of *Licmophora*, a common benthic diatom found attached to sedentary organisms such as hydroids and seaweeds in rock pools. It is frequently flushed into the plankton during rough weather. The thickened cell walls, strengthened with silica, help survival and impart grazing resistance on the shore. True planktonic diatoms have thinner walls. Each cell is 100 μm long.

CHAPTER 7

The productive ones: diatoms and other algae

On occasions the green algae *Halosphaera* appears in the plankton floating near the surface. These bright green spheres 0.5 mm in diameter may occur only in odd ones or twos, but, depending on the warmth, they may increase to hundreds or thousands. Where do they suddenly come from? They are a warm-water species that can be abundant across the North Atlantic, and most probably arrive on our shores on the Gulf Stream in late spring. Around 80% of the total phytoplankton population is composed of diatoms, but *Halosphaera* is not one of them. Placed side by side with a similar-sized diatom, *Coscinodiscus*, the distinction is clear. *Halosphaera* is distinctly green, while diatoms are brown. Our terrestrial lives are dominated by photosynthetic green, but under the waters of our coast is a world coloured a rich golden brown. The second component of phytoplankton is the dinoflagellates, covered in the next chapter.

Diatoms

Diatoms are remarkable. They are found in all environments of the world, but these photosynthesising algae are especially abundant in the marine environment near the surface, where sunlight penetrates into the water. The coastal zone, therefore, is their perfect habitat. They are unicellular organisms that reproduce, usually by mitosis, around every 20 minutes; so, while they may not have a large 'standing crop' biomass compared with, for example, a woodland, their extremely high reproductive rate means that they are the most productive organism in terms of biomass created over a year, exceeding any organism we might find on land. Of total marine primary production, diatoms represent 40%.

In addition, half of all oxygen produced on the planet comes from planktonic diatoms, although this mostly remains within the marine environment and is used by the organisms there. Diatoms exhibit positive phototaxis, moving towards the surface and light by day and slowly sinking at night. Reaching the light is important for their photosynthesis, but the ability to acquire the illumination they need varies with the season. As a measure of this biologists use the term compensation point.

FIGURE 7.2 Collection of phytoplankton after a late spring bloom. Low magnification.

FIGURE 7.3 The large diatom *Coscinodiscus wailesii*, showing both girdle (top right) and valve (bottom left) views. Diameter 400 µm.

This is the point at which the rate of photosynthesis is equal to the rate of respiration – that is, the rate of making energy is equal to that being used and the amount of oxygen produced is the same as the oxygen being used. Most diatom populations reach this point 20–30 m below the surface in summer, while in winter the compensation point is closer to the surface. In general, the ideal depth for diatoms to photosynthesise is actually 2–3 m below the surface because, for most, the level of illumination at the surface is too bright.

These organisms can be found as individual cells or in chains and show remarkable diversity of shape and size. The plastids, or pigment-containing organelles, which are the site of photosynthesis within the cell, are a golden-brown colour due to a pigment called fucoxanthin that is present alongside the chlorophylls A and C. Additional pigments include a variety of carotenoids, which are important material for grazing crustaceans and other consumers. Along with brown seaweeds, such as kelp and seashore wracks, which also possess fucoxanthin, they are classified in the phylum Ochrophyta, meaning ochre-coloured plants. Members of this group, like others in this chapter, have characteristic flagella present at some stage in the lifecycle. This feature

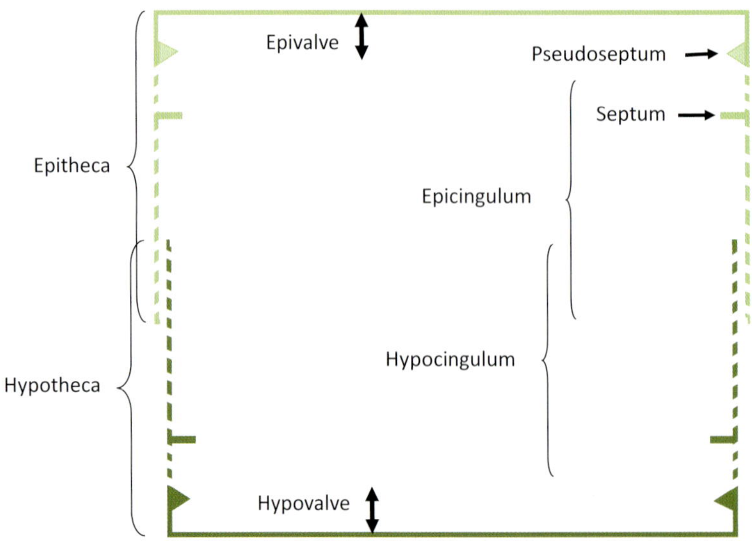

FIGURE 7.4 The structure of a diatom frustule as seen in cross-section. Compare this diagram to the girdle view shown in Figure 7.3, above.

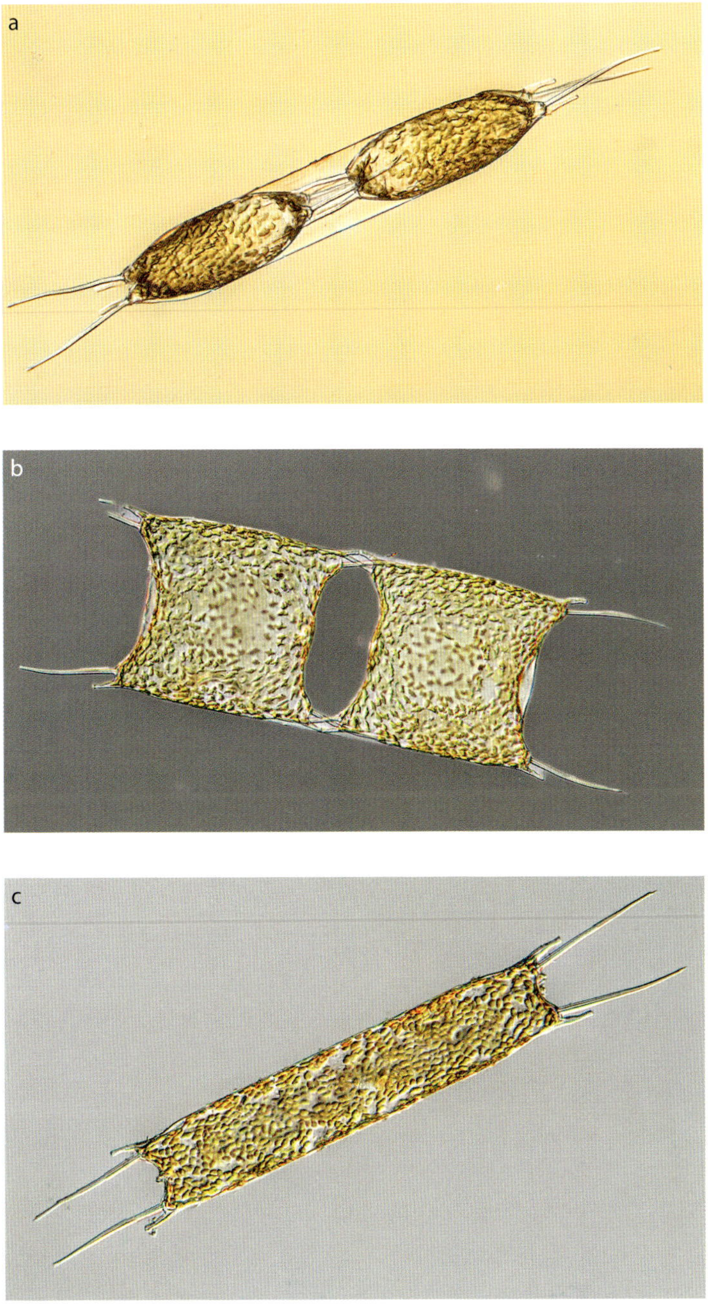

FIGURE 7.5 *Odontella sinensis*. (a) Dividing diatom girdle view and (b) valve view. Length of pair 600 µm. (c) A large mature individual. Cell length 550 µm.

has given them another term: heterokont, which refers to the presence of two different types of flagella: long, forward-moving ones with some stiff hairs, and short smooth ones.

A defining characteristic of diatoms is a cell wall impregnated with silica called the frustule, which has a structure so variable that it is the key to the identification of members of this very diverse group. The adaptation to use silica is probably a result of its widespread presence in the oceans, even if the concentration in the water column changes throughout the year and occasionally a lack of silica may be a limiting factor for the blooming of this group of phytoplankton. Silica is also inert to enzyme action by microbes and even in the guts of consumers is very resistant to

Diatom terminology and biology

Few diatoms have common names. Scientific names in general cross international borders and are understood and accepted by the international scientific community. However, because plankton are so exceptionally mobile in global terms and can thrive wherever the abiotic conditions, such as temperature, are suitable, they can quickly establish populations in almost any part of the world. This can mean that several people, entirely separately, can claim and name organisms as new species. It takes time not just to clarify whether or not an organism is new to science but also to see where it fits with close relatives. Diatoms seem to be the pinnacle of this form of organised chaos, with a significant number of species having multiple names, names that have not been accepted in some quarters or names that are accepted by different groups. For example, the genus *Biddulphia* (named after the British botanist Susanna Biddulph) was first established in 1821 and, while several hundred species have been associated with that name over the years, fewer than 30 are accepted today. Some were changed to *Odontella* (a genus name used in this book) but some now come under *Trieres* instead (see www.algaebase.org). A look at the WoRMS database shows the words 'unaccessed' and 'uncertain' against the names of large numbers of diatoms, including these genera. Synonyms can lead to a great deal of confusion.

digestion. For example, copepods have to mechanically break frustules down before consumption. Using silica also represents a relatively low energy cost for diatoms, enabling quick growth.

Along the shore diatoms as a group can be simply separated into two types: those that float and benthic ones, living on the shore substrate. Of the latter, some live within the surface of sediments, or may be attached to the outer surfaces of seaweed and invertebrates. These benthic forms have thicker silica deposits in the cell wall, possibly to resist grazing by shore herbivores but also, in sediment, to give some protection against the scouring effect of the inorganic particles. The advantage to the floating forms of having a thinner silica coating is that the relative transparency aids photosynthesis and the lower weight provides better buoyancy. Having thinner walls also allows greater flexibility in mitosis, with a quick recovery of the daughter cells. The largest diatoms are found here in the plankton, perhaps for these reasons. One of these, *Rhizosolenia* (*Neocalyptrella*) *robusta*, is just visible to the naked eye. *Coscinodiscus wailesii* and *Odontella sinensis* also grow to a large size of 400 microns or more.

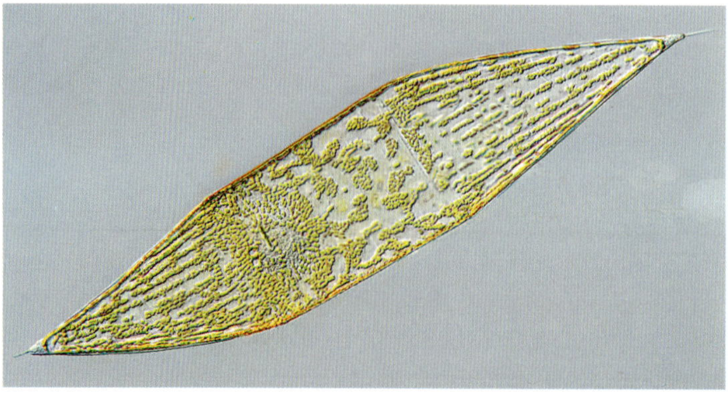

FIGURE 7.6 The large diatom *Rhizosolenia robusta*. Length is just over 1 mm.

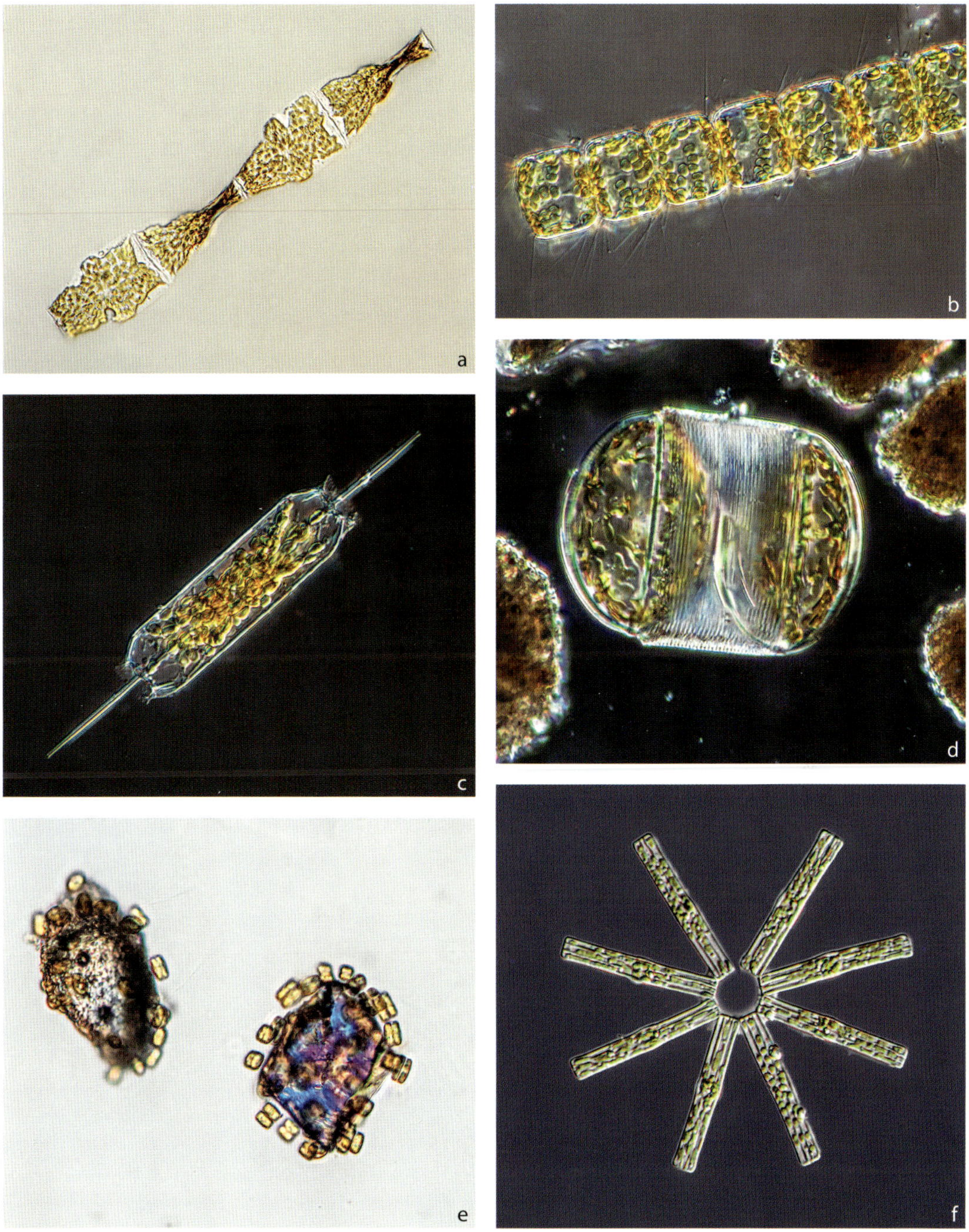

FIGURE 7.7 Diatom variety. Centric: (a) *Helicotheca tamesis*, a twisting chain 400 µm in length; (b) *Lauderia annulata*, always in short chains, common in spring, diameter 80 µm; (c) *Ditylum brightwellii*, a common species 250 µm long; (d) *Podosira stelligera*, two cells connected by common girdle bands, 80 µm long. Pennate: (e) *Delphineis surirella* attaches to mineral grains and enters the plankton at times of disturbance, 50 µm; (f) *Thalassionema nitzschioides* colony, each cell 50 µm.

The frustule comprises two halves that sit inside each other like a Petri dish with the lid overlapping the sides of the base. The larger 'lid' is called the epitheca, while the smaller 'base' is the hypotheca. Each theca is in two parts: a main surface, called the valve, and the sides, or girdle, called cingula. The valve is the most structurally rigid part and the girdle is more flexible, allowing growth and expansion. It is the cingula that differ most between diatoms; they can be simple bands, a double band or have a knobbly bit called a ligula. Under the microscope it is possible for diatoms to present different surfaces to the lens, as described in Chapter 5, providing either a valve view or a girdle view. On the surface, most obviously that of a valve, pores are visible, which are extremely small holes better seen under electron microscopes (see Figure 7.17). These allow gases, nutrients and other materials to pass through between the cytoplasm and the environment. Within the frustule is a cytoplasmic layer that lines the inside, surrounding a central vacuole. The nucleus can be suspended within or to the side of the vacuole and two or more plastids are set within the cytoplasm. In some diatoms the plastids may occupy a considerable space. *Fragilaria* (*Striatella*) *unipunctata* is one such diatom, where there are two half-star-shaped plastids in each cell. At the base of these are tiny, pale, typically round organelles called pyrenoids that act as bodies for food reserves (like oil globules) as well as having roles in aspects of photosynthesis; these are found in all algae. The term chloroplast refers to a specific form of plastid that contains chlorophyll; not all plastids have that particular pigment.

The classification of diatoms underwent a change in 2019 but it retains a split into two groups based on their form – the centrales, usually planktonic diatoms; and the pennales, which tend to be benthic species. These terms describe their symmetry: centric forms are radially symmetrical, while pennate types are bilaterally symmetrical and have a groove or canal running through the middle called the raphe; when viewed at the right orientation, the raphe can be seen to run the length of the diatom. A groove apparently splitting a diatom in half might appear to be a weakness. However, electron microscopy has shown that in section it has a zigzag construction, giving stability. Some pennate diatoms, like *Pleurosigma* and *Gyrosigma*, have a node in the middle

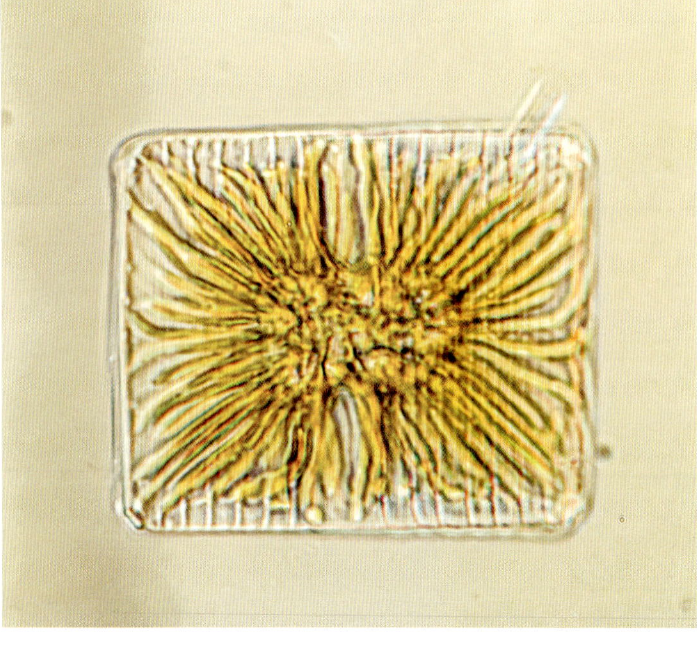

FIGURE 7.8 *Fragilaria unipunctata* is a pennate diatom where pyrenoids are common at the base of the two half-star shaped plastids but pyrenoids are less common in centric forms. *F. unipunctata* attaches to seaweeds and other organisms on rocky shores. It can reach high densities there, attached by a gelatinous thread, but is easily washed into the plankton where it can be abundant. Width 80 µm.

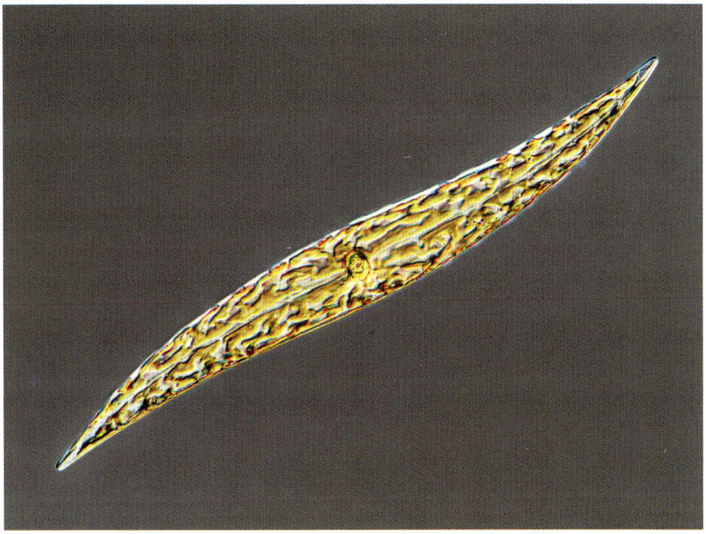

FIGURE 7.9 *Gyrosigma* and *Pleurosigma* are very similar and require high magnification to differentiate them. Both have a sigmoid raphe running down the centre but the plastids are more ribbon-like and complex in the latter species. This example is most likely *Pleurosigma*.

of this groove which produces mucus containing polysaccharides that passes down the groove to be exuded through tiny pores where the raphe opens. As this mucus pours out on to the surface it enables the diatom to glide, allowing a high degree of movement.

Benthic diatoms such as *Navicula* that live on depositing shores would become covered by sediment if they remained stationary, and movement is therefore essential for keeping them in the light. As mentioned above, the mucus they produce contains polysaccharides; these hydrate upon release, creating rod-like structures that push the diatom along the substrate. In some cases,

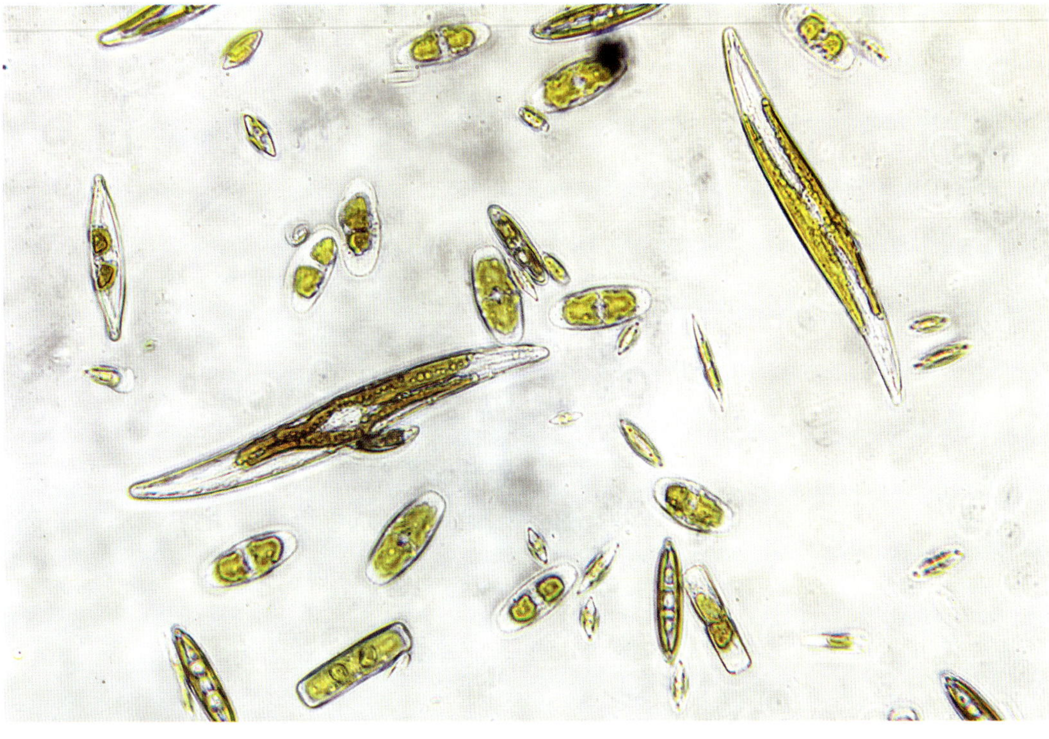

FIGURE 7.10 Diatoms in estuarine mud. Two larger *Gyrosigma* and the rest are various *Navicula* species.

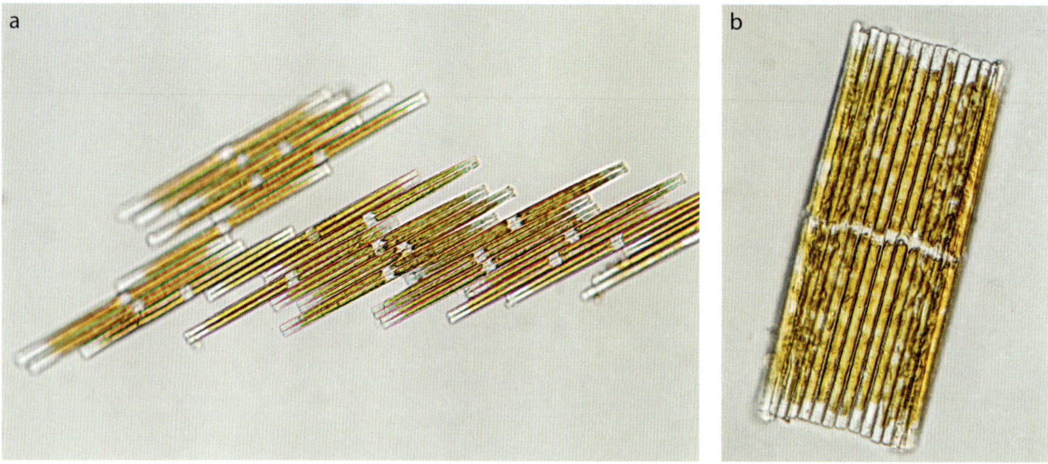

FIGURE 7.11 *Bacillaria paxillifer* colony (a) actively moving by day and (b) collapsed at night.

the mucus coalesces to form distinct stalks. Much of the released mucus builds up in the sediment and plays an important role in stabilising the shore. While *Navicula* is common in sediments, its mobility means that it often becomes washed up into the plankton. If it appears in your samples you will notice that its ability to exude mucus means that it rarely stays still when being viewed under the microscope. In fact, tides washing across the shore will cause a variety of benthic species to enter the plankton.

Bacillaria paxillifer is one of the few diatoms that has an English common name: the Carpenter's Rule diatom. The name comes from this pennate species' colony-forming habit, whereby the individual organisms slide against each other in a fashion resembling a folding carpenter's rule,

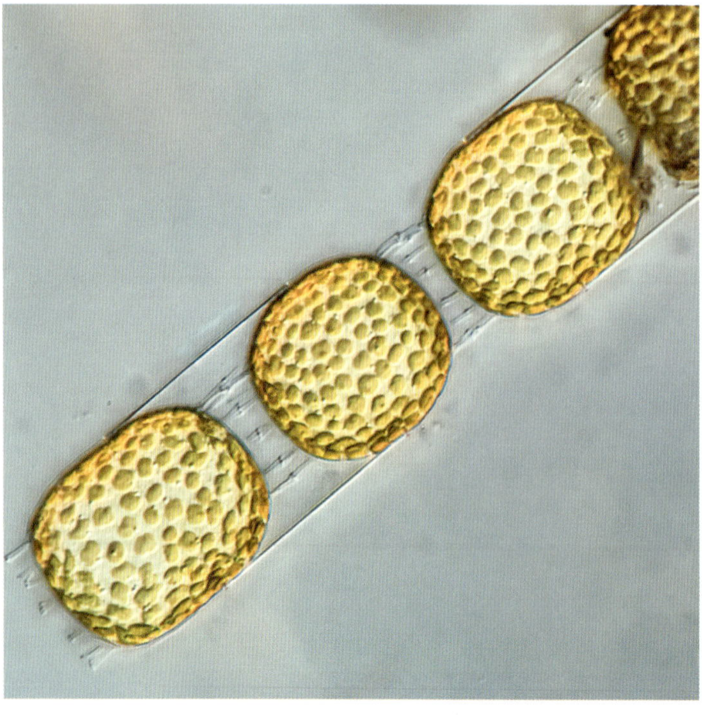

FIGURE 7.12 A Thalassiosirales diatom, *Stephanopyxis turris*, linked by a chitin fibril network. Each individual is 100 µm.

moving quite quickly, if rather randomly, through the plankton. This is a common species and when occurring in large numbers can be very annoying for students of plankton, as they constantly move back and forth, disturbing the rest of the organisms. Interestingly, it is active by day but shuts down at night, forming a tight block.

Although centric diatoms may lack the raphe system, some, like the Thalassiosirales, do have a complex pore structure called the fultoportula, or 'strutted opening'. Instead of mucus secretions they release beta-chitin that forms barely visible external fibrils (see Figures 7.12 and 7.18). These help in buoyancy, but may have further, as yet unknown, functions.

If the conditions are ideal, diatom cells divide into two by mitosis and keep dividing to create a bloom of millions. As well as silica, diatom development also requires nitrate and phosphate. A lack of any of these in the water will cause the population to decline and even disappear. There is typically a bloom of phytoplankton in the spring and as the density of the population rises so these nutrient concentrations will fall as it is used up by the bloom. Seasons have a dramatic effect on diatom populations. On the west coast of the UK, *Odontella* and *Bacillaria paxillifer* begin to bloom in October or November, remain dominant over the winter and then decline in the spring, disappearing by June. By contrast, various species of *Chaetoceros* may be present throughout the year but peak in the summer, along with *Rhizosolenia flaccida*. The long needle-like diatoms such as *Rhizosolenia setigera* are intriguing. For one, they show quite clear movement of the cytoplasm within the frustule. They can also 'migrate' if the nitrogen content in the water diminishes, possibly due to competition with other species. They do this by sinking en masse to find more nutrients and, once these are detected, slowing their descent, taking on and storing the nitrogen, and returning towards the surface. The small diatom *Cerataulina pelagica* can form huge spring blooms sufficient to disrupt zooplankton development and make the water discoloured and quite unpleasant. They have thin, flexible cell walls with a minimum of silica present, contributing to quick growth and an ability to create blooms in a matter of days.

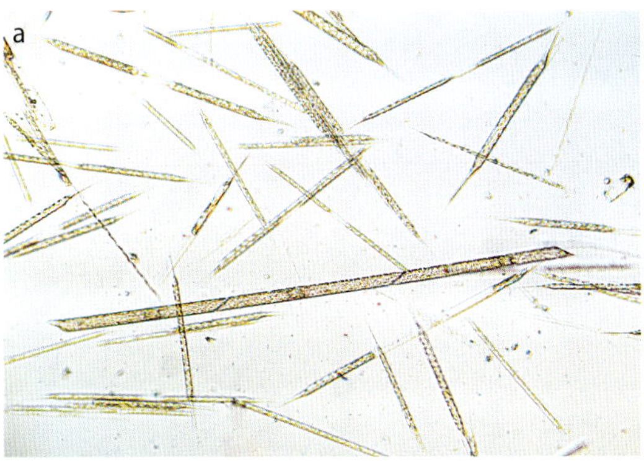

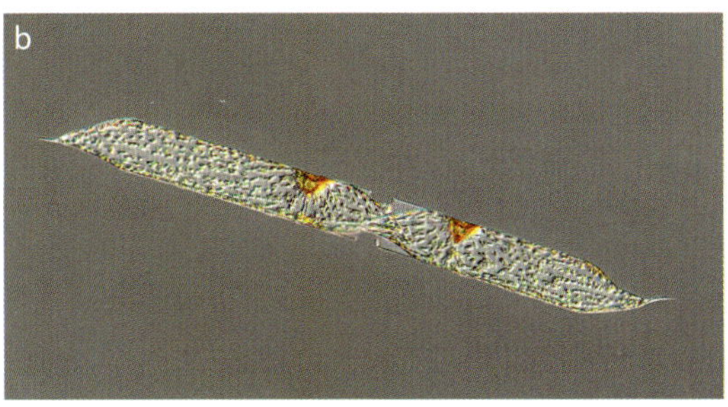

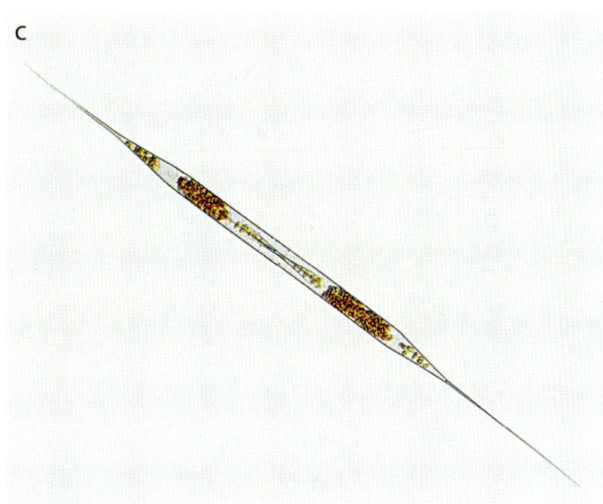

FIGURE 7.13 The centric diatom *Rhizosolenia*. (a) A blooming group. With diminishing nutrients they can sink in the water column in search of food; (b) an individual cell completing division; (c) *R. setigera* dividing. Total length 600 μm.

FIGURE 7.14 Chains and variants of the large genus *Chaetoceros*, a taxon with the largest number of diatom species worldwide. (a) *C. socialis* colony, attached to a central point by threads. Width of colony approximately 700 µm; (b) *C. didymus*, individual cell 30 µm across; (c) *C. curvisetus*, individual cell length 30 µm. Two or more pairs of setae cross over to link them together, along with the secretion of a gel that helps glue them together.

Diatoms also require various trace elements, such as iron and zinc. The presence of the former increases the blooms of *Chaetoceros* and *Coscinodiscus*. Zinc is an important cofactor in the use of the enzyme carbonic anhydrase. Common in most organisms, the form produced by diatoms is unique. In our blood carbonic anhydrase acts as a catalyst in the reaction between carbon dioxide and water, producing carbonic acid. Algal species require carbon dioxide for photosynthesis and in seawater, which is alkaline, it is in short supply. The enzyme is therefore used by the diatoms to produce bicarbonate ions to use as their primary source of carbon.

When diatoms divide during vegetative reproduction the two daughters each inherit one of the parent's thecae and then make a second to match. The new theca is always a hypotheca and, as a consequence, cells become gradually smaller as they continue to divide. This is why there is so much variation in species size and why, to avoid dwindling to a size at which existence is impossible, diatoms need occasionally to pass through an auxospore stage created by sexual reproduction with spermatogonia and oogonia. This is a complex subject already touched on very briefly above. There are several good references that provide detail if required, such as Graham and Wilcox (2000). Possibly the best resource, which includes species-specific photographs of auxospores, oogonia and spermatogonia, is Hoppenrath, Elbrachter and Drebes (2009), but this is unfortunately now out of print.

The maintenance of marine phytoplankton within the illuminated part of the water column depends on a large surface area to produce drag and reduce the chances of sinking. Long hairs, like those of the *Chaetoceros* species, are one useful strategy, but forming chains increases surface area yet further. The hairs of *Chaetoceros* also help to lock the cells together; indeed, *C. socialis* forms

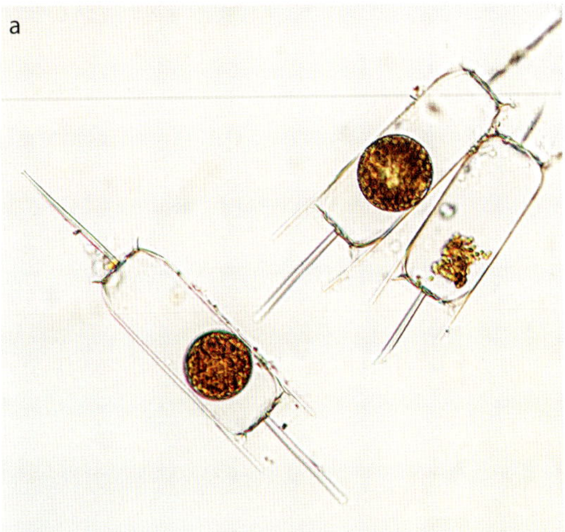

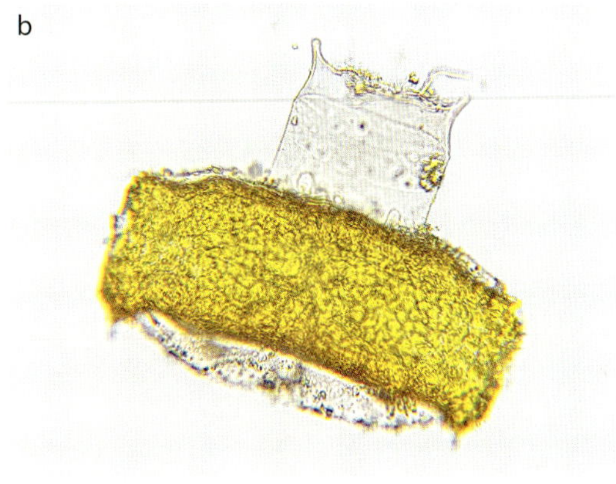

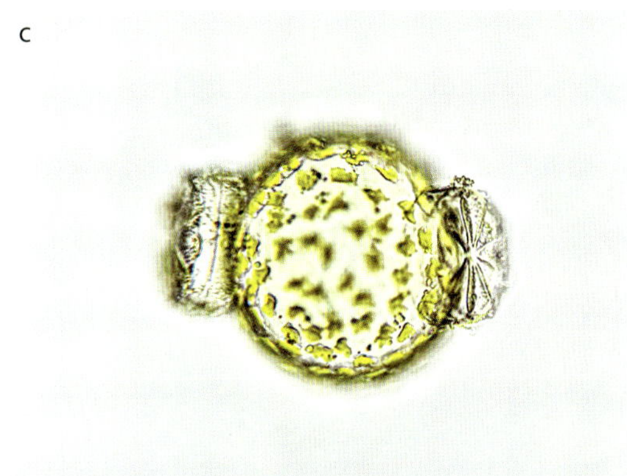

FIGURE 7.15 Restoring diatom size through auxospore formation. (a) An example of an oogonium in *Ditylum*; (b) *Odontella mobiliensis* with an auxospore with the cell (Elizabeth Beston); (c) *Actinoptychus* has two auxospores on the outer edge of the diatom (Elizabeth Beston).

Diatoms at rest

As discussed, diatoms disappear and reappear as conditions change. To bridge difficult periods, which often occur in early summer or autumn, they produce a resting stage (often referred to as a cyst or spore) that sinks to the sea bottom and can survive until conditions improve (see also Chapter 8, Sexual reproduction and resting spores). These contain energy and nutrient stores inside, such as silica, that will be available when the cyst or spore begins to emerge from the dormant state. This spore is quite different to the diatom's vegetative state; the physiological and cytoplasmic changes it undergoes allow it to survive darkness with no photosynthesis, different temperatures and nutrient depletion. Invariably they are simplified (less distinct in their form and having fewer organelles), and thicker and rounder in shape than the vegetative cell. Not all diatoms produce these spores, although our knowledge of this state for many species is still lacking. It is known that diatoms more typical of offshore, oceanic conditions rarely produce resting stages because inevitably they would descend to a depth too extreme to allow them to recover.

In this way, those planktonic diatoms that form spores can survive periods of low nutrient levels that could extend from just a couple of months to perhaps many years. This state also enables diatoms to avoid the heat and strong light of summer.

Re-emergence may be triggered by, for example, periods of turbulent weather along the coast, which will draw benthic material up into the water column and, with it, the resting spores, which germinate quite quickly once they are exposed to light and nutrients. However, it can take several mitotic divisions to regain the diatom's correct size and appearance. As these develop, so species begin to make a reappearance in the plankton and within weeks they may form blooms.

It has been estimated that *Chaetoceros* resting spores can constitute up to 50% of organic matter found in coastal sediments. While sediment can be viewed under the microscope to look for spores, the most effective method for determining the presence of spores is the

one of the most complex chains of all the marine diatoms, with multiple groups linked together. Many diatoms, such as *Odontella* species, link cells with a mucilage secretion exuded from the many pores at the ends of hollow horns that are found on several species in this genus. Collectively these pores are known as areolae, and can be important aids for identification. *Thalassionema nitzschioides* is a small diatom that produces a chain which, by linking at the corners, can create beautiful star shapes. Most diatoms also have the ability to change the contents of the sap vacuole, increasing and decreasing the cell's density to optimise buoyancy.

Salinity will have a noticeable effect on diatom distribution in coastal waters (in fact it, is probably the single most important factor controlling distribution) and a series of samples taken progressively closer to the mouth of an estuary will show a zonation in species. Some, like *Bacillaria*, may increase, as it is a species tolerant of moderately brackish conditions. Even in summer, when it has disappeared from shores on the open coast, it may maintain a small population around the estuary mouth, presumably because its tolerance of lower salinity enables it to take advantage of nutrients arriving from upstream. Saltmarsh, mud and sediment will be typical substrates in this brackish region and many pennate diatom species have adapted to thrive in these harsh conditions.

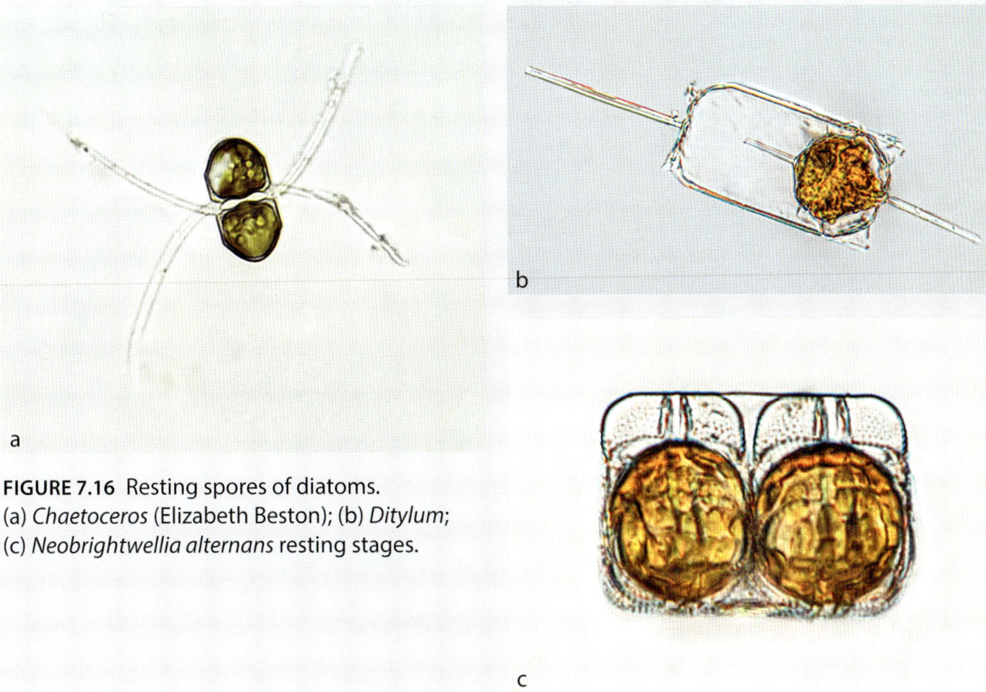

FIGURE 7.16 Resting spores of diatoms. (a) *Chaetoceros* (Elizabeth Beston); (b) *Ditylum*; (c) *Neobrightwellia alternans* resting stages.

use of Plankton Emergence Traps (or PET chamber). Essentially, these traps are cylinders open at the bottom which are pushed into the substrate, where they are held in place. The top of the trap has a net with suitable mesh to retain organisms emerging naturally from the sediment when conditions are favourable. By working in situ, environment conditions such as temperature and nutrients will be natural to the locality and a better determination of *what* conditions favour emergence.

Entomoneis and *Surirella* (both in the same order) and *Navicula* are three such genera that are common here; every tide washing over the surface brings them up into the plankton, before they settle once more. Members of the genus *Navicula*, a highly mobile diatom, are very common at the surface of sediment and are probably the most frequent estuarine diatom to occur in shore plankton. *Entomoneis* has two lobes making up a 'keel' raised above the valve surface, but this can be difficult to see as it usually sits on a microscope slide showing a different orientation. There is a sigmoid-shaped raphe in *Entomoneis* and numerous cingulae, or girdle bands. It can be the most abundant diatom of brackish conditions and so is a good indicator genus. In members of *Surirella* the raphe is positioned along the edge of the valve and, as with *Entomoneis*, can be raised above the valve, which may give both genera the ability to lever their way through fine sediment. Many marine pennate diatoms living in sediment, as we have seen, are relatively mobile in these substrates and, although they have the ability to push their way up to maintain themselves in the light, they will also dive down into the dark where there is material rich in nutrients. Here they switch nutritional mode to heterotrophy and absorb organic carbon from their surroundings. *Surirella* is particularly nimble, being able to change speed or adopt a rolling action. As an adaptation to life

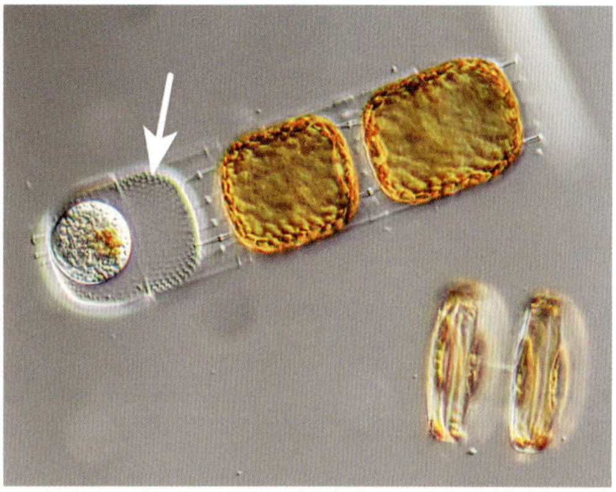

FIGURE 7.17 A short chain of *Stephanopyxis turris* with an arrow indicating barely visible hexagonal areolae inside. These are pores allowing movement of gases and materials. The cell has a possible spore developing inside.

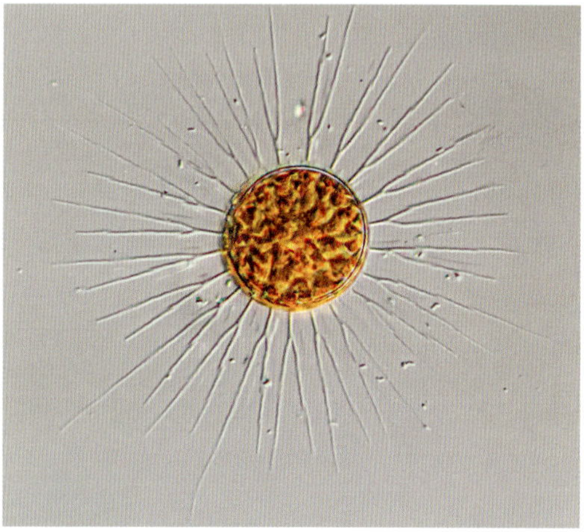

FIGURE 7.18 Chitin fibrils radiating out from a centric diatom. They stretch over 150 um from the valve margin.

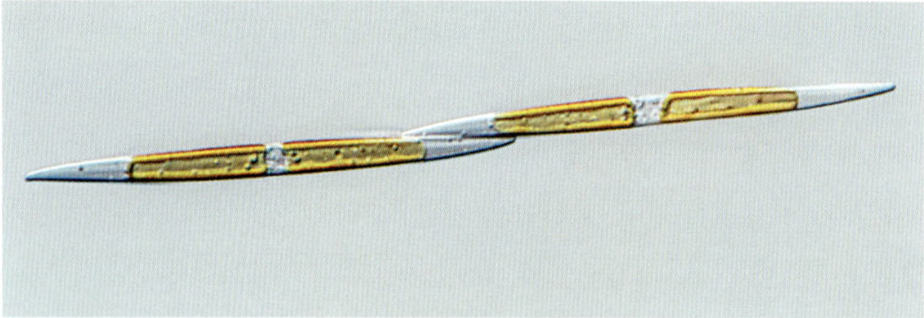

FIGURE 7.19 Two cells of *Pseudo-nitzschia seriata* joining to begin a chain of diatoms. Cell 100 µm long.

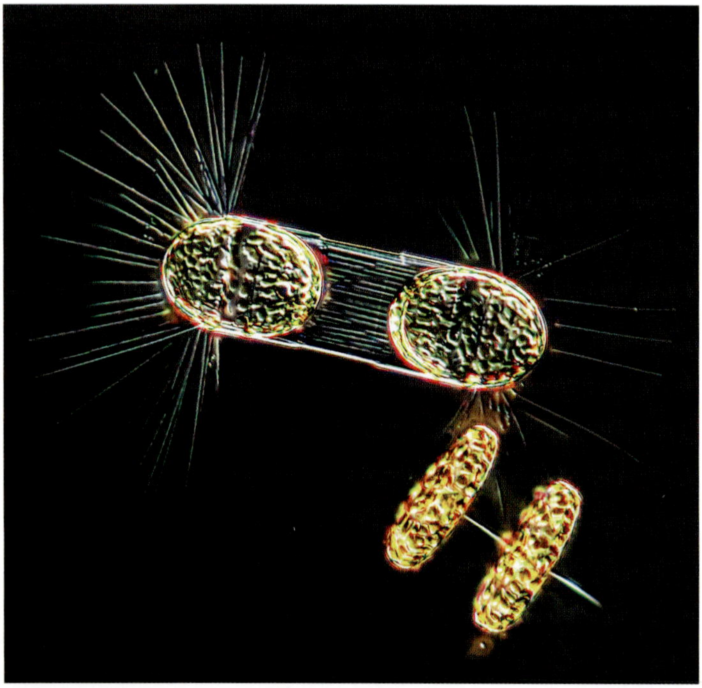

FIGURE 7.20 *Corethron* sp. after division, with long siliceous bristles that produce drag on the water. Two cells of *Thalassiosira rotula* are also present.

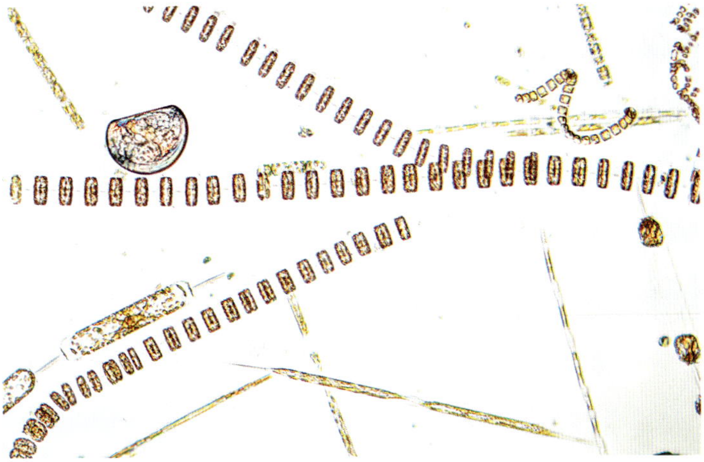

FIGURE 7.21 Chains of *Thalassiosira rotula* diatoms in girdle view. Common in spring, with individuals quickly developing into long chains connected by threads. The number and arrangement of threads varies between species; here they form a single discrete bundle in the middle, whereas in *T. anguste-lineata* there are five bundles of three threads spread out around the valves.

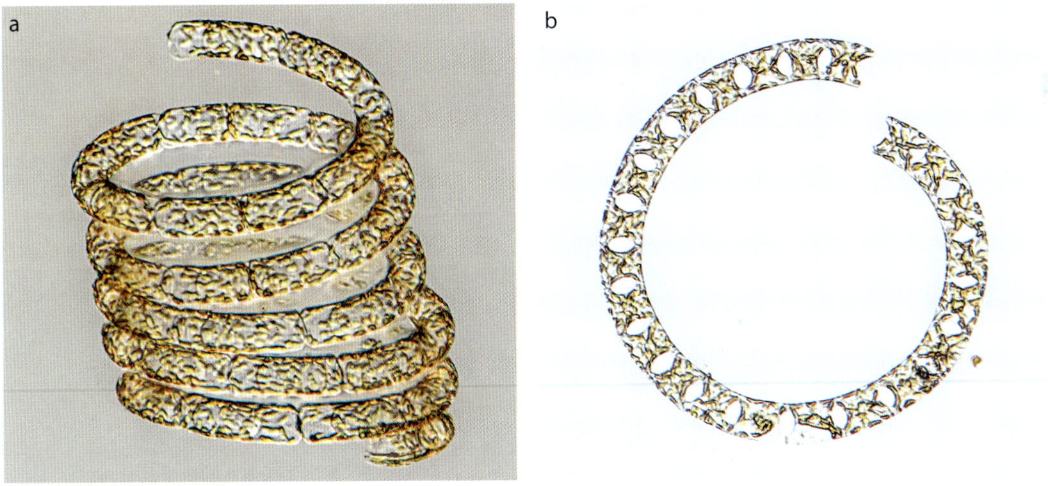

FIGURE 7.22 Spirally coiled diatom chains. (a) *Rhizosolenia striata*, cell 150 µm; (b) *Eucampa zodiacus*, colony diameter 200 µm.

in sand grains and fine sediment, with the abrasion that implies, the frustule of *Surirella* is tough and highly silicified. Indeed, while the *Surirella* found in seashore plankton are often dead, their empty frustules can survive long after cell death because of the high levels of silica in their walls. This, along with the high degree of motility, is quite unusual for a diatom.

Diatoms are not just flushed into the water from mud and sand. *Licmophora* species cover seaweeds and animals on rocky shores (Figures 7.1 and 11.11) and are important food for grazers there. Upon reaching high densities they can easily be washed into the plankton and can be a common sight in samples taken around hard shores. A close relative is the unusual species *Sceptronema orientale*, which, instead of adhering to organisms on the shore, forms an attachment to the harpacticoid copepod *Euterpina acutifrons* as an epizoite. The relationship seems to be neutral, with no obvious gain to the copepod and the diatom uses the animal purely as a substrate. However, it is highly specific to *E. acutifrons* and is not found on any other species. Described in 1983 from Japanese waters, it has, in recent years, been identified in Argentina, the Mediterranean and the Celtic Sea. Clearly, it has a worldwide distribution.

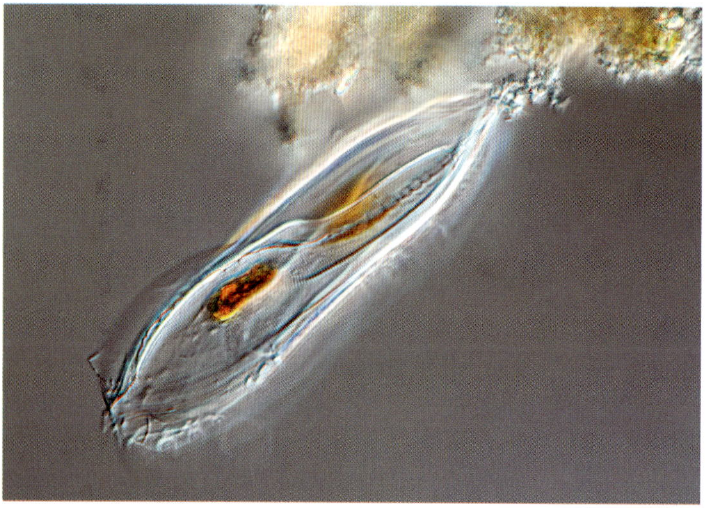

FIGURE 7.23 Valve view of an estuarine diatom in *Entomoneis*. Cell length 200 μm. See Figure 1.5 for girdle view.

FIGURE 7.24 An estuarine diatom in *Surirella*. (a) cell division in *S. gemma*, height 140 μm; (b) empty frustules of *S. gemma*, height 120 μm and (c) *S. fastuosa*, height 100 μm. Both species can be common in shore plankton near mudflats.

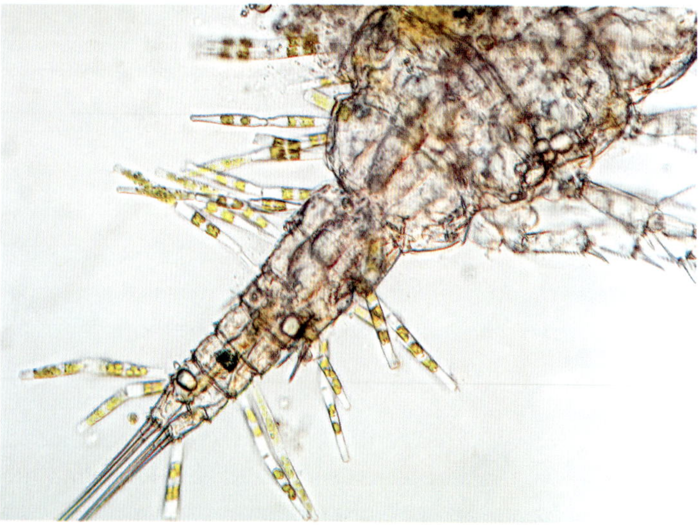

FIGURE 7.25 The epizoic diatom *Sceptronema orientale*, which lives specifically on the harpacticoid copepod *Euterpina acutifrons*.

As the main component of phytoplankton, diatoms are the most important producers in the marine ecosystem. They form the base of the food web, and are consumed by the zooplankton, principally copepods. During their messy feeding they liberate considerable amounts of diatom debris, which, along with their faeces, will sink to the benthos, although much of the soluble carbon becomes dissolved organic carbon to be used by other organisms. Through the consumption of diatoms the copepods make energy available to fish and nektonic organisms. While diatoms are being eaten compounds are released into the water that may indicate to other diatoms nearby that consumers are around.

Diatoms and dinoflagellates, to be seen in the next chapter, can be difficult to identify, especially to species. For the most comprehensive cover of these two groups perhaps the best starting point is *Identifying Marine Phytoplankton* edited by Carmelo Tomas (1997), now available as a free download.

Haptophytes

The haptophytes, which are single cells with two distinct flagella, all have golden brown plastids. They are an algal group living in the phytoplankton and include some important members, such as the coccolithophores, which are characterised by scales (called coccoliths) composed primarily of calcium carbonate that surround the cells. We came across these in the previous chapter, where

FIGURE 7.26 *Phaeocystis globosa* colony. (a) Typical colony in plankton, diameter 200 µm; (b) blooming colonies; (c) a foam-covered beach during stormy weather, due to a *Phaeocystis* bloom.

the giant virus, *Emiliania huxleyi* 86, can be carried by coccolithophores. Ocean blooms can cover expansive areas (see Figure 1.1), as photographed by NASA, and were so common in the past that when they died and sank to the ocean floor they created chalk and limestone deposits, as the scales do not decompose. Of all the algae, these have the single greatest impact on climate change through their ability to remove carbon and sulphur to the benthic ocean depths: today a quarter of the carbon that makes this journey originates from the coccoliths.

Along the coast, however, coccolithophores appear only rarely. Haptophytes here are represented by the genus *Phaeocystis*, which lacks the coccoliths, and when it does occur it can be in such abundance that the bloom becomes a nuisance. However, they can be found in small numbers at almost any time of the year. Within the plankton they exist in small colonies of around 50–100 individuals completely engulfed in a thick gelatinous ball. When highly magnified, the paired flagella on the cell are barely visible. The mucilage is probably a food reserve produced when light is available, and is also a reservoir of minerals such as manganese, iron and phosphate. It also helps with buoyancy. Individual cells are small and probably of a poorer nutritional value than diatoms, but they still represent an important food source for crustaceans and protist predators in the zooplankton.

The nuisance they present results not from any toxicity but the large amounts of gelatinous material they produce. In the spring of 2016, a particularly intense bloom occurred on the west coast of the UK such that a plankton net would rapidly fill with mucus making it impossible to sample; the beaches and coast were covered in slime, the rotting smell from which deterred tourism and swimmers. Even fish reject them. Rough weather makes the problem worse, whipping up the mucilage and covering the shores in thick foam. In parts of the North Sea *Phaeocystis* spring blooms can be a common occurrence.

Phaeocystis increases when silica is exhausted, often after there have been blooms of diatoms, which use silica to build their frustules. They can even survive low phosphate conditions and water rich in dissolved inorganic nitrogen. Blooms are not directly poisonous but high densities can lead to the creation of anoxic (low in oxygen) conditions.

Tiny stars: silicoflagellates

Particularly delightful single-celled, silica-shaped stars and members of the phytoplankton are the silicoflagellates. They reached their peak in the oceans of the Miocene and many microfossils have been described from 120 million years ago. Today there are just a few species to be found and seeing one under a medium-power objective is like looking through a telescope to view a star in the night sky. If you are not looking carefully they can be easily missed, at only 50 microns across. Zoom in with a high magnification and their simplicity is amazing. Normally, one or two are present in most collections.

What usually turns up in a plankton sample is, however, the external skeleton, which can survive for months or probably years after the death of the cell. Consisting of a tubular silica structure (Figure 7.28), it is considered to be somewhat like a basket holding and supporting the cytoplasm. The structure shows a plasticity of form that can vary, particularly in the number of spines. The regular construction of the most common species, *Dictyocha speculum*, is a hexagon with a spine at each corner. Within this there is an apical ring with supporting bars radiating out to the hexagon. The inner ring can have additional developments; in one subspecies, *D. speculum* var. *septenaria*, an extra spine is present. *D. speculum* var. *aculeatus* has the normal spine at the

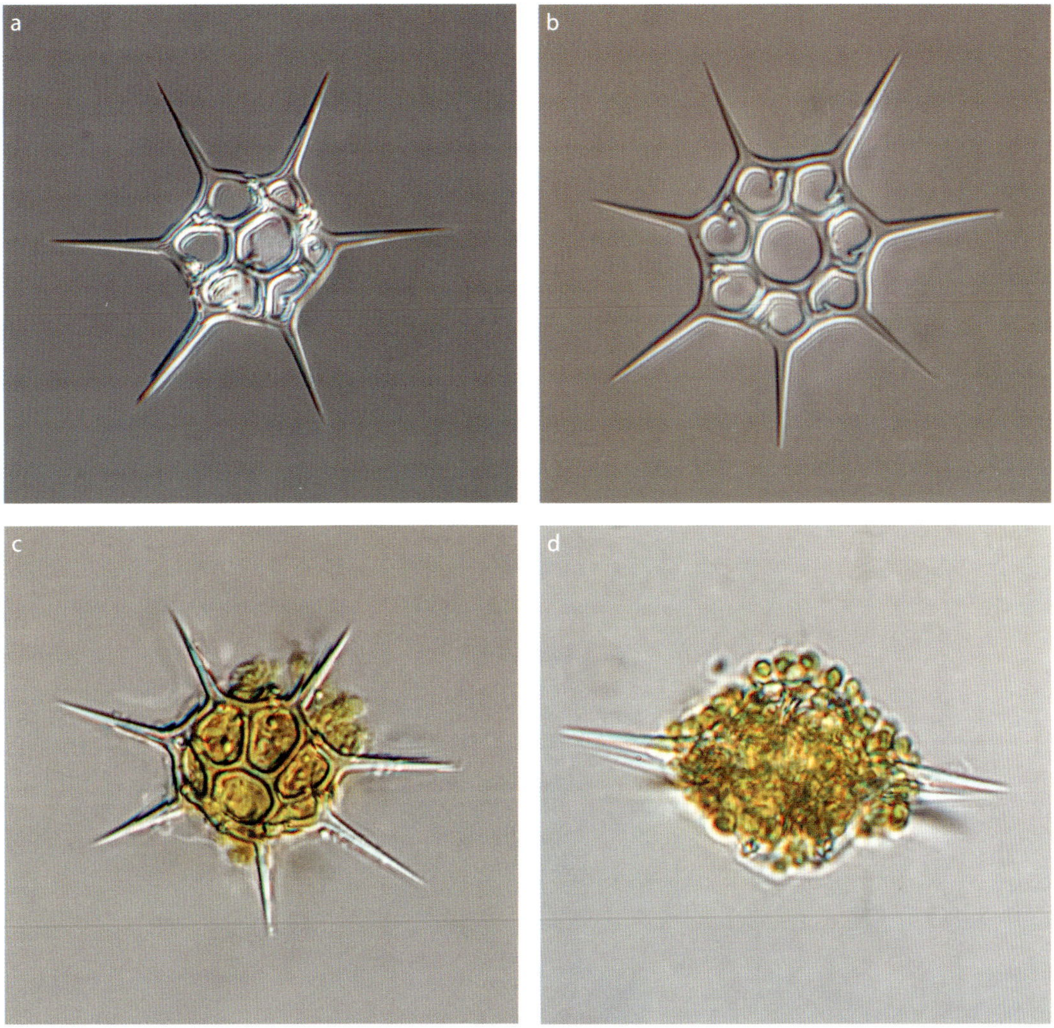

FIGURE 7.27 The silicoflagellate *Dictyocha speculum*, width including spines 60 µm. (a) The external skeleton; (b) *D. speculum* var. *septenaria*, with an extra spine present; (c) a living specimen, front view and (d) side view.

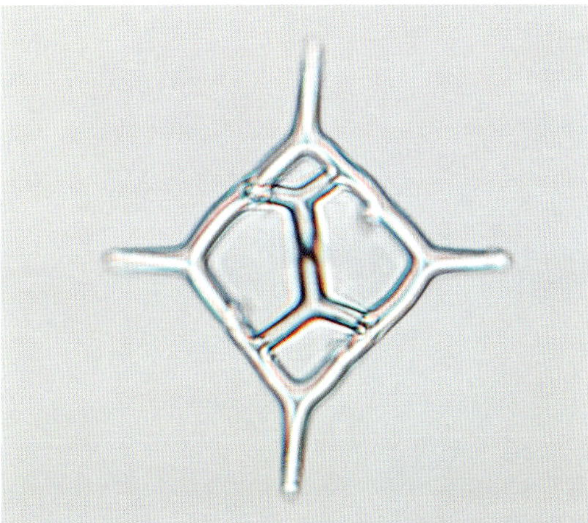

FIGURE 7.28 The skeleton of the rarer silicoflagellate *Dictyocha fibula*. Width including spines 40 µm.

top and bottom, but the pairs to the right and left are short stumps. All spines are shortened in the subspecies *D. speculum* var. *brevispinus*.

The skeleton thus creates a perfect matrix on which to hang the cellular material, which looks quite disorganised under a light microscope. Essentially, the cell is lobed, with a criss-cross of connecting cytoplasm wrapping around the silica and possesses a flagellum which provides only minor adjustment in the water, possibly to angle the cell for maximum exposure to light. The plastids are located on the outer tips of the lobes, with the nucleus occupying the space at the very centre of the cell. One interesting fact is that the cytoplasm can produce pseudopodia possibly to extend the plastids into the light. The photosynthetic pigments are similar to those of diatoms, including fucoxanthin, so that they appear green speckled with brown. The presence of fucoxanthin puts them squarely within the Ochrophytes; like the diatoms, they are important producers in the phytoplankton.

Dictyocha speculum is a cold-water (boreal) species and the most common around the UK, particularly in the autumn. A second species, *D. fibula*, is much rarer but does occur and is a lover of warmer water, around 18–20°C. Both are capable of forming blooms, but most recorded examples are of the latter species. With warming seas we are likely to see blooms of both species increasing and occurring further north. Importantly, silicoflagellates are a significant food source for juveniles of the important copepod *Calanus*.

Halosphaera

We met this uncommon and interesting organism at the head of this chapter, where we noted that its colour distinguished it from much of the rest of the phytoplankton. *Halosphaera* is a member of the division of green algae called Chlorophyta. This taxon is believed to be a modern representative of some of the earliest algae and, as such, is considered a 'living fossil'. They are unicellular and possess flagella, with *Halosphaera* having four. The lifecycle consists of two main phases, starting with a single motile form with flagellae that are several times the length of the cell. This is approximately ten microns across and, with luck, it is possible to see them,

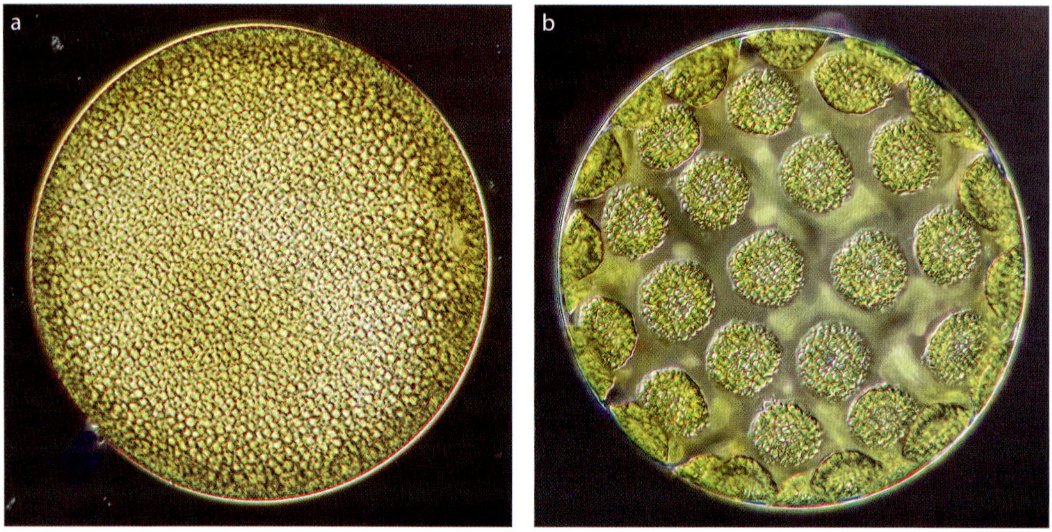

FIGURE 7.29 *Halosphaera*. (a) A developing cell, 160 µm; (b) a mature cell, 220 µm.

although they spin around and are difficult to photograph. In the second phase they grow until the cell reaches 400–500 microns in diameter, and become less mobile. Individuals are now covered in small rosettes of chloroplasts. A resistant coat surrounds them, thought to consist of a biopolymer that can fossilise in sediment. This second phase, called a phycoma, is a cyst, and over some weeks it matures to produce small round bodies inside that are the new flagellate stages. When developed they can be released through slits in the coat. The phycoma contains a high lipid content that provides buoyancy, so it tends to float in the top few metres of the water column. While, like diatoms, *Halosphaera* does photosynthesise, it is, like silicoflagellates, also a phagotroph, meaning that as part of its feeding strategy it will ingest organic matter and small organisms by engulfing them.

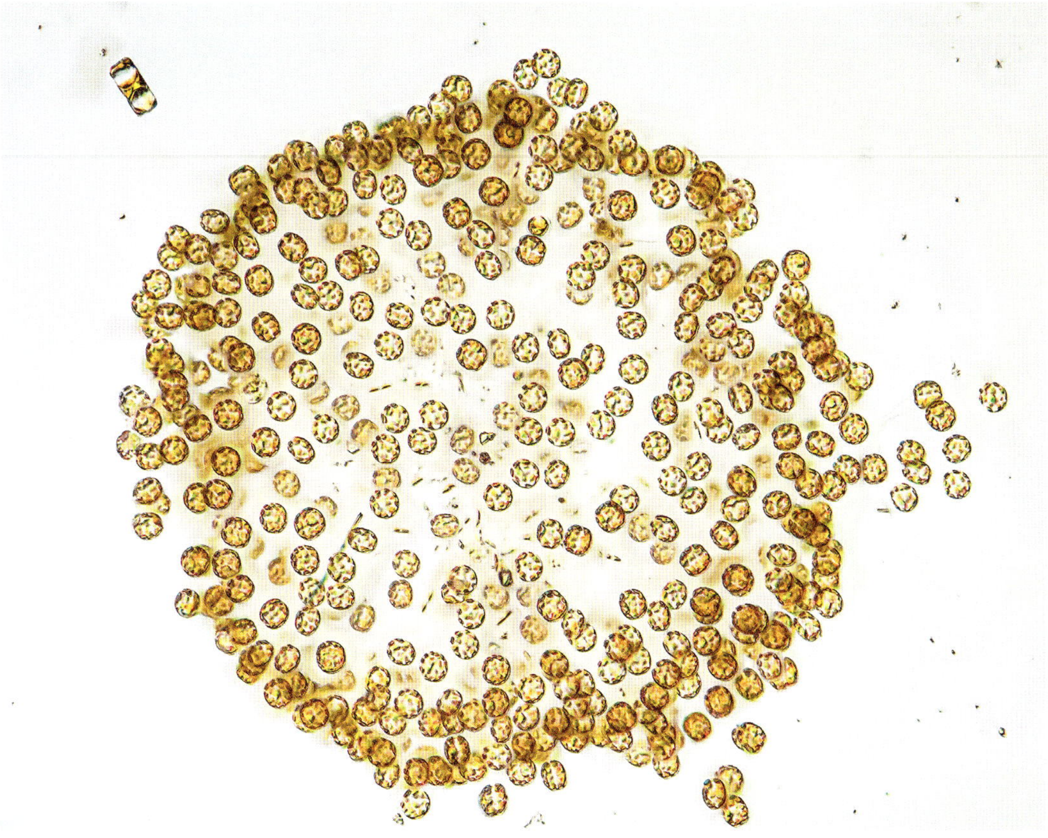

FIGURE 7.30 An unusual diatom colony of the species *Thalassiosira subtilis*. Occasionally it occurs in the plankton, where this species forms irregular colonies embedded in mucus.

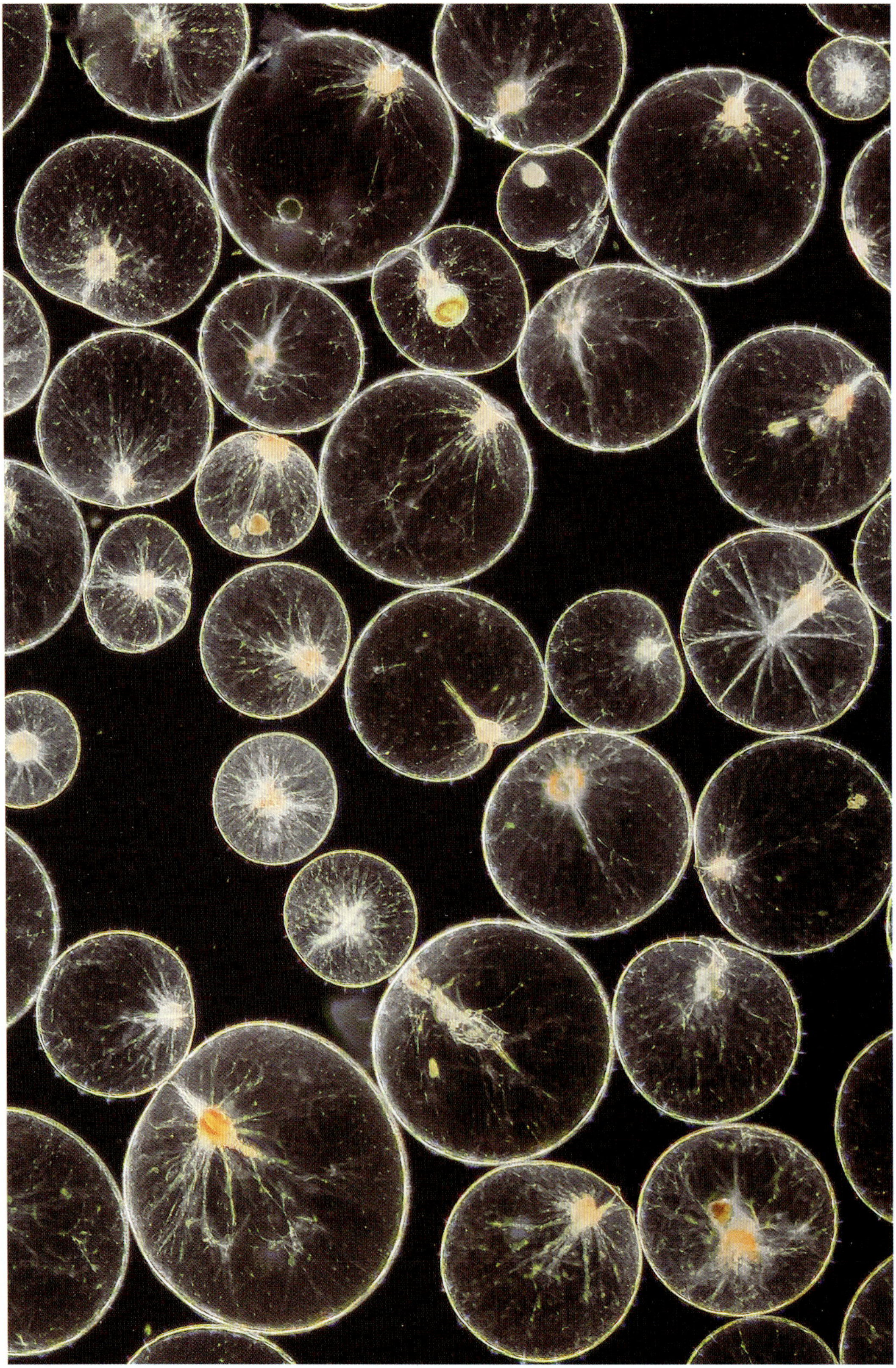

FIGURE 8.1 *Noctiluca scintillans*, an unusual naked dinoflagellate found worldwide.

CHAPTER 8

Plant or animal? Dinoflagellates

> Last night the water within the harbour was splendidly luminous. No trace of light, however, appeared on the smooth surface but when agitated, it blazed.
>
> (Gosse 1856)

The second most important component of the phytoplankton is the phylum Dinoflagellata, consisting currently of 4,000 species worldwide. This group of single-celled organisms shows the greatest diversity in the marine environment, especially along the coast, where they will be invariably present in plankton samples taken from the shore. Different species peak in different seasons and, along with diatoms, they contribute hugely to productivity. Sampling in estuaries shows steadily diminishing numbers upstream, although once into fresh water there are some distinct species to be found. Dinoflagellates are one of the many groups within what are known as the protists or protozoans. These unicellular creatures probably exhibit the greatest range of nutrition types of any taxon. There are those that carry out photosynthesis as part of the phytoplankton, while others are symbionts and parasites or predatory species with the ability to engulf prey. Just to complicate matters, there are a surprising number living as heterotrophs (consuming organic matter) that may also have plastids with the ability to photosynthesise. Despite being single-celled the size variation is quite astounding, from those a few microns across – almost impossible to see even with the compound microscope – to *Noctiluca*, large enough to be visible – just – to the naked eye.

Morphology

Dinoflagellates vary in form but they all have some similarities to diatoms in that there are two parts to the body, the anterior and posterior, which are not always both visible under a microscope. The Greek word *dineo* means to whirl, which individuals can do in both sedate and manic fashion due to the presence of two characteristically different flagella: one long and vertical, the other shorter and positioned across the cell. The former is the most obvious one when seen under the microscope, as it moves like a spiralling ribbon.

The cell is covered with a membrane more like a pellicle, meaning a thin layer composed of protein and other organic materials. Within this layer, called the amphiesma, there are minute sacs or alveoli. Two types of dinoflagellate are categorised based on this amphiesma. In one type, the alveoli are enlarged and full of a polysaccharide, particularly cellulose, forming 'thecal plates'. These can be visible under the microscope and have patterns particular to species, which assists in identification. Dinoflagellates with these plates are known as thecate, or armoured, forms, while those lacking the cellulose plates appear to be naked and are known as athecate, or unarmoured, forms. Athecate dinoflagellates are in general minute, and a fine mesh net of 10–20 microns is needed to collect them. One of the largest of this type, however, *Akashiwo sanguinea*, can be picked up occasionally in 50-micron nets when there is a coastal bloom in late summer. One dinoflagellate genus, *Symbiodinium*, is unique in being somewhere between thecate and athecate, and is a symbiont typically associated with cnidarians such as sea anemones.

Both groups have a girdle or groove around the middle called the cingulum, like the band around diatoms. The anterior, or epitheca, is above this and the posterior, or hypotheca, below.

FIGURE 8.2 The armoured dinoflagellate *Tripos* showing one of the two flagella. Length 240 µm.

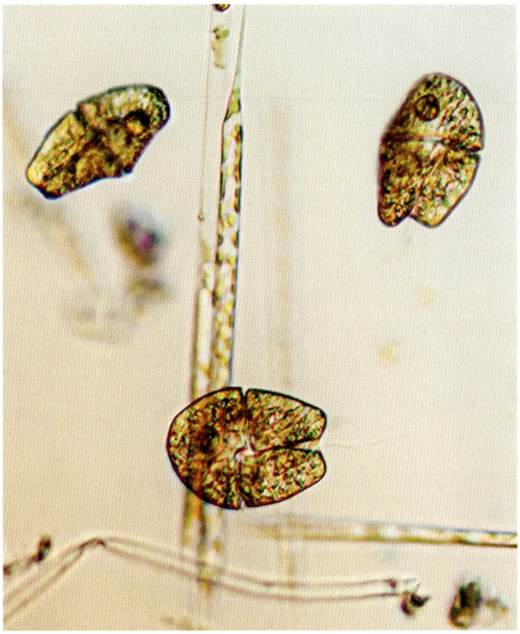

FIGURE 8.3 *Akashiwo sanguinea*, a large 'naked' dinoflagellate. The flagellum of the specimen at the bottom is just visible to the right. Length 70 µm.

A further groove, the sulcus, runs perpendicular to the cingulum from the hypotheca just into the epitheca. The girdle or cingulum is not always complete and on different species may be in a different position around the sulcus. Non-thecate forms tend to have a rounded top, or apex, but at the top of thecates there is a canal opening, the apical pore complex. This has several small plates around it and is probably involved with cell division. The pusule is a mysterious organelle at the centre of the cell, usually in the hypotheca (see Figure 8.4c). It consists of a cell membrane folded multiple times to create tubules and sacs. It is not thought to change shape, move or contract, but does have an opening to the outside into the flagellar canal. Only the heterotrophic marine dinoflagellates have a well-developed pusule, leading those studying them to think that its purpose is to do with osmoregulation and excretion or some form of water uptake.

Momentum is provided by flagella. Within the cingulum lies a smaller, coiled flagellum lying at right angles to the sulcus. Its whiplash action propels the organism at some astonishing speeds, around 200–500 microns per second. Within the sulcus itself there is a longitudinal flagellum attached to provide the main means of forward propulsion. It is particularly flexible, allowing the dinoflagellate to suddenly stop and reverse its swimming direction by changing the position of this flagellum. The ability to control direction is important for dinoflagellates, as they swim towards light (positive phototaxis), and it helps individuals avoid intense predation. However, generally they lack any eyespots that would guide them, and it is not clear how they judge different angles of illumination. While some seem to have light pigments under the sulcus, others seem to be able to use their plastids to detect light. Either way, their progress towards the light is rarely direct.

As with diatoms, some means of preventing sinking is helpful. Members of the genus *Tripos* are horned forms. These projections increase the organism's surface area to provide additional drag and slow the rate of sinking. The greater the size of the horn, the greater the drag, but the

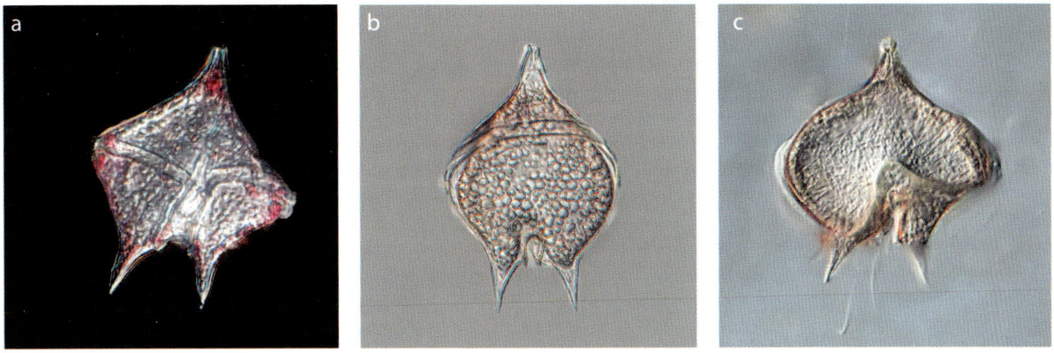

FIGURE 8.4 A *Protoperidinium* dinoflagellate. Most species are around 50 microns, with the largest peaking at 150 microns. They whirl around like spinning tops with a flagellum visible below the hypotheca. (a) A ventral view showing the cingulum around the centre with the epitheca above. The groove going down the middle is the sulcus. (b) Dorsal view: the lines on either side of the cingulum are the girdle lists. (c) A different example showing the large pusule and the flagellum coming from the sulcus. Length 140 µm.

chances of damage are also increased in stormy weather going into winter. Those species that have a pair of long horns sweeping up and backwards, such as *Tripos muelleri*, can be seen under the microscope with broken horns or breakages through the cingulum. In consequence, the very short-horned species *T. furca* and *T. tripos* can become the more dominant forms later in the year. *T. fusus* is long and curved, with a flagellum emerging from the cingulum at the centre, where it is widest. The epitheca continues into a long horn, while the hypotheca has a long horn and a very short, inconspicuous third one. *Tripos* species can be the fastest of the dinoflagellates and some, like *T. pavillardi*, form short chains that together allow yet greater speed.

Within the surface of the cell membrane are clusters of structures, their function not fully understood. Suggestions include defence or assistance with feeding. Three different types

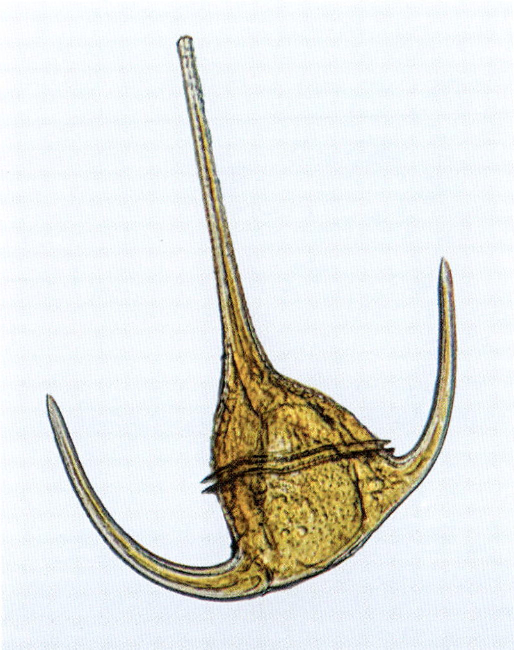

FIGURE 8.5 *Tripos tripos*. In the centre is the girdle or cingulum, bordered by a pair of toughened bands, the girdle lists. The central apical horn is attached to the epitheca, while the two smaller side horns come from the hypotheca. Length 300 µm.

are found, all of which can be fired or ejected to the outside of the cell. The first type, called a trichocyst, is present in all dinoflagellates and consists of projectile rods made from a crystalline protein. Produced in the amphiesma, they are positioned perpendicular to the surface and are discharged if exposed to a change in temperature or water turbulence; in fact, any external disturbance can cause them to eject. As they project out from the membrane they become long and thin, lengthening by up to eight times the initial size, creating a form of jet propulsion. This may be an escape response if threatened by predators. The second form, nematocysts, should not be confused with those body parts of the same name found in cnidarians, such as the sting cells of sea anemones. Only a few dinoflagellate species have them and they are modified trichocysts. Finally, there are mucocysts, sacs under the surface that release rods of mucus, again when disturbed. The latter can be found in species of *Gymnodinium*.

Nutrition and feeding

Dinoflagellates are important photosynthetic forms in the phytoplankton, with plastids just visible as small round organelles. Chlorophyll A and C are the main pigments, but a unique form of xanthophyll, a carotenoid called gyroxanthin, is found in several genera.

Mixotrophic species also exist in this group. These can carry out photosynthesis but will also exhibit phagotrophy, engulfing prey items. One such is *Gyrodinium britannicum*, a species approximately 80 microns long with very thin thecal plates and the cingulum running around the midpoint. Another is *Dinophysis*, a fairly complex, laterally compressed dinoflagellate with the cingulum located very close to the apex and the sulcus extending outwards, creating a sail-like frill around one end of the cell, called lists. This has orange and red plastids, but uses phagotrophy to obtain supplemental nutrients by engulfing prey.

Many dinoflagellates need to feed in this way in some form or another to obtain nitrogen. Some consume only small amounts of material, while others rely entirely on consuming other dinoflagellates, small polychaete larvae, nematodes, large protists and even small fish (especially fish larvae). To obtain prey they use several methods. They may engulf it whole by taking the prey into the cell and creating a vacuole within which it is digested; this is the method used by the naked forms. Alternatively, some sort of feeding tube or appendage can be present. *Noctiluca*, for example, has a tentacle that is used for seizing food and transferring it to the mouth (see Figure 8.15). As we will see below, this species can grow to a substantial size on a diet of particulate matter, diatoms, copepods and fish eggs and larvae. Members of the genus *Protoperidinium* – very common warm-water species during spring and summer in the shore plankton that are replaced in late summer and autumn by *Tripos*, possibly the most diverse dinoflagellate taxon and another mixotroph – have plastids but feed by extending their cytoplasm out between the thecae in a manner like pseudopodia to collect material, a method called pallium feeding.

Some obtain nutrients not by directly predating but in other ways. Small dinoflagellates have a large surface area to volume ratio and can absorb dissolved organic carbon through their cell membrane, along with inorganic phosphorous and nitrogen, a method known as osmotrophy. The smaller surface area to volume ratio of larger species means that they cannot make use of osmotrophy and their growth rates will consequently be slower, as absorbing and metabolising the carbon they need is a slower process.

Many dinoflagellates also have close relationships with other organisms that can help in their feeding. Some are parasitic or semi-parasitic forms that live in algae, invertebrates and fish. Others,

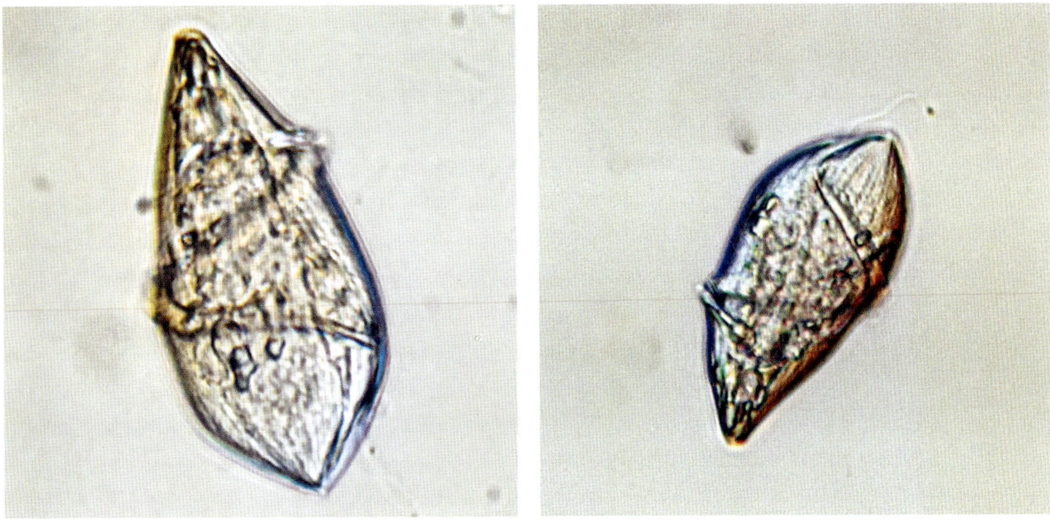

FIGURE 8.6 Two views of *Gyrodinium britannicum*. Length 80 µm.

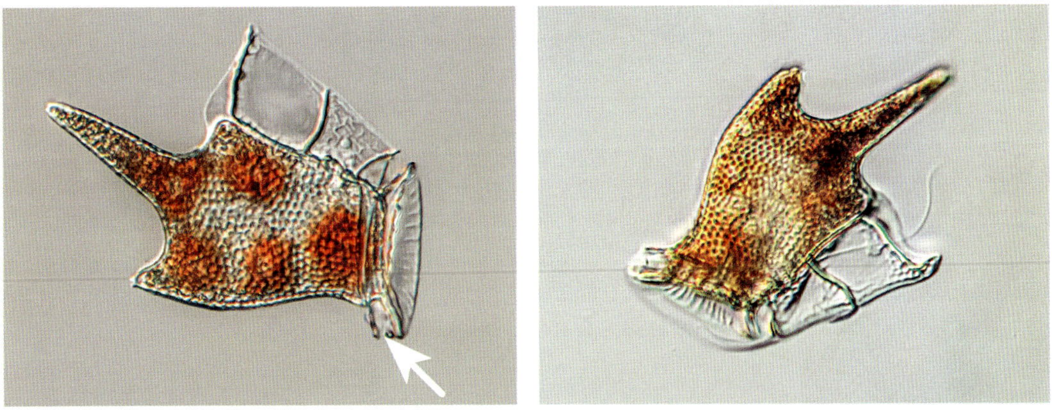

FIGURE 8.7 Two views of *Dinophysis tripos*. Note: the arrow indicates the cingulum and the lists. The 'wing' at the side of the cell is the sulcus lists. Length 90 µm.

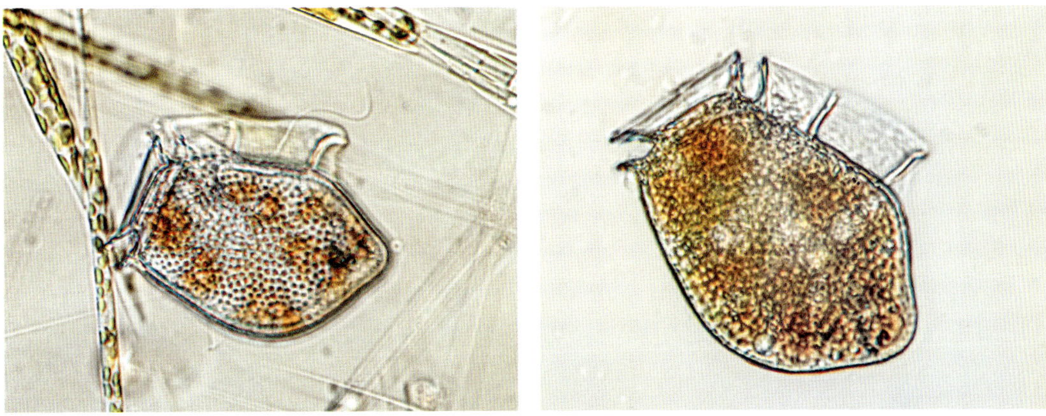

FIGURE 8.8 Two views of *Dinophysis acuta*. Length 80 µm.

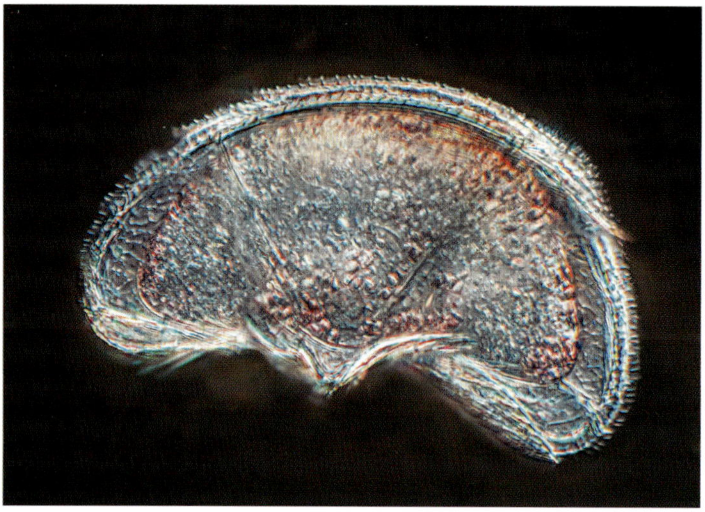

FIGURE 8.9 *Protoperidinium pentagonum* is a large, kidney-shaped species. Width 90 µm.

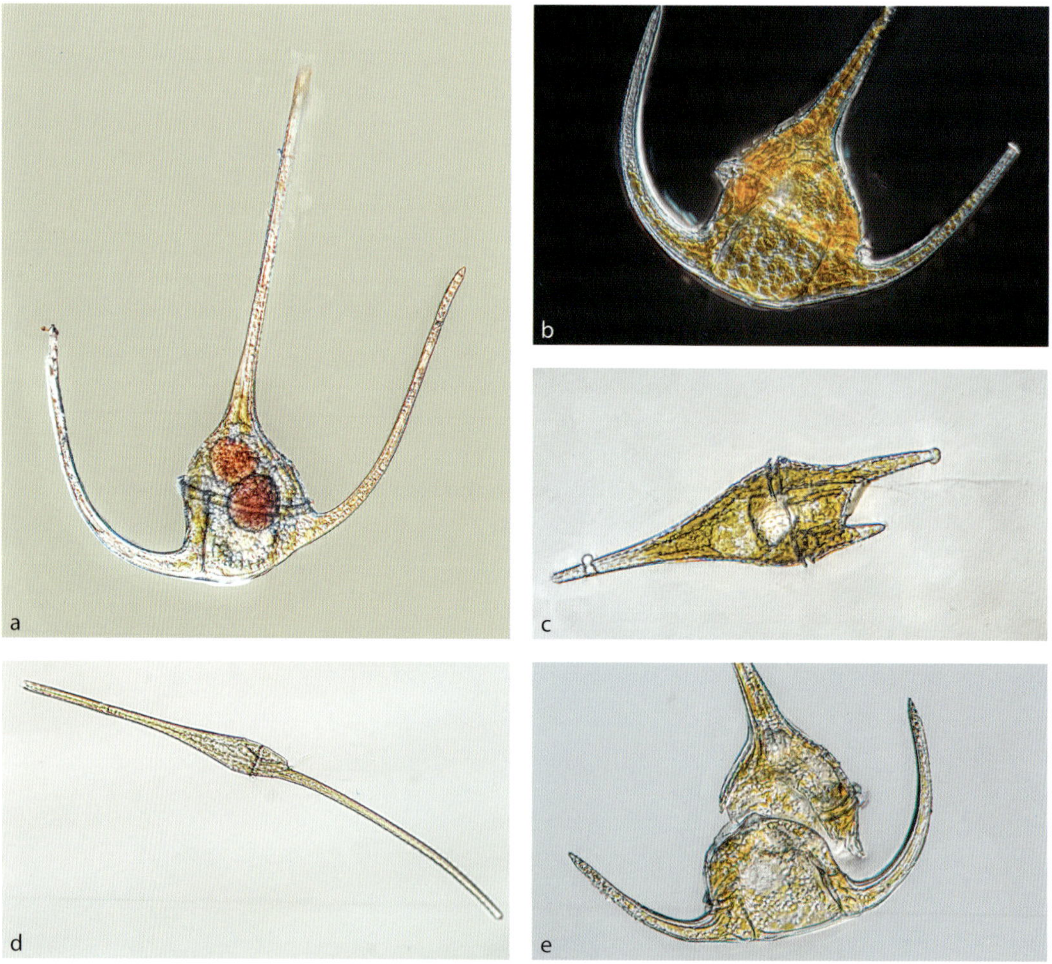

FIGURE 8.10 Various examples of *Tripos*. (a) The two red vacuoles are food that has been consumed, length 300 µm. (b) The cell is full of chloroplasts for photosynthesis, length 220 µm. (c) *T. furca*, note flagellum, length 150 µm. (d) *T. fusus*, length 400 µm. (e) Asexual reproduction in *Tripos* with the cell dividing.

FIGURE 8.11 The colonial Snakelocks Anemone *Anemonia viridis*. Note the green tentacles where the dinoflagellate *Symbiodinium* is located in the tissues.

including the genus *Symbiodinium*, a group of photosynthesising dinoflagellates that exhibit symbiosis and associate with cnidarians (jellyfish and sea anemones), where they live within the gastrodermal cells, are symbionts. These are also referred to as zooxanthellae and include other groups, such as diatoms and some algae. As part of these associations the host provides inorganic carbon, nitrogen and phosphate as well as a protected refuge in which to live. In return, the dinoflagellate can produce carbohydrates from their photosynthetic activity. It appears that this partnership is far more complex than first thought. Recent research has shown that what were once considered as a simple two-species relationships almost certainly involve an assemblage of organisms: zooxanthellae, viruses, endolithic algae and fungi, bacteria and Archaea.

Toxic tides

The pigment gyroxanthin, mentioned above, is prominent in the phenomenon known as toxic red tides, frequently associated with agricultural run-off. When high nutrient levels and a warming temperature coincide, sudden blooms occur in which *Gymnodinium catenatum* and several other species produce the compound saxitoxin. This is known to block the sodium–potassium pump, causing paralysis by stopping nerve impulse transmission. It is concentrated up the food chain as shellfish consume plankton and larger predators, including humans, in turn consume the shellfish. In humans the resulting condition is called paralytic shellfish poisoning and can lead to total paralysis and death.

Dozens of different species of dinoflagellates have been associated with shellfish poisoning. Primarily, these are the photosynthetic forms found along the shore and in estuaries. The toxins can seriously affect animals, causing liver and nerve damage, while *Gymnodinium breve*, for instance, produces brevetoxins that in humans may cause gastrointestinal problems, which can be very unpleasant but are rarely fatal. In some cases rough weather has blown foam from a bloom of

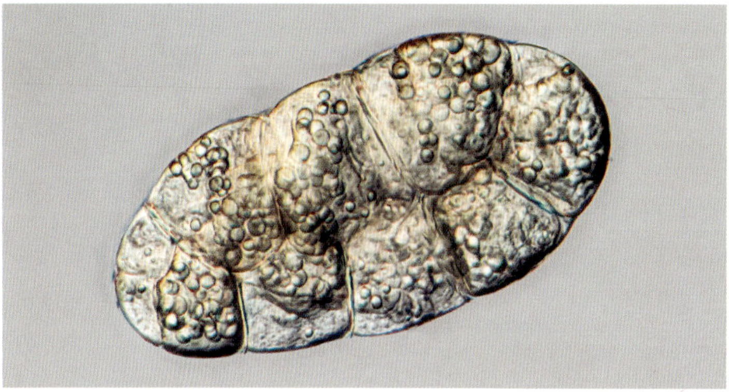

FIGURE 8.12 The genus *Polykrikos* is unusual as these dinoflagellates form colonies, consisting of 2, 4, 8 or 16 zooids with fused sulci. Individual cells are similar to *Gymnodinium* and may contain chloroplasts, nematocysts or trichocysts. Rarely abundant, *Polykrikos* is a coastal genus. Colony size 100 µm.

G. breve inland, where it has contaminated food. This species is also a particularly common cause of red tides worldwide. The plastid colour of these and most dinoflagellates is golden-brown, like those of diatoms, despite the fact that they lack the fucoxanthin, and accounts for the colour of these blooms.

Noctiluca: light of the sea

The quotation from Gosse's *Tenby* at the beginning of this chapter relates to *Noctiluca scintillans*, a real exception among dinoflagellates. It is naked, with empty alveoli under the cell membrane. Individuals tend to look like a disc under the microscope but are more of a squashed ball in shape and can be quite large, up to 2 mm in diameter. More typically, they are around 400–900 microns across. There is a dip or invagination on the ventral surface, the oral pouch, from which emerges a tentacle that traps food to transfer to the mouth. There is also a small single flagellum, but movement is limited and rotational movement absent, as there is no transverse flagellum. The bulk of the cell consists of vacuoles and a network of cytoplasmic strands that criss-crosses them.

Noctiluca has a preference for warm conditions, around 15°C or more, so different areas of Europe will see peak coastal blooms at different times. In the Baltic they usually appear in early summer,

FIGURE 8.13 A night view of the shore, showing disturbed water glowing white with bioluminescence produced by the dinoflagellate *Noctiluca scintillans*. (Hushaan @fromtinyisles)

Secrets of buoyancy

Buoyancy is achieved by the use of smaller, 'lighter' ions in the vacuoles, such as potassium rather than magnesium or calcium. This small difference is sufficient to reduce sinking in the water column. In fact, by moving and exchanging ions, *Noctiluca* can adjust buoyancy to allow a slow rise or fall. When looking down a microscope at samples in a dish it is too easy to focus on what is towards the bottom. Decanting or shaking a sample can cause air to be trapped within some organisms, such as water fleas, meaning that they will present towards the top of the water. *Noctiluca* is so buoyant, even without air bubbles, that it too usually floats to the surface. Therefore, always give the surface a quick check with the naked eye to see what is caught in the surface tension before focusing in on what is within the sample: water fleas, cypris larvae, diatoms and, possibly, *Noctiluca*.

FIGURE 8.14 *Noctiluca scintillans* photographed with DIC lighting to show the details of the cytoplasm and the large vacuoles. Diameter 550 µm.

but on the Atlantic side of the UK they occur when the temperature of the sea is at its maximum, in late summer. Blooms continue through the autumn and, with climate change, we are seeing them extending into the winter. In addition, they are becoming more difficult to predict. When conditions are right for them – warmth and sufficient nutrients – the cells reproduce asexually by mitosis. As the numbers increase along the coast spring tides bring them on to the seashore. A night visit to an open shore on a rising tide when a *Noctiluca* bloom occurs can be spectacular. As the water strikes the land the disturbance of the *Noctiluca* causes it to release cold blue light. This bioluminescence is produced by other species, including *Tripos* and *Protoperidinium*, but they produce a less startling display than *Noctiluca*, as the latter is so much larger and so abundant.

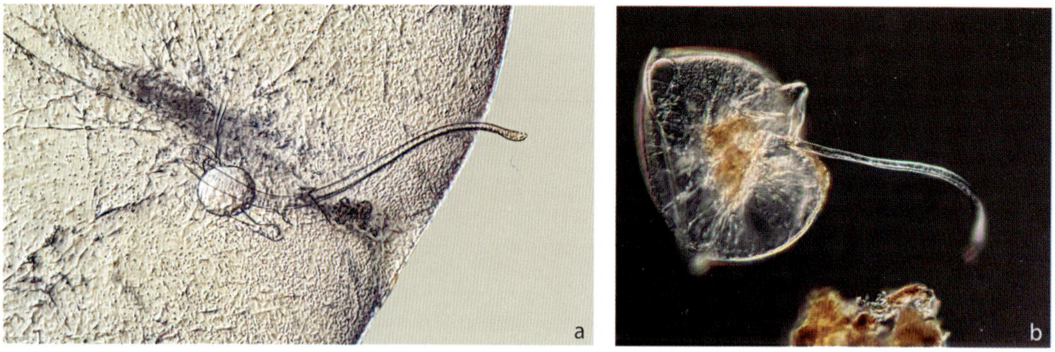

FIGURE 8.15 *Noctiluca scintillans*. (a) A closeup of the 'mouth', nucleus and tentacle; (b) an active individual that has extended the tentacle to feed.

Bioluminescence is almost certainly an adaptation to reduce attacks by predators by putting them off with a sudden spark of bright light. To produce light, dinoflagellates have specialised microstructures called scintillons (hence the name, *N. scintillans*) that are less than half a micron across containing the protein luciferin and the enzyme luciferase. The enzyme catalyses the oxidation of luciferin and the reaction emits light. Any sudden environmental disturbance causes a rapid change in pH in the scintillon, from 8 to 6, which precipitates the reaction. Therefore, landing on the beach has a similar agitating effect to attempted predation. Interestingly, studies have shown that there is a diurnal rhythm to the functioning scintillons, in that they only work at night. This is probably an adaptation to reduce energy consumption, as the light is not strong enough to be seen by day. As well as appearing on the shore, *Noctiluca* has a worldwide distribution across the oceans. Bioluminescent night blooms can be seen in summer from Australia to the Mediterranean. My most vivid memory of bioluminescence was on the shores of Morocco.

Sexual reproduction and resting spores

Dinoflagellates bloom and increase their numbers by mitosis, as described above for *Noctiluca*. Dinoflagellate chromosomes are unique. Other organisms will pass through cell resting periods (when no division is taking place) with the DNA uncoiled within a nuclear envelope, where DNA replication can take place. Dinoflagellates' chromosomes are permanently formed and pass out of the nuclear envelope through microtubules (microscopic tubes) at the time of division. When cell division takes place thecate forms usually separate the plates at an oblique angle across the anterior and posterior sections. For example, in *Tripos* the anterior plates pass to one daughter cell while the posterior ones pass to the second. After division the missing plates are then synthesised.

With environmental stresses such as low nitrogen or declining temperature come the need to produce genetic variation, via meiosis, and perhaps a resting stage (see above). Typically, this happens in autumn. There is a considerable amount of variation across the group and most species have never been studied. Meiosis in *Noctiluca* will produce flagellated gametes that resemble the species *Gymnodinium*. For many species there are two phases of meiosis. After the first stage the gametes (which can be the same or slightly different in size) each have a flagellum. When the two fuse both flagella remain so that a biflagellate cell can move through the water until it creates a resting stage, or cyst. This can remain in the plankton or sink, but it has no means of movement and requires an obligatory period of several months of encystment before it can 'germinate'. The stimulus needed to bring the resting stage to an end requires both light and oxygen and usually this occurs in springtime. After emergence a second stage of meiosis will usually occur to finally reform the vegetative cells.

Not all cysts are produced via the sexual route, but their production will always be instigated by poor conditions, including a lack of nutrients (for instance, dinoflagellates have a particular need for vitamin B_{12}), low light levels and reduced temperature. Cysts or spores are known to survive for many years in sediment or trapped in crevices on rocky shores. Research in recent years is showing that coastal sediments around the world harbour huge numbers of resting spores and with improving techniques to extract them for study our knowledge is increasing. The cyst wall consists of cellulose and the highly resistant molecule dinosporin, which is a biopolymer, or large organic compound. These are combined with mucus in multiple layers to form the wall. Inside are large stores of lipids and red-pigmented organelles, thought unique to dinoflagellates (see Figure 8.16b).

Smaller creatures are much better able to cope with turbulence than larger. Storms in late autumn and winter reduce populations of larger species, but all sizes will be engaged in the formation of resting spores. They do not drop to the bottom quickly, and oils inside the spores will aid buoyancy; thus, by remaining in the plankton, resting stages will be dispersed along the coast as well as out to sea or deposited on the shore for later germination. Turbulence reduces cell division, but once calm conditions are restored mitosis will continue for those vegetative cells that have survived and so the presence of dinoflagellates continues through the winter in moderate numbers. Prolonged and severe winter storms will be the main reducing factor, sometimes eliminating populations for a month or more, along with very occasional heavy infestation by chytrid fungi.

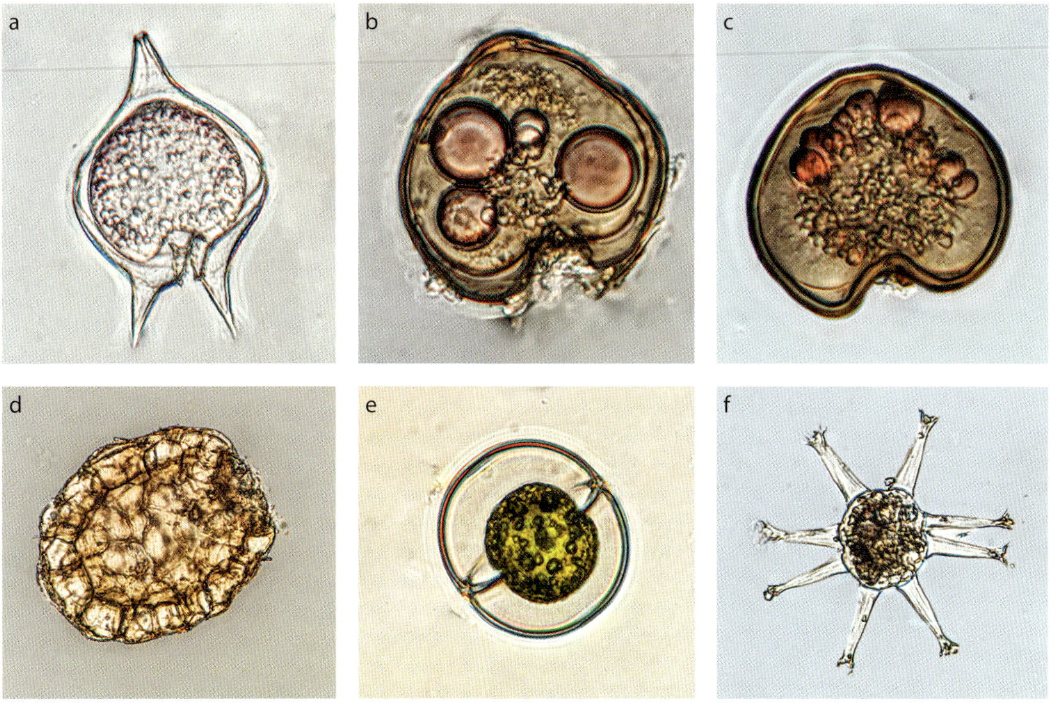

FIGURE 8.16 Resting spores. (a) *Protoperidinium* with a developing spore; (b) and (c) *P. oblongum* spore showing red-pigmented organelles with lipid and other stored food; (d) *Polykrikos* resting spore; (e) *Pyrophacus* resting spore; (f) undetermined species resting spore. Lengths of 80–100 µm.

FIGURE 9.1 The beautiful 'radiolarian' *Acanthometra pellucida*. Diameter, including spicules, 450 µm.

CHAPTER 9

Ciliates, tintinnids and other strange protists

Sometimes referred to as protists or protozoa, the subjects of this chapter are the single-celled consumers of organic matter, bacteria and other protists. They have a huge variety of structure and form and from the perspective of marine plankton we will divide them into four groups: the choanoflagellates, foraminifera, ciliates and the radiolaria.

The choanoflagellates

Only the very fortunate planktologist manages to find these in a sample, despite their apparent abundance. They are less than ten microns in size, so the only option is to place a sample under ×40 magnification and look very carefully for tiny blobs before increasing the magnification – if you can. In my time of looking I have seen them on only a few occasions but they are difficult to discover and the collar (the 'choano' in their name) is impossible to see under a light microscope. Choanoflagellates are an important part of the nanoplankton, consuming bacteria suspended in the plankton and, in turn, being consumed by microplankton. This gives them an important role in moving energy up the food chain. Many are stalked and sessile, some are free-swimming and many are colonial.

The single round or ovoid cell has a flagellum that faces out through a long, widening collar constructed of microvilli. As the flagellum moves so it creates a swirling current pushing nearby bacteria into the villi, where they become trapped and moved down to the base to be ingested. Under the microscope often the only way of detecting these organisms under high magnification is by seeing the swirling of bacteria. They are found in many watery environments and marine species usually have a lorica, or protective casing, around them. This is constructed with silicate rods that are transported out of the cell and glued in place.

Choanoflagellate cells are remarkably similar in arrangement to the choanocytes (digestive cells) found in sponges. This has led to the theory that they form a close link between unicellular and multicelled organisms, perhaps representing the evolution of single-celled organisms and their amalgamation into multicellular organisms. Those who disagree with this theory argue that they may be a regressive stage of sponges. Either way, they are very similar.

The foraminifera

The foraminifera, or forams for short, constitute the most significant group in the phylum Granuloreticulosa. They are essentially amoebae that have several specialities: they produce a spectacular shell, or test, in which they live, as well as extensive pseudopodia (extensions of the cytoplasm) for collecting prey. Of the 4,000 marine species worldwide, just 40 are known to be zooplanktonic. A high proportion of those found in shore plankton are benthic species flushed into the water by turbulence. All are heterotrophs, consuming any suitable-sized particle, such as bacteria, fungi, dinoflagellates and diatoms, as well as dead organic matter. It is also thought that many forams can live for several years.

Most people are acquainted with the amoeba, a naked single-celled organism that changes its shape as it glides around watery substrates by using cell extensions called pseudopodia with which it can engulf food such as bacteria. There is a wide range of modified amoebae in the protists that make coverings, and forams are just one such group, which can be further subdivided by the type of test they make.

Organic tests are flexible coverings made from a secreted mix of mucopolysaccharide and protein. This enables the organism to change its shape quite rapidly when in search of food. Agglutinated tests are made from secretions, as before, with the addition of materials such as diatom tests, sponge spicules and silt particles embedded in the mucopolysaccharide. It is further made rigid by incorporating calcium and iron ions. Some species are particular about the materials they add, but others will just use what is available. Finally, the calcareous test is a mixture of calcite (calcium

Foram tests

Most chalk and limestone is made from foram shells that sank to the bottom of the sea during the Cretaceous period, 145 to 66 million years ago. The proportion of coccoliths (seen in an earlier chapter) to foram tests in these rocks varies considerably in different parts of the world. The former dominate the cliffs around Dover in southern England, whereas the Great Pyramid of Cheops is made from rock that was originally foram shells. Their fossil record goes even further back, to the Lower Cambrian period, 540 million years ago. They are still abundant today and found in a wide range of environments, particularly in the sea.

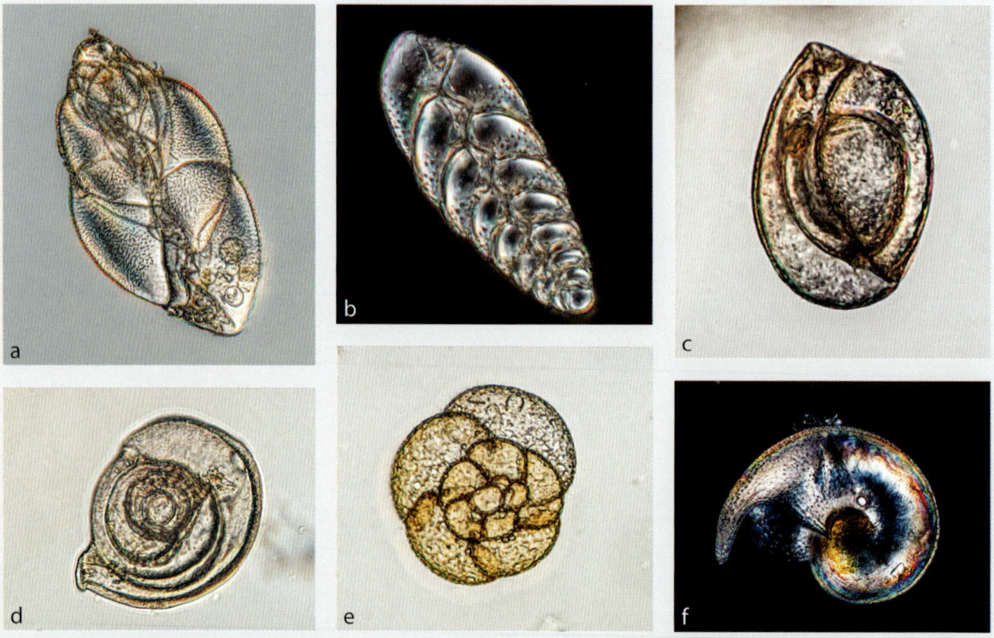

FIGURE 9.2 A selection of foraminiferan tests to illustrate the variety of species. (a) genus *Guttulina*, 120 µm; (b) family Bolivinitidae, probably *Bolivina britannica*, 140 µm; (c) and (d) family Miliolidae, 80 µm; (e) family Discorbidae, possibly *Rosalina*, 90 µm; (f) undetermined species, 110 µm. Sizes are length or diameters.

FIGURE 9.3 Foram tests are often incorporated into the tubes of the Sand Mason Worm *Lanice conchilega*. An enlarged view of the tube side shows a number of species, including the central white specimen *Elphidium* sp., 1 mm diameter.

carbonate) and an organic secretion, creating a rigid structure. The arrangement of calcite crystals produces a distinctive appearance. These too can be further subdivided. One of the most beautiful is the *porcelaneous* test, made by species including *Elphidium*, which looks like white porcelain in reflected light. *Hyaline* tests are glass-like and perforated with many tiny holes. *Microgranular* tests have a coarse or gritty surface. This last form can be present in huge numbers and on death accumulate in the benthos to create a dense ooze. That formed by *Globigerina*, found offshore and into deeper water, especially in the Mediterranean, is a particularly famous example. *Globigerina* is a millimetre or less in size and very spiny, which helps with feeding, but it is a southern species and a very unlikely occurrence in northern Europe.

In the neritic environment and on the shore foram shells are used by a wide range of other animals for shelter and as a place to lay eggs, including copepods, nematodes and polychaetes. A number of creatures incorporate tests into protective covers of their own. The best example is the Sand Mason *Lanice conchilega*, a species of marine worm common on the lower shore of sandy beaches, which secretes a protein tube with the open end just visible above the surface and with sand grains and often large numbers of foram tests studded across it.

Living and feeding

The foram test invariably has more than one chamber, the amoeba adding more chambers as it grows, although it will usually live in the most recent addition, near the entrance. Since the amoeba is inside the test it needs to emerge in order to move or feed. Forams have the ability to produce very long and fine extensions of the pseudopodia called reticulopodia. Some tests have fine perforations over the surface and, while the bulk of the reticulopodia emerge through the main chamber entrance, other, finer extensions pass through these tiny holes to reach the outside of the test. The bulk of the cell remains inside the test, while the reticulopodia are extended beyond it to form a large network around the organism. Some of the reticulopodia stay stretched across the test, maintaining and cleaning it. Reticulopodia are unique in that, internally, they have a bidirectional flow of protoplasm. By extending and contracting the network the foram can move, quite quickly, particularly in search of food. Once detected, the prey will be engulfed into a vacuole, from which it is passed back into the cell within the test to be digested. This mode of feeding categorises them as phagotrophs, whether carnivore or herbivore. The nucleus always remains within the test and in some cases there may be more than one.

Large forams and most planktonic species have algal cells living as endosymbionts inside their cell, exchanging nutrients and helping with test formation. The foram has multiple chambers that help foster the symbiotic relationship, where the algae are enclosed within vacuoles produced from the host membrane. A wide range of algal species have adapted to the forams, from diatoms (such as *Navicula* and *Nitzschia*) to dinoflagellates (*Symbiodinium* and *Gymnodinium*), the green alga *Chlorella* and the unicellular red alga *Porphyridium*. Those hosting diatoms are often not specific about the species they will form relationships with, and those may come and go, with one species replacing another. They may also mix diatoms with dinoflagellates. The forams adapt to the presence and needs of a photosynthesising symbiont by spreading the cytoplasm containing the symbiont vacuoles out to produce a larger surface area, making the shell thinner on their side and moving closer to the light. Planktonic forams are even known to push the cytoplasm containing the symbiont through the small test perforations to receive better illumination by day, bringing them back in by night.

Within the plankton

While the number of species living in the plankton is low, forams are abundant in coastal sediment, both sand and mud along the coast and into brackish estuaries. On rocky shores they take shelter in microhabitats. *Elphidium crispum* is common creeping over seaweeds, while *Allogromia*, one of the species that has a single chamber and strengthens the test with iron and calcium ions, can be found inside holdfasts of *Laminaria* (kelp). *Haliphysema* lives around the edges of rock pools where the pink fringe weed *Corallina* grows. Brushing a fine net (a tea strainer with a good handle will work) back and forth through the weed at the side of a pool may yield very small sea spiders and the occasional foram like *Haliphysema*. They collect the spicules of sponges to cover the cell and are quite distinctive. *Cribrostomoides*, by contrast, adds sand grains randomly.

These are just a few of the species living on the shore that can, for a while, exist as tychoplankton. If their tenuous grip on the substrate is lost they will be washed away by waves to become a component of the shore plankton. Many will be dead, but living ones do occur and after turbulent weather they will be particularly abundant. Sometimes the diversity is quite astounding. Identification is difficult and one of several sources is Haynes (1973).

Foram reproduction

The details of reproduction are understood in only a few cases, but it is known that there are differences between the species. Typically, reproduction begins with a zygote that is naked, looking like an amoeba with no shell. The first thing it does is to secrete a basic test, called the proloculum. As it grows and matures within this test, in some cases adding new chambers, it reaches an asexual or vegetative phase called the agamont. Young amoebulae, or gamonts, form by mitosis. As these mature and enlarge they produce a new test with a large proloculum, adding debris to the outside to produce a flotation chamber. As this floats off into the plankton it repeatedly divides to produce and release gametes. Once these have been fertilised, a swimming zygote is formed to complete the cycle.

Radiolaria: Polycystina and Acantharia

The classes Polycystina and Acantharia, both belonging to the phylum Radiozoa, are exclusively marine and can be common in the shore plankton. Although both are referred to as radiolarians, Polycystina is the only true form, although the acantharians are very similar in structure. They vary enormously in size, from 50 to 500 microns, and are intricately beautiful planktonic amoebae. These were the species that particularly inspired Ernst Haekel in producing his book *Art Forms in Nature* (1899–1904). Species of Polycystina are most likely to be regularly seen.

Acanthometra pellucida is a spectacular species in the Acantharia that is common in the neritic zone and is one of the most exquisite under the microscope. When a bloom occurs in late summer, it can begin to appear on the shore and in samples. This is also one of the largest species within Radiolaria. The characteristic features are the 20 perfectly arranged spines that create a frame or skeleton on which the cell is based. Each spine is a crystal of strontium sulphate. A central capsule encloses the cytoplasm, within which are located a number of symbiotic algae, the zooxanthellae that we have come across before. Several species may be present; they are not host specific, and one mature *Acanthometra* cell can accommodate thousands of individuals. Most organisms that possess symbionts pass them through the generations in gametes, but the reproduction of radiolarians is via a flagellate stage that contains no symbionts, and the process by which they regain symbionts in the next generation is unknown.

The cell's cytoplasm extends outwards into pseudopodia, forming reticulopodia (as we saw with the foraminifera, above), and this outer region is called the cortex. These fine extensions hold out the photosynthetic algae as far away from the spines as possible to trap as much light as possible.

FIGURE 9.4 A bloom of *Acanthometra pellucida*; some have consumed diatoms. Each individual approximately 450 μm.

FIGURE 9.5 A single young specimen of *Acanthometra pellucida*, 250 µm across spicules.

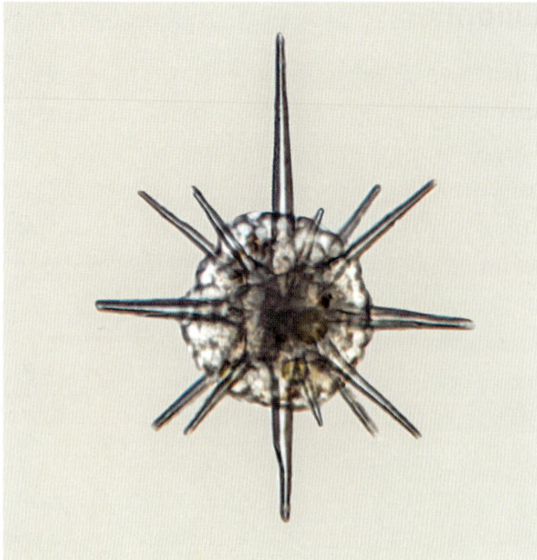

FIGURE 9.6 *Acanthostaurus* species, height of spicules 130 µm.

There is a nucleus in the centre, but in many species the vegetative cell's nucleus is divided up into a number of small ones. The phylum name, Radiozoa, means 'ray feet', and comes from the axopods that radiate out from the cell. At the centre of the cell is a complex of microtubules forming the axoplast from which the axopods radiate outward as groups of microtubules covered in flowing cytoplasm. A finer form of these, lacking the microtubules, are called filipodia and, combined with the axopods, produce the means to collect food. Microplankton, such as bacteria, diatoms, ciliates and larval crustacea, become caught on their surface and then are engulfed into the cell. Kinetocysts (unique to Radiozoa) and mucocysts, which we saw in dinoflagellates, are found on the axopods and can catch prey with threads and mucus. Sometimes the prey is too large for this and so, once a vacuole has been created in the axopod, digestion occurs in situ before the vacuole is moved into the central core.

Although the basic body pattern is the same, organisms in Polycystina lack the strontium sulphate spines. Instead, they have numerous variable rods of silica spicules. Outside the central capsule there is a layer called the calymma, which is a flotation device. In some genera, like *Acanthometra*, the calymma looks like an organised mass of bubbles, while in others it is more random. The bubbles are cell vacuoles filled with oils and low-density ions. During gales when there is severe turbulence they can release some of the oil or replace the ions with more dense forms which causes the cell to sink to less stressful conditions. Over time the cell replaces the oil and ions so that it rises in the water column, back to the light. Although they lack any central means of locomotion, the calymma can allow them to slowly migrate through different currents, facilitating dispersal.

Mitosis is the method of vegetative reproduction, but after successive mitotic divisions the cell ends up with a polyploid nucleus. When this happens the nucleus breaks into small flagellated individuals called swarmers that disperse in the water before losing their flagella and growing into adults. Direct formation of gametes and sexual reproduction is rare and, if it does occur, then they tend to fertilise themselves.

Ciliophora: the phylum of ciliates

Ciliates are an astonishingly diverse and complex group of protists. With over 12,000 known species and plenty still to discover, they are found in most habitats. In the marine environment they thrive in the benthos and plankton, on both rocky and sedimentary shores, adapting well to saltmarshes and estuaries with their brackish water. In these places some create sedentary colonies, while others are fast-moving individuals. Individuals living in the surface of shore sediment, particularly soft mud, will easily become swept up into plankton samples from the benthos. They are variable in size, from around 20 microns up to half a millimetre or more; despite being unicellular, they can exhibit quite complex body forms and structures. The pellicle, or cell membrane, is unique and referred to as an alveolar membrane system as it consists of vesicles and an underlying fibrous layer or cortex.

The classification of ciliates is an ongoing problem and they are extremely difficult to identify to genus, requiring considerable specialist knowledge. Around mud and brackish marshes some occasional spectacular species can suddenly appear, requiring rather specialised resources for help with identification, but in this book we will limit the number of species and groups described to those most likely to occur in general seashore plankton:

- Haptorida: e.g. *Didinium* spp.
- Peniculida: e.g. *Paramecium* spp.
- Suctoria: e.g. *Tokophyra* spp. and *Ephelota* spp.
- Peritrichia: e.g. *Vorticella* spp.
- Choreotrichida: e.g. the tintinnids
- Heterotrichida: e.g. *Stentor* spp.
- Euplotida: e.g. *Euplotes* spp.

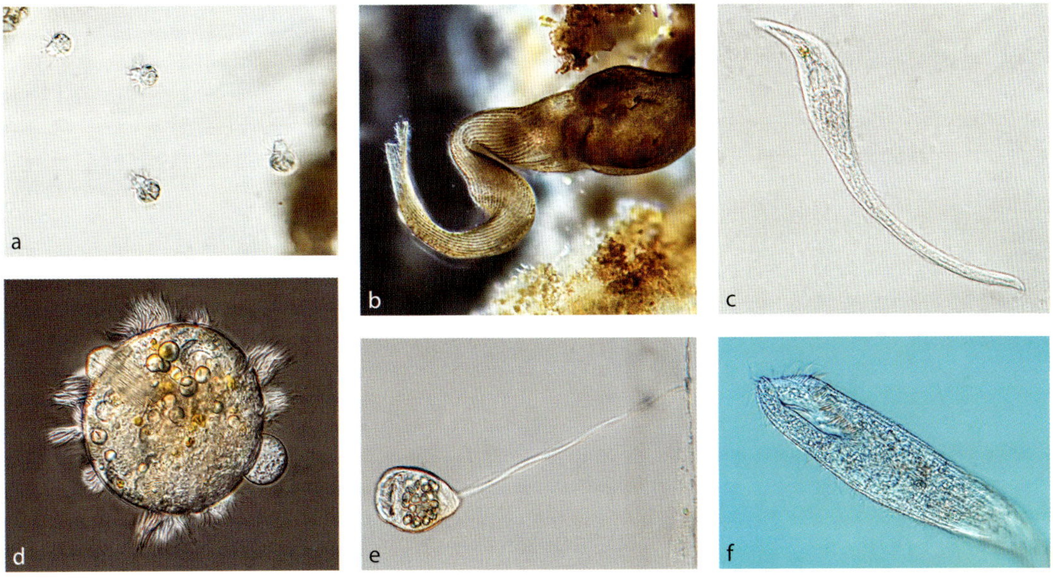

FIGURE 9.7 Examples of ciliates. (a) A group of the mixotroph *Mesodinium*, 15 µm; (b) Haptorida, a large ciliate, 950 µm; (c) a Haptorid *Trachelocerca* sp., 550 µm; (d) the predator *Didinium* in the process of dividing, 90 µm; (e) stalked peritrich *Vorticella* sp., cell diameter 100 µm. Note the myoneme for contraction in the stalk. (f) Undetermined species, possibly *Paramecium*, showing the oral groove, 100 µm.

Cilia

Some species have compound cilia, where the kinetosomes are grouped together, creating broader and longer structures known as cirri or membranelles, as found on *Euplotes*. Cilia are essentially a shorter version of flagella, but their movement is different and there are variable patterns of kineties across the group that create quite different behaviours. In the case of *Euplotes*, these membranelles can be used to creep along substrates.

When a cilium beats it describes a cone shape rather than beating in a single plane. It produces a power stroke towards the back of the creature to drive it forward, folding in on itself on the recovery stroke. The disturbance in the water that the cilium produces stimulates neighbouring cilia to beat, so forming a coordinated wave. Locomotion can be a spiral side-to-side movement in the forward direction. If the ciliate meets debris or another organism they pause and even reverse slightly, as if gauging what to do, in what appears to be a kind of trial-and-error behaviour. Some, like *Didinium*, are particularly adept at reversing and manoeuvring. It has been discovered that in *Paramecium* the cell membrane has a resting and an 'action', or depolarised, potential. When resting, the cilia beat to the rear so that the organism goes forward, while in the depolarised state the cilia reverse and the movement is backward.

The cilia covering the cell of a sessile ciliate are presented differently to those found in errant (more mobile) forms as they are not being used for movement. Instead, sessile ciliates have contractile protein structures called myonemes around the cell. *Vorticella* is attached to a surface by means of a stalk within which a myoneme can be seen spring-like when the ciliate is extended. A *Vorticella* under the microscope can very suddenly disappear from view as the myoneme contracts, shortening the stalk and pulling the cell close to the substrate. Myonemes are also present in the funnel-shaped species *Stentor*, where they are located within the pellicle so that, when contraction takes place, the entire cell seems to collapse.

The cilia

Ciliates possess two important characteristics. For one thing, they are unique in having two types of nucleus: micronuclei and macronuclei. We will consider these when we look at reproduction. The second feature is the presence of cilia, or hair-like structures, used for feeding and locomotion – although some species are sessile. The cilium has several components, including basal bodies, or kinetosomes, as well as a mix of two microtubular roots and a fibrous root. Together this is called the infraciliature. The microtubular roots create the anchor for the cilium and give it support. The cilia, the collective term being called somatic ciliature, are arranged in rows called kineties and the pattern that these rows form is used in species identification. Most ciliates have a 'mouth' area called a cytostome, around which may be specialised oral ciliature to help with feeding.

Feeding

Ciliates have no plastids for photosynthesis and are either filter feeders or phagotrophs. The collection of suspended material such as bacteria and particulate matter requires special oral cilia. These compound cilia lie in a surface depression to one side of the cytostome called the adoral zone of membranelles (AZM). In both *Vorticella* and *Stentor* its movement produces a water current that draws food items into the cytostome. The AZM in *Euplotes* operates when it is walking, drawing suspended matter off the surface and passing it under the cell into the cytostome. Once food has entered the cytostome a vacuole forms around it and enzymes are then released into the space to begin digestion. Soon after the vacuole is produced the pH inside is changed to enable digestion as the new organelle then travels around the cell. During the journey the digested material is absorbed into the cell cytoplasm before the vacuole arrives back at the exterior surface, bursts and releases any waste.

FIGURE 9.8 *Euplotes* with the large membranelles visible, cell width 80 µm.

FIGURE 9.9 A marine *Stentor*, 200 µm.

FIGURE 9.10 Part of a colony of the peritrich *Zoothamnium alternans* within a *Rhizosolenia* spp. diatom bloom. A myoneme for contraction is visible within the stalk and is similar to that seen in the solitary *Vorticella*. Length of colony 350 µm.

Capturing and ingesting food is more complex. Prey consists of other protists, diatoms, pieces of algae and suitably sized small invertebrates. Incredibly, some species, such as *Didinium*, can consume prey the same size or even bigger than themselves. In ciliates that capture large prey the cytostome has adapted to have both trichocysts and microtubules, creating a structure called the cytopharynx. It is easy to forget, when looking at such structures, that we are dealing with a single, albeit complex, cell. When trichocysts are fired out at the prey they stick to it. The cytopharynx then pulls the food into the cell, which opens up a huge space to accommodate the size of the prey.

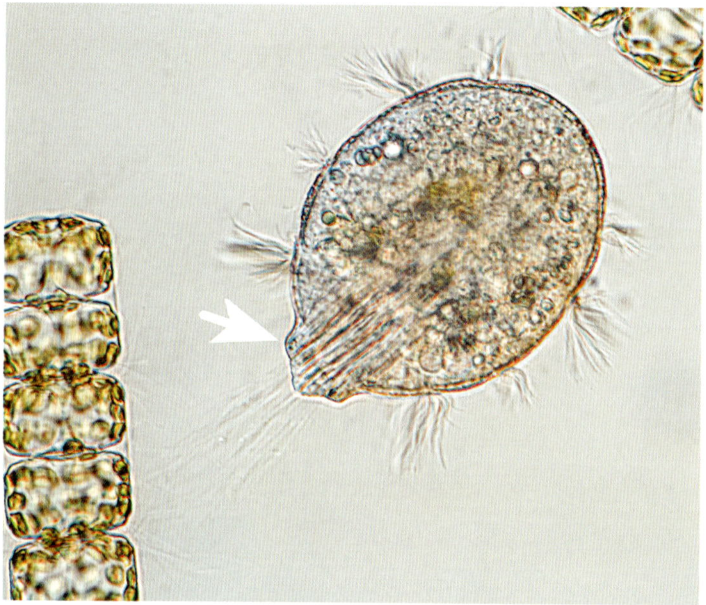

FIGURE 9.11 *Didinium* sp. with diatom chain of *Lauderia*. The arrow marks the cytopharynx and discharged trichocysts. The ciliate normally has two bands of cilia, but this individual has four as it is about to divide into two new individuals. 90 µm.

FIGURE 9.12 The difference between the two suctorian genera *Tokophyra* and *Acineta* is so subtle as to make identification at this magnification difficult. The latter has a lorica covering the cell and usually a short stalk. Possibly, this is the former genus; it can be fairly common in the late spring. Total length 200 µm.

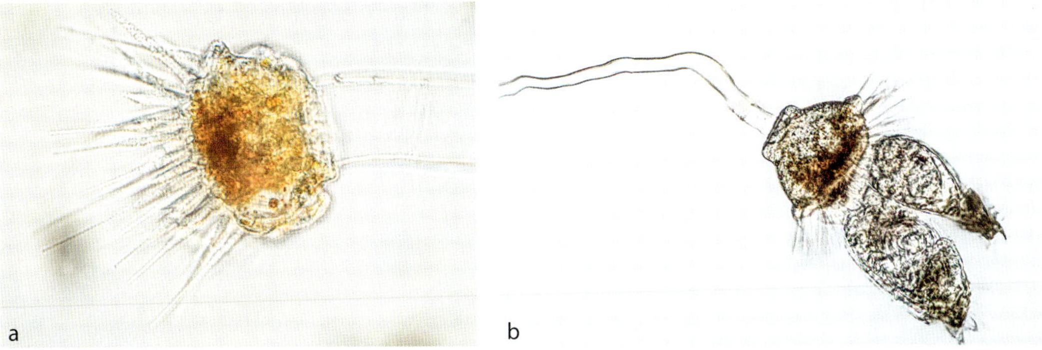

FIGURE 9.13 *Ephelota* sp. (a) A close-up view of the cell, width 90 µm; (b) two *Synchaeta* rotifers that have been caught by the tentacles and immobilised with toxins and digestive enzymes.

A third, more complex, feeding mechanism is found in members of the group Suctoria. This is an unusual group of sessile ciliates that do not have cilia in the adult, although kinetosomes do occur, left scattered within the pellicle. They mostly have a stalk that is secreted from the scopula, the name of the front region in the developing young ciliate when the stalk is first forming. The cell at the end of the stalk possesses generally immobile tentacles, although some may have a slight lateral 'twitch'. In *Tokophyra* and most suctorians the tentacles have a knob at the tip called the haptocyst. When a prey item touches a haptocyst it discharges a barb that contains enzymes causing paralysis, preventing it from swimming away. A temporary tube forms inside the tentacle and the prey is slowly sucked inside to form a food vacuole. In some species the tubular tips contain toxicysts that not only paralyse the prey but also inject enzymes so that digestion begins immediately, with products flowing into vacuoles. *Ephelota* has these types of tentacle in large numbers, ending in what appear to be sharp tips on which the prey is impaled. As a robust species, they appear to be able to immobilise quite substantial prey. Suctorians may appear to be poorly adapted for the plankton, but the stalk needs only a piece of debris, organic or mineral, for attachment and these organisms may in consequence be common in early summer shore plankton.

Ciliate reproduction

As we have already learned, ciliates have two types of nucleus, the micronuclei and the macronuclei. The micronucleus is used in reproduction, synthesising DNA and ensuring that equal numbers of chromosomes enter the daughter cells. In contrast, the macronucleus has multiple sets of chromosomes (called hyperploidy) that are used in the general operation of the ciliate. The macronucleus varies a great deal in ciliates, but the nuclear envelope always remains intact during cell division, although the size does reduce. After cell division a great deal of reconstruction of all the specialist parts, such as the cytostome, is needed. Asexual reproduction is through mitosis. Suctorians undergo a process of budding, where the region near the tentacles is drawn inwards to create a bud on either side.

Sexual reproduction is by a process of conjugation, where a pair fuse to exchange micronuclei, ensuring a mixture of genetical material. Appropriate individuals need to meet; that is, ones that have not come from the asexual division of the same parent organism. The individuals are called conjugants and they attach near the cytostome. The micronuclei then undergoes two divisions, which is equivalent to meiosis. Three daughter micronuclei from each disintegrate and become incorporated into the cytoplasm of the conjugant while the remaining micronuclei carry out mitosis, and these will be the gametes. This is just the beginning of the process, which is long, drawn-out and complex but will result in each conjugant producing four daughter cells with reshuffled genes.

Differences are apparent between species; for example, *Vorticella* produces two conjugants of different size, a microconjugant and a macroconjugant, and invariably genetic material goes only one way, into the latter individual. In *Ephelota* the microconjugant is completely absorbed into the macroconjugant.

Tintinnids: the little gems

Tintinnida is a huge ciliate group, with around 1,500 species. This makes them the largest taxon within the Ciliophora and they are believed to represent some of the most advanced forms. They are almost entirely marine and make up one of the most important components of the plankton. So abundant are they that, when collecting samples with a 50-micron mesh net from the shore, there will almost certainly be tintinnids present on any day of the year, and possibly they will dominate the sample. There have been occasions when I have withdrawn my net from the water thinking that it is full of sand, only to find that each grain is actually a rounded tintinnid. They quickly die and in a sample dish look like uniform-shaped minerals lying at the bottom, although once under the microscope some will probably prove still to be alive.

The 50-micron mesh, while collecting a good range of tintinnid species, has been shown to severely underestimate the biodiversity and population density of the group, reminding us of the importance of understanding your aim in taking samples. For a general look at micro and mesoplankton 50-micron nets are ideal, but if you want more precise information on a group such as tintinnids a targeted approach is needed because, in this case, the organisms are likely to be substantially smaller than the mesh size.

FIGURE 9.14 A pair of tintinnids carrying out conjugation. They have linked in the centre where the two cells have emerged from their lorica. Length of pair 120 µm.

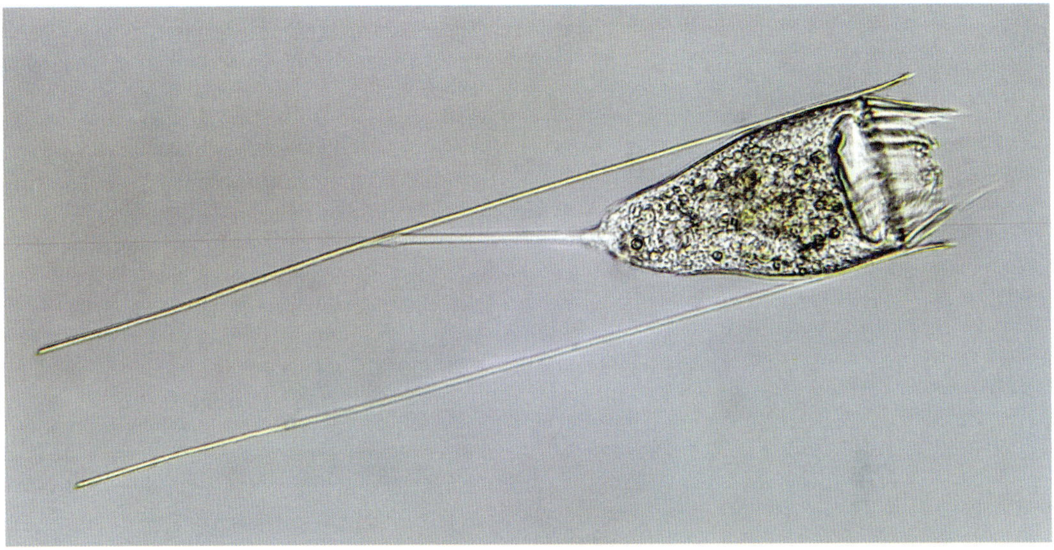

FIGURE 9.15 *Eutintinnus* in its hyaline tube. Tube length 90 µm.

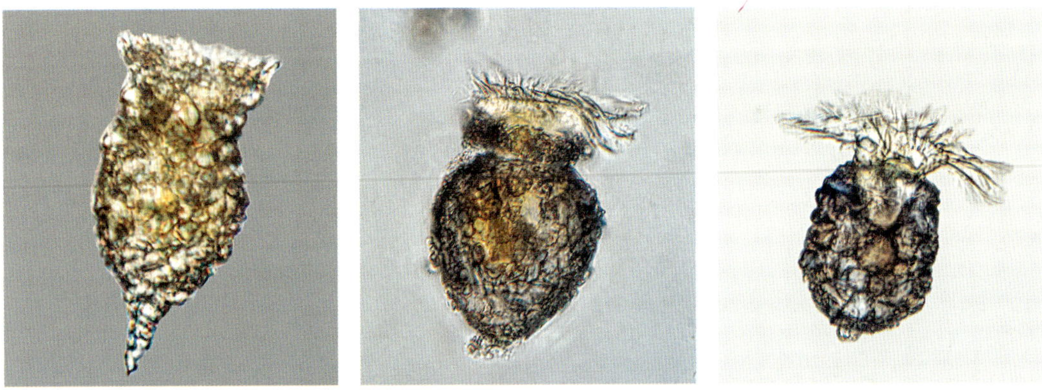

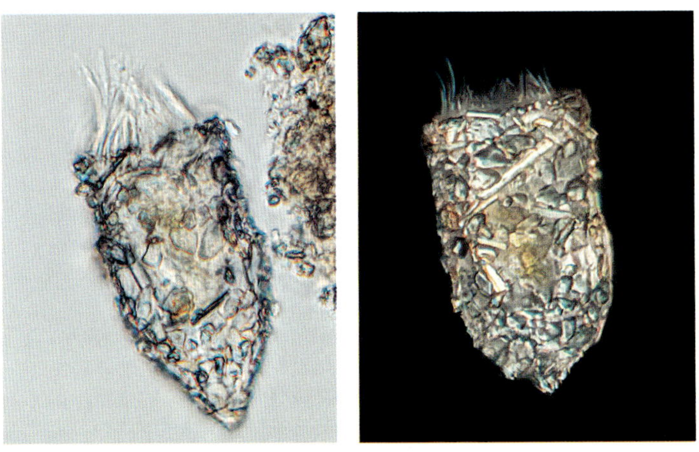

FIGURE 9.16 Five different species of encrusted loricate tintinnid. Note the typical bell shapes and ciliated crowns. All approximately 60 µm.

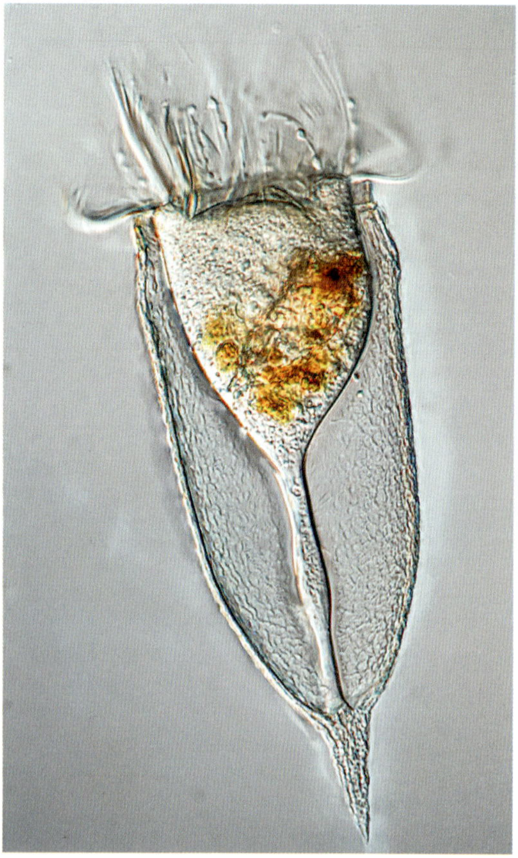

FIGURE 9.17 *Favella* sp. with crown extended out of the wine-glass-shaped hyaline lorica. Note the membranelles, or compound cilia, in the crown. Lorica length 220 µm.

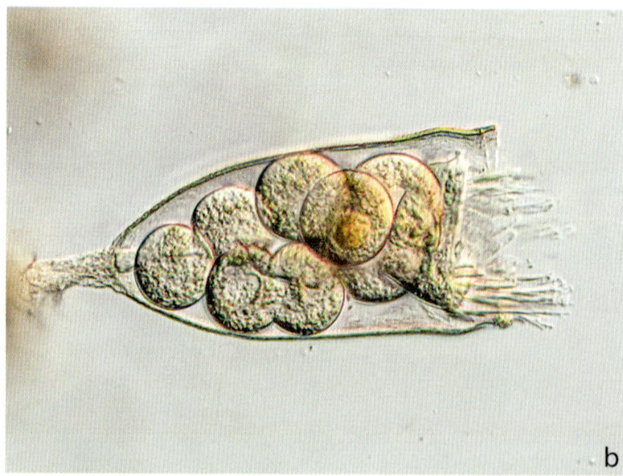

FIGURE 9.18 The endoparasitic dinoflagellate *Euduboscquella* sp. with *Favella* as the host. (a) The dinoflagellate inside the host indicated with an arrow. (b) Large sporocytes (macrospores) produced by the endoparasite. Lorica length 170 µm. (c) Small sporocytes (microspores) with the host now dead. Lorica length 180 µm.

Tintinnids' resemblance to sand is due to the tough coat, or lorica, that they secrete around their cell. This is commonly bell-shaped, giving this group its name: tintinnid is from the Latin for bell. Many of the small species stick tiny minerals to the lorica so that they are camouflaged and protected. The shape and degree of mineral covering can help in species identification; however, as with all ciliates, this is not straightforward. The lorica is made from proteinaceous material, but the actual chemical composition has eluded scientists for years. It was thought to contain chitin, although this is now largely dismissed.

The lorica is open at the anterior (the mouth of the 'bell' shape), where feeding takes place, and occasionally, with species such as *Eutintinnus*, the posterior end is also open. The cell is anchored by a fine, contractile stalk to the lower end of the tube. Myonemes, like those present in the genus *Vorticella*, enable the organism to withdraw back into the lorica. When the cell extends the shape becomes long and thin, with a deeply incised mouth, the buccal cavity resembling a funnel. Around the top of this a crown of membranelles, compound cilia, is extended just outside the lorica for feeding.

The membranelles, when active, can also enable rapid swimming. As we saw with some other protists, the beating can be suddenly reversed, enabling backward locomotion as well. A fresh plankton sample with tintinnids should be viewed under low magnification (e.g. ×5) to see their smooth gliding action (compared with the jagged, erratic movement of crustaceans).

Just below the cilia crown is a lobe of cytoplasm which is the region of lorica secretion, helping to determine the shape. Running down the side of the cell are sparse rows of cilia, with some near the buccal cavity. Most of the structures and organelles can be seen in those species that do not produce an encrusted lorica, such as *Favella*, one of the larger genera.

Prey or predator? Symbiosis or parasitism?

Tintinnids form an essential link between the nanoplankton, which they feed on, and crustacea such as copepods and larval fish, which consume them. Although copepods are sometimes abundant, they do not seem to have any strong effect on tintinnid numbers, whereas the availability of tintinnids' main food sources does. Estuaries, for example, have very high bacteria levels and, of all the coastal areas, they can have the highest tintinnid populations.

Larger dinoflagellates will consume them while smaller dinoflagellates will parasitise them. An example of the latter is the dinoflagellate genus *Euduboscquella* that gains entry to the host, usually a *Favella* species, by being ingested as a spore, where it becomes endoparasitic. They differentiate into a new phase without flagellae and a modified amphiesma ring that is thick and fibrous. The lifecycle is complex and two possible spores (or sporocytes) may then be produced by the parasite – either a small number of macrospores or many microspores. Finding these parasites is difficult and requires a substantial bloom of tintinnids to increase the chance of locating them. It is thought that parasitism has a negligible effect on populations.

As consumers, tintinnids have a particularly high metabolic rate. Nanoplankton such as bacteria, cyanobacteria and small flagellates are primary food sources for tintinnids, but their diet can be quite diverse, including diatoms and dinoflagellates. The size of their prey is determined by the lorica diameter, and the most efficient food source has been found to have a diameter 25% that of the crown diameter – that is, 6.5–8.5 microns. Hence, they cannot consume large diatoms or ones that form chains, such as *Odontella* or *Chaetoceros*, respectively. Diatom blooms have a detrimental effect on tintinnid numbers, which sharply decline.

One interesting association between tintinnids and diatoms is represented by the members of the genus *Eutintinnus*, which has a clear lorica and is open at both ends. It can form a symbiotic association with the diatom *Chaetoceros tetrastichon* to the extent that, even when the tintinnid dies, the lorica remains on the diatom. The tintinnid helps the diatom to reach the light and receives some photosynthetic metabolites from the diatom in return.

Environmental factors

While tintinnids can completely dominate the plankton, with, at times, many thousands in a sample, different species will peak at different seasons. *Favella* is common over the shore in late spring and summer. The much smaller *Tintinnidium* species can peak after a diatom bloom. In fact, generally as diatoms decline so tintinnids increase in density. Other factors influence tintinnid numbers too. A drop in salinity will cause an increase, possibly through reduced competition, as organism diversity is lower. The salinity of seawater is typically 3.5% (mainly sodium chloride) but near the mouth of an estuary this can reduce to nearer 3%. The tintinnid density may then rise

If you encyst…

Like other inhabitants of the plankton, tintinnids will sometimes form cysts enabling them to survive conditions unsuitable for growth. Temperature and light levels are known to be key factors in initiating the process of encystment. As cysts form so they descend to the bottom sediments. Turbulence in the spring or autumn may cause them to rise up into the water column again, where an increase in temperature and light will trigger the emergence of a new stage from the cyst. It is unclear what proportion of cysts will eventually develop, but sediments hold a huge cyst bank of a wide range of planktonic organisms. Only a limited number of cysts have been identified and linked to tintinnids, although it is generally thought that they all have the ability to encyst. This process is important on the coast to maintain high densities of tintinnids.

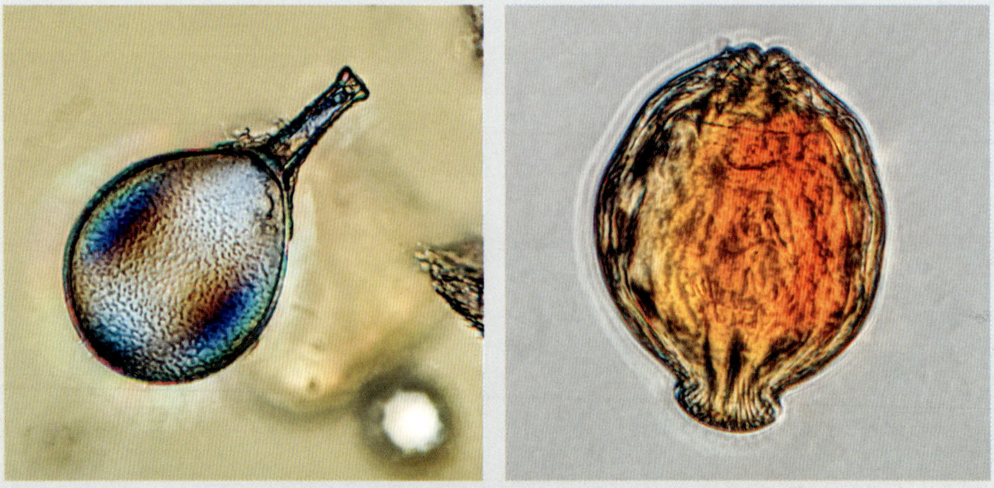

FIGURE 9.19 Two examples of probable tintinnid cysts. Average length 60 μm.

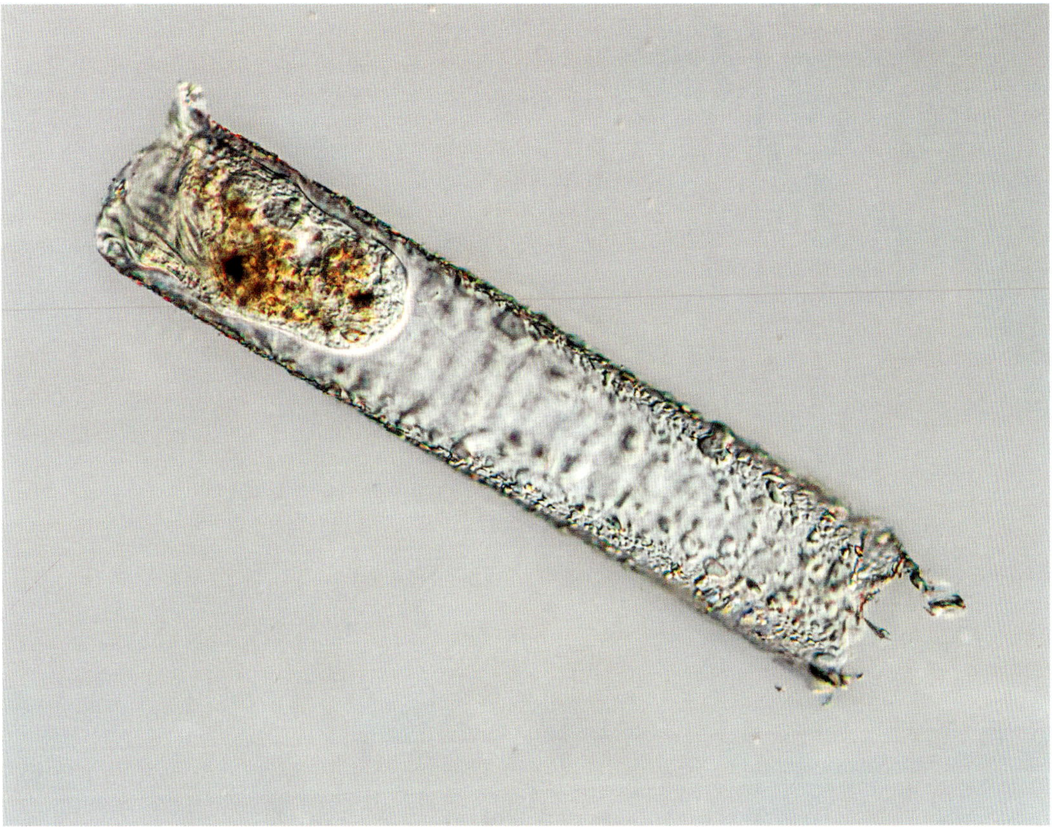

FIGURE 9.20 The tintinnid genus *Leprotintinnus* with a lorica open at both ends and encrusted with minerals. The organism is visible inside on the left.

to a peak before declining as one moves up the estuary into fresh water. Studies on tintinnids in estuaries show that, although they can deal capably with salinity changes, the tidal flow is more of a challenge as it can flush them into unsuitable conditions, and probably has the greatest effect on numbers, with densities fluctuating considerably over a matter of days.

As a result of the sheer numbers of species and their high densities in the plankton, protists of all kinds play a prominent role in the microbial food web as both consumers – of bacteria, particulate matter and so on – and consumed. This is discussed further in Chapter 19.

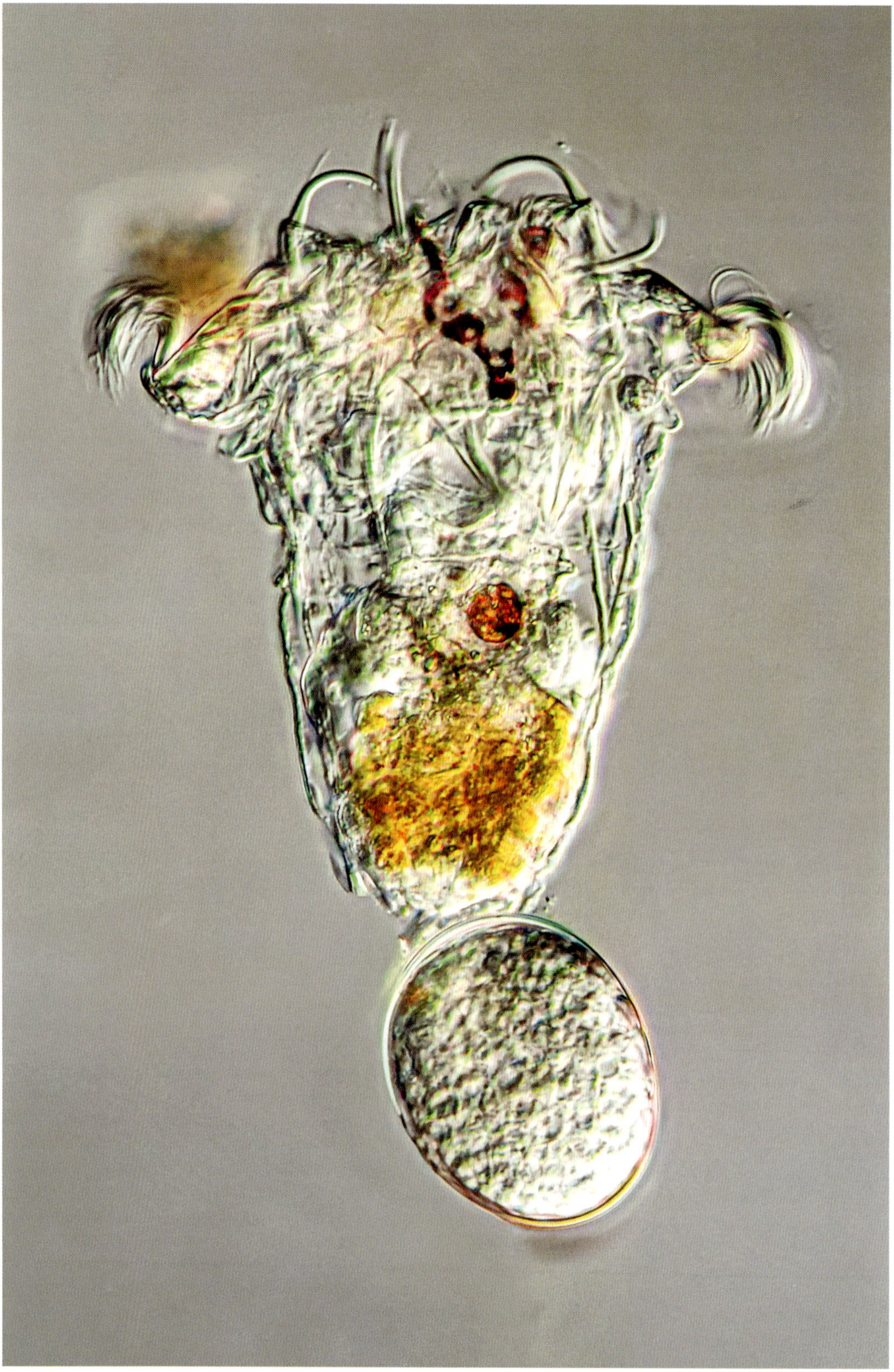

FIGURE 10.1 *Synchaeta*, the marine rotifer. Female with an egg attached. Length 400 µm.

CHAPTER 10

Rotifers, gastrotrichs and tardigrades

> It is a beautiful creature, lively in its movements, and rich in its organisation … clear as glass, perfectly colourless except the distended stomach and the large red eye.
>
> (Gosse 1856)

Henry Gosse had a particular love for rotifers and wrote regularly about them. Marine species of rotifer had been recorded as being bioluminescent years before he carried out his work and he tried many times to demonstrate this, but with no success – it is probable that these early records were in fact of the dinoflagellates that were abundant in the same habitat. In the early part of the nineteenth century rotifers, because of their size (usually less than one millimetre), were considered to be protists (in fact, some protists were called rotifers), so no wonder there was confusion. Rotifers were not the only organisms to be confused with protists; for years, so were the gastrotrichs. Now, like the rotifers, they have been given their own phylum. Tardigrades, or water bears, are yet another group of unique species with their own phylum. But first we will consider the rotifers.

Manic wheels of life

The word rotifer means 'wheel bearer'; another term for this organism is 'wheel animalcule'. These names are in reference to the corona, an area on the animal's head that has large numbers of cilia that in many species look like a pair of rotating wheels. It is the corona that produces fast locomotion as well as the water current necessary for feeding. Structurally, rotifers are unusual and complex. They originated in the marine plankton before evolving to inhabit every habitat on Earth, leaving just a few genera in the plankton. Most species, nowadays, are found in fresh water, but they are also found in moss and soil, where they are common and easily observed. Marine rotifers are more exclusive and completely dominated by the genus *Synchaeta*. Around 25 species are known worldwide and all are very difficult to tell apart, which has resulted in quite limited scientific study. They can suddenly appear in early summer above sheltered sedimentary seashores to dominate the zooplankton and a month or two later disappear just as suddenly. Populations of *Synchaeta* are strongly influenced by temperature and different species are believed to have their own optimum thermal ranges. Within the Mediterranean a succession of different species can be found from early spring through to the autumn, although heat stress in late summer can cause complete loss of populations.

The family Synchaetidae is part of the major Monogononta group of rotifers and displays typical rotifer biology. They are divided into a head, body and foot. The corona comprises a circle of cilia around the head and another around the mouth, although in *Synchaeta* the corona is on appendages, distinct small tufts on either side of the head. Within the head is a brain and on the outside sense organs such as sensory bristles and pits, along with a sensory dorsal antenna and a pair of similar lateral antennae on the side of the body. The entire shape of the rotifer's body is cone-like, tapering to the foot, which terminates in a pair of small toes. The latter connect to adhesive glands that secrete a substance that allows temporary attachment to a substrate and in some species a crawling action.

FIGURE 10.2 Philip Henry Gosse's engraving of *Synchaeta* that he titled 'The Luminous Syncheta'.

Inside the body is a complete gut with specialised compartments, including a stomach. The mouth is a small aperture at the top of the body, just beneath the cilia ring. Within the mouth is the mastax, a fairly complex set of internal jaws that in *Synchaeta* are quite slender. There is a particularly well-developed hypopharynx muscle, which gives the jaws a sucking action. The top cilia create a water current in opposition to the tufted cilia on the side of the head and the combined effect is to bring material into the mouth. From the pharynx the food passes through the oesophagus down to the stomach, where there is a pair of gastric glands. *Synchaeta* is primarily an algal feeder, mainly consuming diatoms up to 50 microns across. Larger diatoms are avoided in part because of the mucus that they produce, which can make their consumption difficult. A bloom of large species such as *Odontella sinensis* and *Bacillaria paxillifer* can therefore have a negative effect on *Synchaeta* population growth.

Under the microscope amazing detail of the internal organs of live *Synchaeta* is visible because the epidermis is so thin and transparent. An external cuticle is secreted by the epidermal cells, which also have a thin skeletal layer, providing flexible support to the body. The epidermis is one of the unique features of rotifers, in that it is syncytial tissue. A syncytium is a fusion of multiple cells that lack dividing membranes and appears like one giant cell with many nuclei inside (our skeletal muscles are examples of syncytial tissue). In rotifers this nuclei number is fixed at somewhere between 900 and 1,000. A further feature of rotifers is that the cells of their body exhibit eutely:

that is, cell constancy. After they have grown to maturity the number of cells is fixed and no more can develop; mitosis completely ceases, so cells cannot be replaced. Any growth in size that does occur is due just to cell enlargement.

As well as the gut, other organs may be seen through the transparent epidermis. A pair of fine tubes called the protonephridia run down each side of the body, terminating in flame bulbs. These are bunches of fused cilia that form a flagellum-like structure which constantly beats, appearing like a flickering flame inside the body. These are the excretory and osmoregulatory organs.

There is no body musculature as such, although beneath the epidermis lie a small number of circular and longitudinal muscle bands. The body cavity acts as a hydrostatic skeleton. Contracting muscle bands will cause some slight change in body shape, but these contractions mainly act to extend or shrink the foot and corona.

The usual rotifers found in shore plankton are female *Synchaeta*; males are present only occasionally. Females have one small ovary, along with a gland to produce yolk. The number of eggs the mother produces is determined by the fixed number of nuclei in the ovary, usually between 10 and 50. Reproduction follows alternating phases of parthenogenesis and sexual reproduction. In early summer, when they are reaching their peak densities, parthenogenesis dominates, where amictic females produce asexual diploid eggs. These can sometimes be seen inside the body, but later are attached to the outside and eventually dropped to develop and hatch within the plankton. These can produce more amictic females or mictic ones with the ability

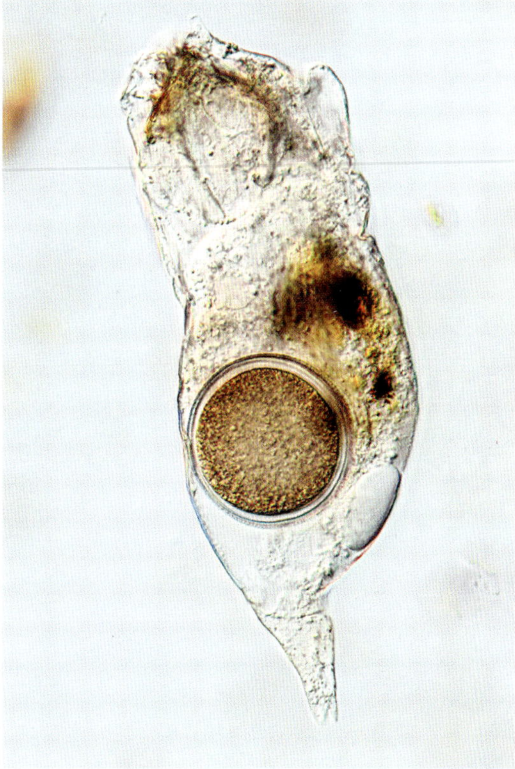

FIGURE 10.3 A recently dead *Synchaeta* sp. with a resting egg inside. This drops out of the plankton to remain in the benthos until conditions improve. Length 300 µm.

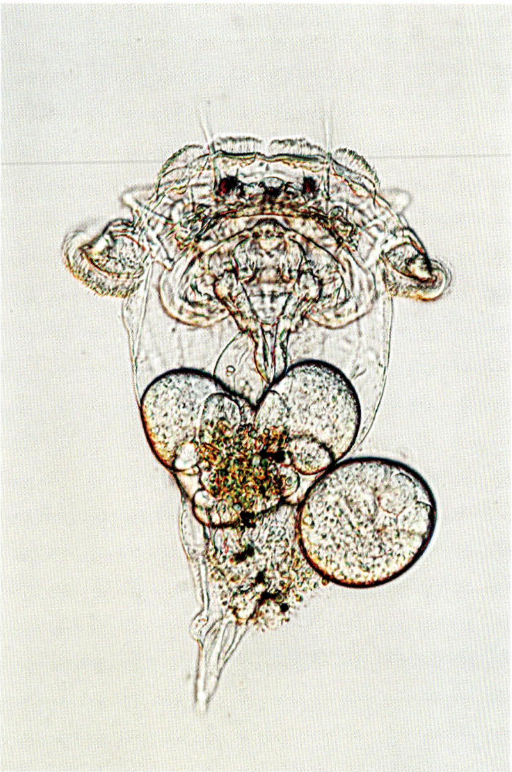

FIGURE 10.4 Female *Synchaeta* sp. with three eggs visible inside the body below the mastax. Length 350 µm.

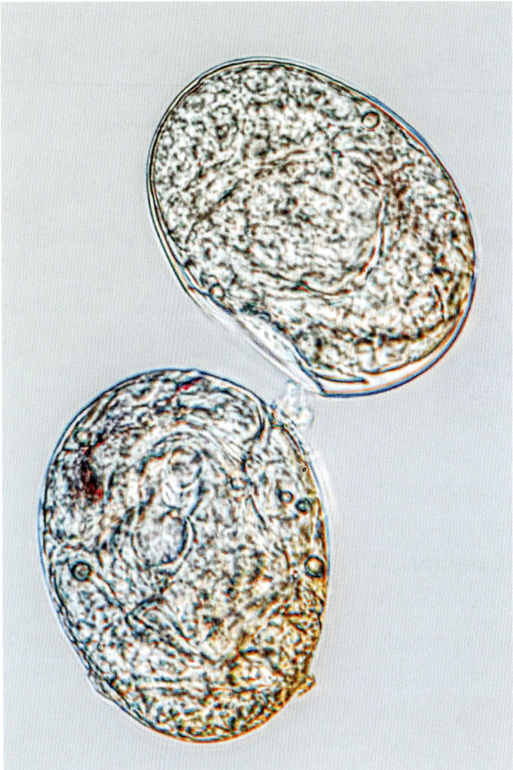

FIGURE 10.5 Developing *Synchaeta* sp. eggs. Average diameter 50 µm.

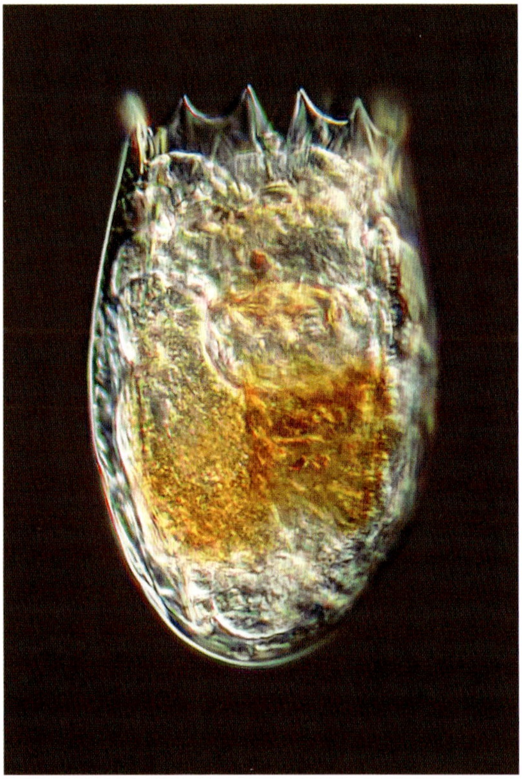

FIGURE 10.6 The estuarine rotifer *Keratella*. Length 300 µm.

to produce haploid eggs that, if left unfertilised, will develop into males. The males are much smaller than the females and the pear-shaped testes may be seen with developing sperm inside. If the haploid eggs are fertilised, then the diploid state will be restored. A third type of egg may be produced, the resting egg, with a tough exterior. These eggs, with a high lipid content, are often adorned with a fine covering of short spines that enables them to attach to debris. These are produced when environmental conditions become unsuitable and descend to the benthos. With improving environmental factors the resting stage is broken to produce amictic females.

On the coast *Synchaeta* is the genus most likely to be found in the plankton, but around estuaries and coastal marshes *Keratella* may appear. It can exist in a salinity down to 0.4% and occurs in pools on saltmarshes and mudflats as well as in the plankton. The two genera are quite different and so identification is straightforward. *Keratella* has no foot and there is a covering over the body called the lorica. Six spines adorn the head and there is a single red eye placed centrally.

Marine gastrotrichs

Gastrotrichs are among the smallest multicellular creatures on earth. They are widely distributed wherever water is present, occupying an important niche among debris and in tiny spaces between sand grains. In the marine environment they favour sheltered sedimentary shores where they can rapidly move around these interstitial spaces, having the appearance of a ciliate protist, probing and active. Although they have circular and longitudinal muscles they use the cilia that cover the body for their locomotion.

Scraping the top few millimetres off the surface of a mixed sand and mud shore should reveal plenty of examples, with backwater habitats usually being more productive for the planktonologist; clean, exposed sand will not supply sufficient food for a thriving population of gastrotrichs. The surface of these sediments is easily disturbed by even a gentle breeze as the tide goes out, often taking gastrotrichs into the plankton. Although there are a few known species that are specific to the pelagic way of life, most shore plankton samples will pick up those that have been flushed off the sediment.

The gastrotrich most likely to appear in the shore plankton is *Turbanella hyalina*, in the order Macrodasyida, which contains the largest number of marine gastrotrich species. All gastrotrichs have a through gut with a mouth and anus. Being so small, they require no circulation or gaseous exchange system and the outer cuticle can be quite thin, with no chitin present. They are all long and relatively flexible, which allows them to move between the sand grains. The body can be covered in scales or spines and even hooks. Some of these are adhesive tubes that release a glue to temporarily stick the animal to grains and other material. *T. hyalina* has a number of these tubes visible, protruding from the body (see Figure 10.7). True planktonic species tend to have very few adhesive glands. At the front of the head are many sensory bristles, probably to assist in finding food. Gastrotrichs feed on bacteria, protists and fragments of detritus, which are swept into the mouth by cilia, with the pharynx acting as a suction pump. At a length of around 200 microns, *T. hyalina* is one of the smaller species, while the usual gastrotrichs encountered in sediment will be at least double this – with one-millimetre examples possible. Most gastrotrichs are hermaphrodites. Eggs are glued to a sand grain, where they are then fertilised by sperm either from the egg-laying individual or from another individual. There is no larval stage; rather, juvenile gastrotrichs hatch from the eggs.

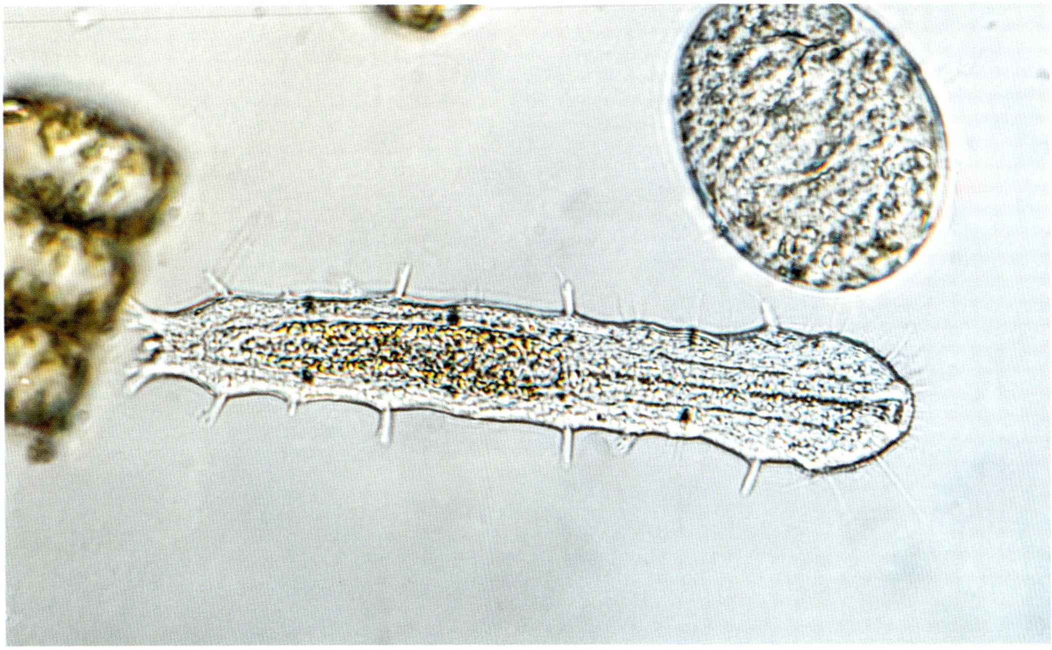

FIGURE 10.7 The common gastrotrich *Turbanella hyalina*, found in sediment surfaces, will occasionally appear in shore plankton. To the left is a small diatom *Lauderia* chain; the large sphere is a *Synchaeta* egg. Gastrotrich length 160 µm.

Entoprocts and trochophores

This chapter has seen some curious microscopic creatures, multicellular organisms with unique features. For example, members of the phylum Entoprocta are sessile, often living in colonies of zooids attached by stalks to rocky shores. They produce both lecithotrophic and planktotrophic larvae that live in the plankton. However, the larvae appear to be a type of trochophore, a first-stage larva common to many groups, including molluscs and annelids (see Chapter 14). The phylum Sipuncula (meaning 'little pipe'), a type of marine worm, are the same, living on the shore and producing several forms of trochophores, one of which goes on to produce a unique oceanic type, the pelagosphaera, found only in this phylum.

Marine tardigrades

Unlike land and freshwater species, marine tardigrades are a poorly understood group. Many tardigrade species are believed to live in the oceans, both as plankton and in sediments of the abyssal depths. Intertidal habitats are also inhabited by tardigrades, from fine mud in estuaries to coarse gravel and rocks. Here they live among algae and in crevices, including around barnacles. On the latter they will not only feed on detritus but as semi-parasites. They have been found living on the gills of sessile rocky shore organism such as mussels and inside crabs. Inevitably, these shore tardigrades find their way into the plankton, generally as a result of waves dislodging individuals, although I have found an individual caught in the shed exuvia of a barnacle; presumably they were on the barnacle when the exoskeleton was shed into the plankton.

They are often referred to as water bears because of their loping, bumbling gait and their appearance: the body is tubular with four pairs of short, unjointed legs that are hollow extensions of the body. Toes at the end are tipped with claws. The outer cuticle contains chitin and they reach maturity through moulting, up to six times. Reproduction is both asexual and sexual, with eggs being laid into the shed cuticle. To feed, tardigrades pierce cells with a stylet and draw out

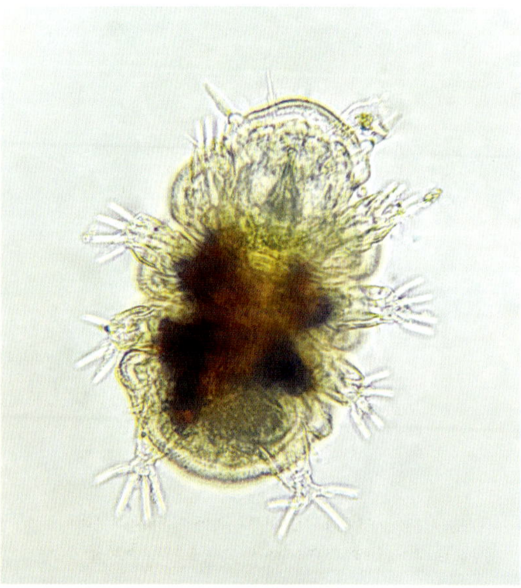

FIGURE 10.8 *Orzeliscus* sp., a marine tardigrade. Length 250 µm. (Elizabeth Beston)

the contents. They are well known for their amazing ability to survive extreme conditions and can suspend their metabolism when necessary, entering a 'tun' state. Tardigrades have their own phylum, and much debate exists about their ancestry, which seems to align with Arthropoda. They have been placed in this chapter for convenience, but also because of the phenomenon of eutely, which they share with rotifers: that is, when maturity is reached the cell number is fixed and no more mitosis takes place.

FIGURE 11.1 The Compass Jellyfish *Chrysaora hysoscella* in open water. Diameter of umbrella 30 cm (Isabella Tuiran).

CHAPTER 11

Predatory jellies

There is a significant amount of gelatinous life in the plankton, giving rise to a separate term, 'gelatinous plankton'. Primarily, this refers to two phyla, Cnidaria and Ctenophora, both of which use the plankton for food and dispersal.

Introducing the cnidarians

This is a phylum of stingers including sea anemones, jellyfish and hydroids with a worldwide species number exceeding 11,000. They are a diverse and successful group. Much of this is down to two principal reasons: first, they can rapidly produce colonies by asexual reproduction; secondly, their lifecycle involves two phases, medusoid and polypoid. This phenomenon, where two visibly different phases occur in a species' lifecycle, is known as 'alternation of generations'. The medusoid form, or medusa, is jellyfish-like organism that swims in the plankton. The polyp is usually attached to a substrate, such as rock or seaweed, like a sea anemone. Cnidarians are radially symmetrical, a quality that is quite obvious in both medusa and polyp, whereas the relationship between the forms is less clear to the casual observer; simplistically, the former is an upside-down version of the latter.

The polyp is a sac with an opening, surrounded by tentacles, at one end. The body is constructed of two cell layers, the endoderm and ectoderm, which sandwich a jelly-like layer called the mesoglea. The cavity within the sac forms the gut, although it is incomplete because there is only one way in and out. Food is collected by the tentacles, which contain stinging cells called cnidae or nematocysts. A very basic nerve network is present, but nothing that could be described as a central nervous system. There is also a lack of any head structure (no cephalisation). All of these

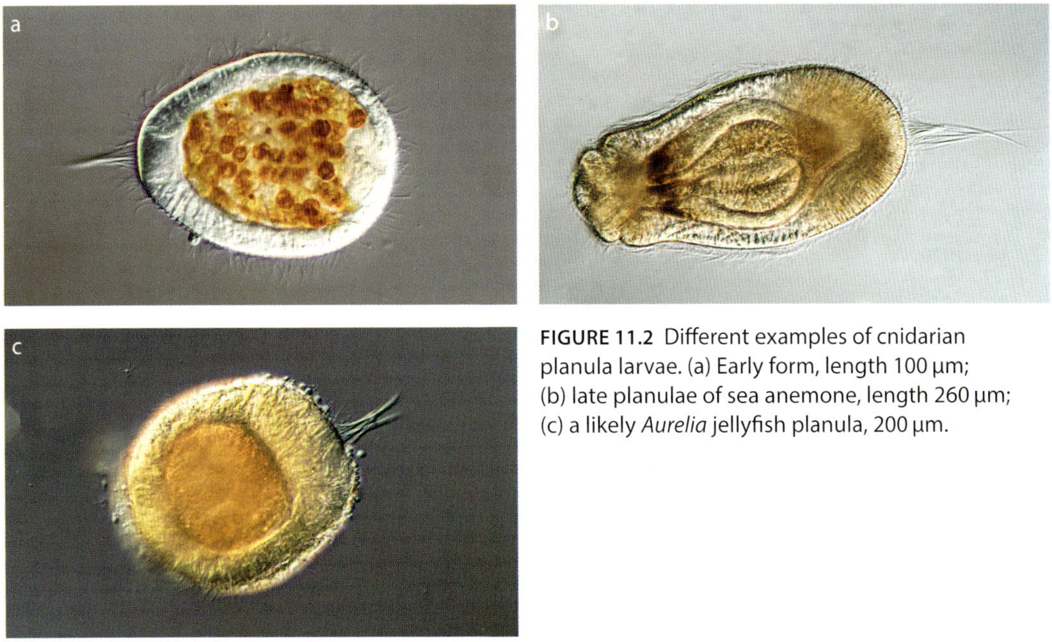

FIGURE 11.2 Different examples of cnidarian planula larvae. (a) Early form, length 100 µm; (b) late planulae of sea anemone, length 260 µm; (c) a likely *Aurelia* jellyfish planula, 200 µm.

features are characteristic of this phylum, as is the production by all members of this group of a larva, the planula.

Sea anemones: Anthozoa

Having said all the above, sea anemones typically miss out the medusa stage of the lifecycle. The enlarged polyp also has internal membranes or septa that divide up the gastric cavity, a feature not found in other cnidarians. Asexual reproduction produces internal buds in some, such as the common Beadlet Anemone *Actinia equina*, and longitudinal cloning in others, like Snakelocks Anemone *Anemonia viridis*, which can fill rock pools with clones (Figure 8.11). Sexual reproduction also occurs, and most sea anemones release zygotes and planulae into the water. The planula larval stage, covered in cilia and with a tuft of longer ones at the front, is a delight to watch under the microscope as they glide slowly around the slide. Some species produce a lecithotrophic planula that carries sufficient yolk to sustain it without searching for food and spends only a day or so in the plankton. By contrast, others are planktotrophic and have little in the way of nutrient stores. The planktotrophic stage can last for weeks, ensuring good dispersal but increasing the chance of being eaten. Ultimately, the planula transforms into a tiny juvenile anemone that settles on the shore.

Mating occurs in only one sea anemone species, *Sagartia troglodytes*, which is found on most coasts of Europe among kelp holdfasts or coarse sediments. The female sidles up to a receptive male and pushes her pedal disc against his to create a space into which they both share gametes for fertilisation. They remain in this position for several days while the eggs develop and hatch into the planula.

In the tube anemones (Ceriantharia), which live within a mucus tube in soft sediment, the planula quickly forms a second larval type, the antipathula. In the early stage it looks like a toy rocket, with a bulbous body and four ciliated pod-like 'legs'. At some stage old larval tissue is ejected through the mouth and a slimmer, late-stage antipathula, is born. The 'legs' steadily increase in number and length to become tentacles, at which stage it is called a cerinula larva. Now it looks very 'spidery', which has given rise to the alternative name arachnactis larva. The body (the new polyp) and tentacles elongate to transform into a juvenile anemone that burrows into the mud.

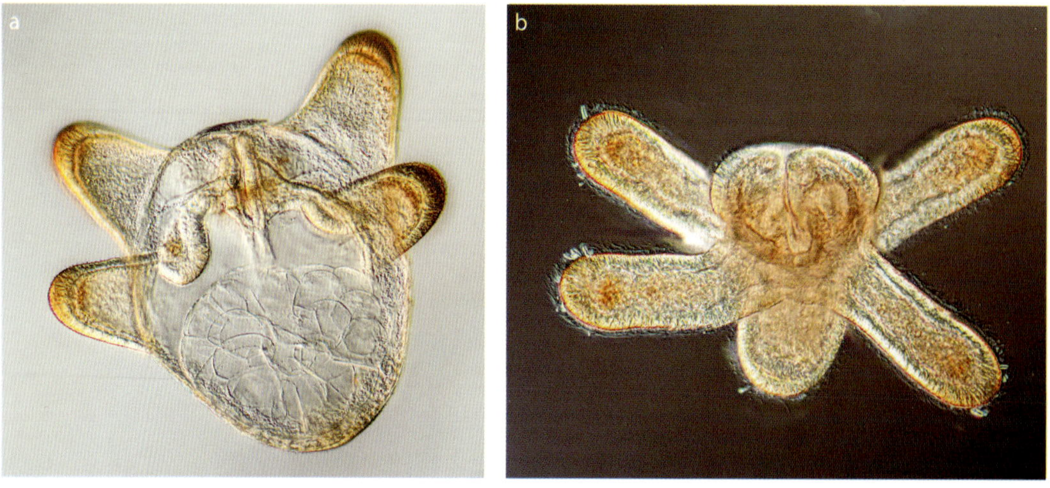

FIGURE 11.3 Antipathula larva of a tube anemone, probably *Cerianthus lloydi*. (a) Early form and (b) the slimmer, late antipathula form. 600 µm across, in both.

Jellyfish: the Scyphozoa

These are the most prominent and visible members of the plankton, with many exceeding a metre in length. In Europe the Lion's Mane Jellyfish *Cyanea capillata* is the largest, with a bell around half a metre wide and trailing tentacles up to 30 m long covered in batteries of cnidae, or sting cells. These hang down from the bell, which has an outer layer of cells called the epidermis. The bulk of the bell is gel-like, filled with mesoglea, with the organs at the bottom. Circling the inside of its rim is the ring canal, which also connects to the middle, where the gastric pouch lies. Beneath this pouch is the mouth, around which are four oral arms that help to push food into it. The gonads are sited around the base of the gastric pouch. The margin has a number of notches in it, so that the edge looks slightly lobed, and within each notch lies a sense organ called the rhopalium. Typically, these are statocysts, which determine the orientation of the jellyfish, but they

FIGURE 11.4 Lion's Mane Jellyfish *Cyanea capillata*. (Mariusz Smenzyk)

FIGURE 11.5 Moon Jellyfish *Aurelia aurita*. (Chris Hardy)

may also have a pigmented ocellus, which detects light. Using specialised epidermal cells around the edge of the bell, they can contract to expel water from underneath to propel them forward and, usually, upwards to the light. When they stop they sink slowly, increasing the chance of the tentacles beneath encountering prey, which can be immobilised by the sting cells or collected in mucus and passed to the mouth. With some minor modifications this body structure is typical of all jellyfish and they also exhibit alternation of generations, which we met above.

Another *Cyanea* species is the Blue Jellyfish, *C. lamarckii*. This also has long trailing tentacles but is smaller, at around 20 cm across, and, as the name suggests, is a deep blue. Around the mouth are many more tentacles and the bell's margin has 16 lobes with rhopalia in the notches. *Cyanea* are cold-water species common in the arctic. Of all the jellyfish likely to be encountered in temperate regions, they probably give the most painful sting. Their reproduction is similar to that of the *Aurelia* species, described below.

The Moon Jellyfish *Aurelia aurita* is also known as the common jellyfish because of its frequently huge abundance, not just in Europe but across the world. They are one of the smaller species, being between 20 and 30 cm across the umbrella. It is an inshore species and, as well as being common along the coast, it may be found in estuaries, where it can tolerate very low salinity. It lacks the long tentacles found in most jellyfish but along the margin of the umbrella, which has eight notches containing rhopalia, it has short regular ones trailing down. These contain cnidae used for both catching prey items, such as fish larvae, and protection against predators. Most food is collected on mucus streams that glide across the umbrella surface, assisted by the action of cilia, and move prey to the mouth and gastric cavity in the centre. Almost anything of a suitable size will be eaten, including larvae, diatoms and other medusae from hydroids. The most distinctive bodily feature are four C-shapes in the bell, the gonads. Oxygen and carbon dioxide simply diffuse in and out of the body, but when water oxygen saturation becomes very

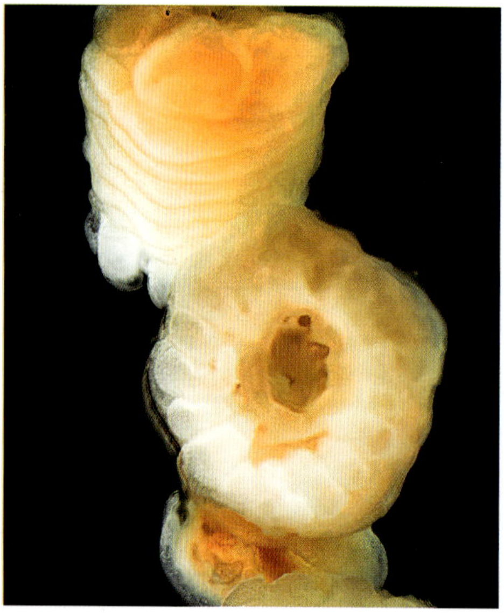

FIGURE 11.6 Lateral and top view of the scyphistomae of Moon Jellyfish *Aurelia aurita*. Note the layers, each one developing into a medusa, called an ephyra. Diameter 2 mm.

FIGURE 11.7 Early-stage ephyra of Moon Jellyfish *Aurelia aurita*. Diameter 2 mm.

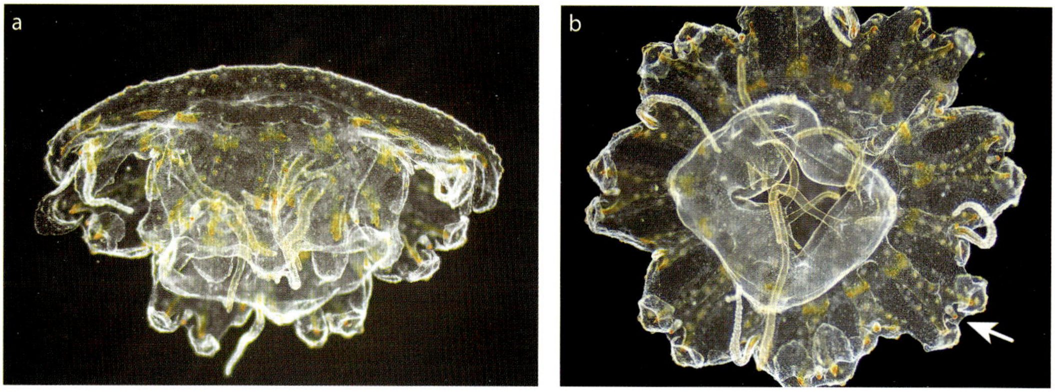

FIGURE 11.8 Late-stage ephyra, probably of Mauve Stinger *Pelagia noctiluca*. (a) Side view and (b) ventral view. Note the brown spots, which are groups of sting cells. The arrow points at the rhopalium. Diameter 6 mm.

low feeding is reduced. It is thought to be a species advantage as it reduces competition with other moon jellyfish, especially when they are in high densities. Consuming food would increase the demand for oxygen. The jellyfish's body is high in polyunsaturated fatty acids and is eaten by a wide range of species, particularly turtles and sunfish.

Members of *Aurelia* have separate sexes and are sexually mature at as little as 6 cm in diameter. Fertilisation is internal, with the eggs developing embryos and being brooded in a chamber until they hatch to become planula larvae. These are released through the autumn and soon after the adults begin to die. This can happen when they are a few months old, although many can migrate into deeper offshore waters where they may survive until the spring. The planula develops and by early spring settles on the lowest regions of the shore, where it becomes a simple polyp called the scyphistomae. The function of this is to produce, through strobilation, a series of larvae called ephyra that form at the polyp tip. These ephyra enter the plankton and over the next few months feed and disperse, steadily enlarging until the final phase, when they mature into an adult jellyfish. This lifecycle is usual to most species of jellyfish. Ephyras suffer a high mortality rate as they are thought to be consumed by sessile seashore feeders such as barnacles, tunicates and mussels.

It has been discovered that, during its lifecycle, *Aurelia* has the ability to reverse ageing and revert to an earlier stage of development. A number of related species (some hydroids and flatworms) can do this too. All these species have pluripotent stem cells that can differentiate and replace damaged or ageing cells, returning the organism to a more 'youthful' individual.

Hydroids and the hydromedusa

Adult scyphozoan jellyfish are large and conspicuous, making them an unlikely find in the plankton net. By contrast, the marine hydroids produce hundreds of different forms of tiny medusae, a small jellyfish-like stage called a hydromedusa. Alternation of generations, as in jellyfish, is usually present in the lifecycle of hydroids, except that the dominant phase is reversed. For them it is the asexual polyp stage that survives the longest, usually living as a colony attached to rock or seaweed on the shore. These will produce and release the hydromedusae that enter the zooplankton to feed and disperse. It is these that possess the gonads for sexual reproduction and will lead to the formation of planula larvae to settle on the shore and produce new polyps.

Larger jellyfish

Of the larger species, the Compass Jellyfish *Chrysaora hysoscella* is the one most likely to be encountered in northern Europe. At over 30 cm across, the margin has 32 lobes, rhopalia in between and 24 tentacles. The most distinctive characteristic of this species are the 16 V-shaped brown markings on the top of the umbrella. It is a protandrous hermaphrodite: that is, the male gonads mature before the female ones, and fertilisation is internal. The planulae are released a few days later, during summer and into winter. The ephyra are produced from the scyphistomae in spring, with maturation to adult by the summer. They are omnivorous, catching zooplankton with cnidae on the tentacles and passing food items to the arms, which hang below the bell, and then to the mouth.

The Mauve Stinger *Pelagia noctiluca* is a warm-water species and is more likely to be seen out at sea rather than along the coast. This one, too, is a painful stinging jellyfish and stinging cells are not limited to the tentacles, as small warts on the umbrella are full of cnidae. The ephyra shown in Figure 11.8 is likely to be of this species, as there are so many brown warts of sting cells. It is a species that has not often been seen in groups, although, in recent years, with warming seas, it has started to appear around the UK. In November 2024 a huge bloom of them occurred in the Irish Sea, with large numbers being stranded on the shore. As this species spends much of the time at the surface in the open Atlantic, strong winds can blow them into groups and on to shore. At night they are known to descend into deeper water, returning to the surface layers soon after dawn. The name *noctiluca* may be familiar, as we encountered it among the dinoflagellates in connection with bioluminescence – and this jellyfish also glows in the dark. Like all the jellyfish, they are predators on zooplankton. Sexes are separate, and the eggs and larvae are retained in the tentacles before release.

The Barrel Jellyfish *Rhizostoma pulmo* is unusual. To begin with, it is in a different family group and lacks the long trailing tentacles of many other jellyfish. The mesoglea in the umbrella

FIGURE 11.9 Mediterranean Jellyfish *Cotylorhiza tuberculata*. (Atanas Zaro)

FIGURE 11.10 Barrel Jellyfish *Rhizostoma pulmo*. (Jose Manuel Santos Alfonso)

is very thick and gelatinous, so this jellyfish may reach 30 kg in weight. The general colour is white, with a dark pigmentation around the margin. The four large arms underneath the bell have many openings and pits, into which microscopic plankton are drawn when it feeds. The pits are often occupied by a parasitic crustacean called *Hyperia galba* (see Figure 13.52). The sexes are separate and the gonads are different in colour, the male being blueish and the female brown. In northern Europe they can be common in summer on western shores; as winter approaches, they are known to move into deeper water to avoid turbulence. The sting to humans is moderately painful and has been likened to a burning sensation.

Cotylorhiza tuberculata, the Mediterranean Jellyfish, is also known as the Fried Egg Jellyfish because of its appearance (see Figure 11.9). It is a warm-water species abundant across the whole of the Mediterranean. A dumpy little jellyfish approximately 15 cm in diameter, it has a weak sting that does not particularly affect humans. It has the typical jellyfish lifecycle, beginning with the planula larvae in spring and early summer, which seek a hard surface on which to create the scyphistomae. During the summer ephyra are shed, later developing into the medusa. Once matured in the autumn, sexual reproduction occurs, producing large numbers of new planulae. An important feature of this species is the presence of dinoflagellates (zooxanthellae) living within the mesoglea. They provide the jellyfish with metabolites from photosynthesis for energy in the form of fatty acids, while the host provides protection. Without this symbiotic relationship it has been shown that the medusa cannot grow.

Jellyfish medusae can take months to develop and during this time can reach high densities, peaking throughout the summer. The bell is very susceptible to damage by boat propellers and fishing nets. As autumn approaches there is the increased chance of gales and strong wind causing water turbulence that will damage medusae. If the damage is not too extreme repair is possible, but a mass stranding caused by wind and tides will soon reduce the population.

FIGURE 11.11 Colonial polyp stage of *Obelia* sp. Note that the hydrocaulus ('stem') is covered in benthic diatoms (*Licmophora* sp.). Height 1 cm.

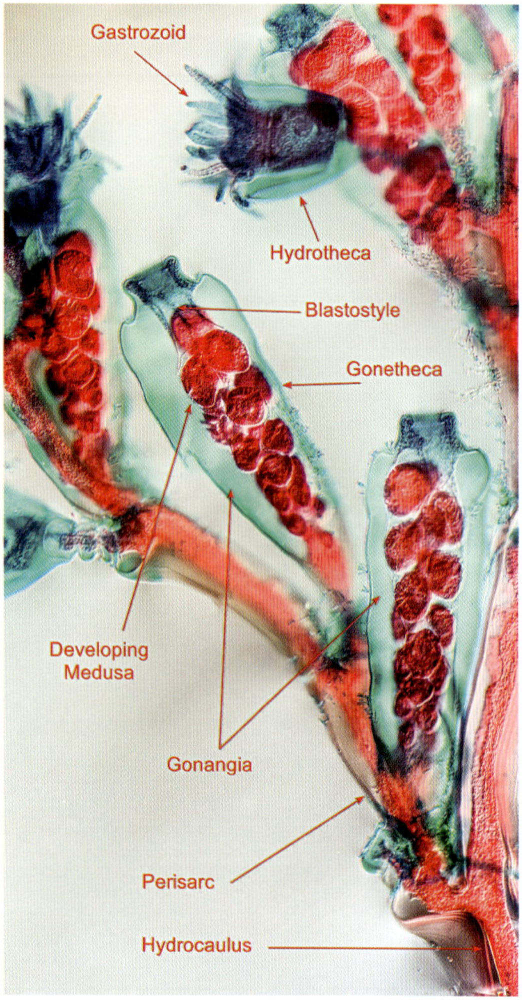

FIGURE 11.12 Microscopic view of part of an *Obelia* colony, showing the feeding polyp, the gastrozooid, and the asexual gonangia that produce the medusae. The specimen has been selectively stained.

Hydroids exhibit a wide diversity of form. A simple one is *Obelia*. It is also an abundant and widespread species to be found, standing erect from the surface, on most sheltered shores where there are suitable substrates for the colony to develop: kelp plants, rocks or shells, or even the surface of other animals, such as sea squirts. Many colonies may grow together to form a carpet a centimetre or so thick. Each colony consists of a root-like structure, the hydrorhiza, growing across the substrate surface and sending up occasional 'stems', the hydrocaulus. In *Obelia* the polyps grow alternately along the hydrocaulus in a zig-zag fashion, but other species produce an elaborate branching fan or have minimal stem and bunches of polyps. On the exterior of the stem is a protective layer, the perisarc, made of a mix of chitin and protein. The apical tip is the growing point for the colony and below this the specialised hydroid polyps, or zooids, are produced for the various functions, three types in all: gastrozooids for feeding, dactylozooids for defence (filled with batteries of sting cells) and gonozooids, also known as gonangia, for reproduction.

In Figure 11.12 a small section of the hydrocaulus is shown, with several mature growths branching from it. The feeding polyp, or gastrozooid, has sting-cell-laden tentacles for collecting food. The hydrotheca is a protective coat within which the gastrozooid can be withdrawn. Smaller buds develop behind these to produce the asexual gonangia, which extend into long vase formations. Surrounded by a gonetheca coat, there is a stalk through the middle (a blastostyle) that form the medusae by mitosis. Those at the top of the vase are the most mature and when ready they are separately expelled into the water, where they immediately contract to move, jellyfish-like, through the plankton. Under the perisarc there is a tube of tissue, the coenosarc, that transports nutrients from the gastrozooids across the growing colony, especially for the formation of medusae.

When released, the medusae require time to develop tentacles and gonads. They feed on microplankton, collected by the manubrium in the centre to be digested in the stomach and distributed through the radial and ring canals (transport structures). At the base of each tentacle is a statocyst. The mesoglea in the bell is the gelatinous bulk of the medusa, as in jellyfish, and below it, when fully mature, lie the four gonads. The gonads are really a collection of gametes and after fertilisation a planula larva is released. This is non-feeding and may only remain in the plankton a few hours or a day before finding a suitable benthic substrate to establish a polyp colony.

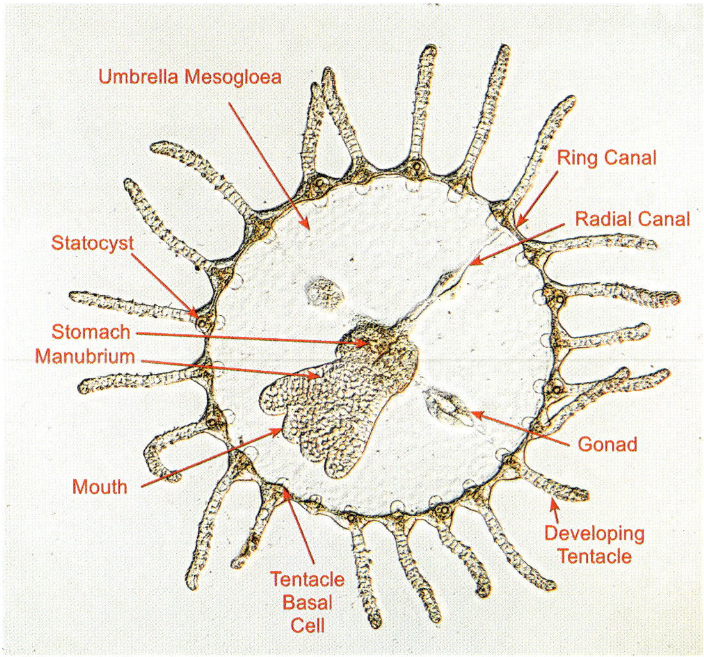

FIGURE 11.13 A maturing hydromedusa of an *Obelia* sp. Diameter 2 mm.

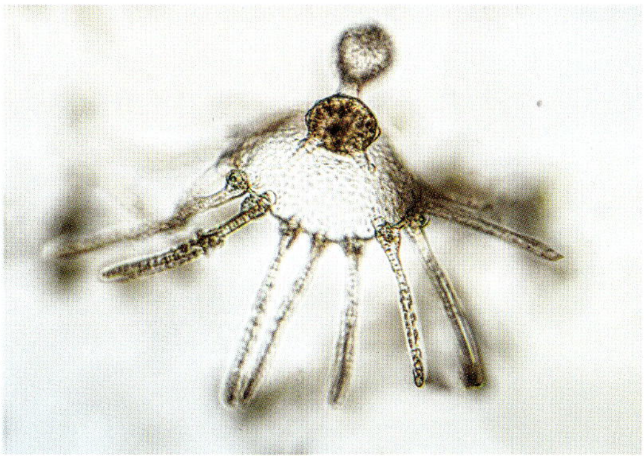

FIGURE 11.14 An active young *Obelia* sp. hydromedusa feeding with the manubrium extended. This is a tubular structure with the mouth at the end. Width of photo 650 µm.

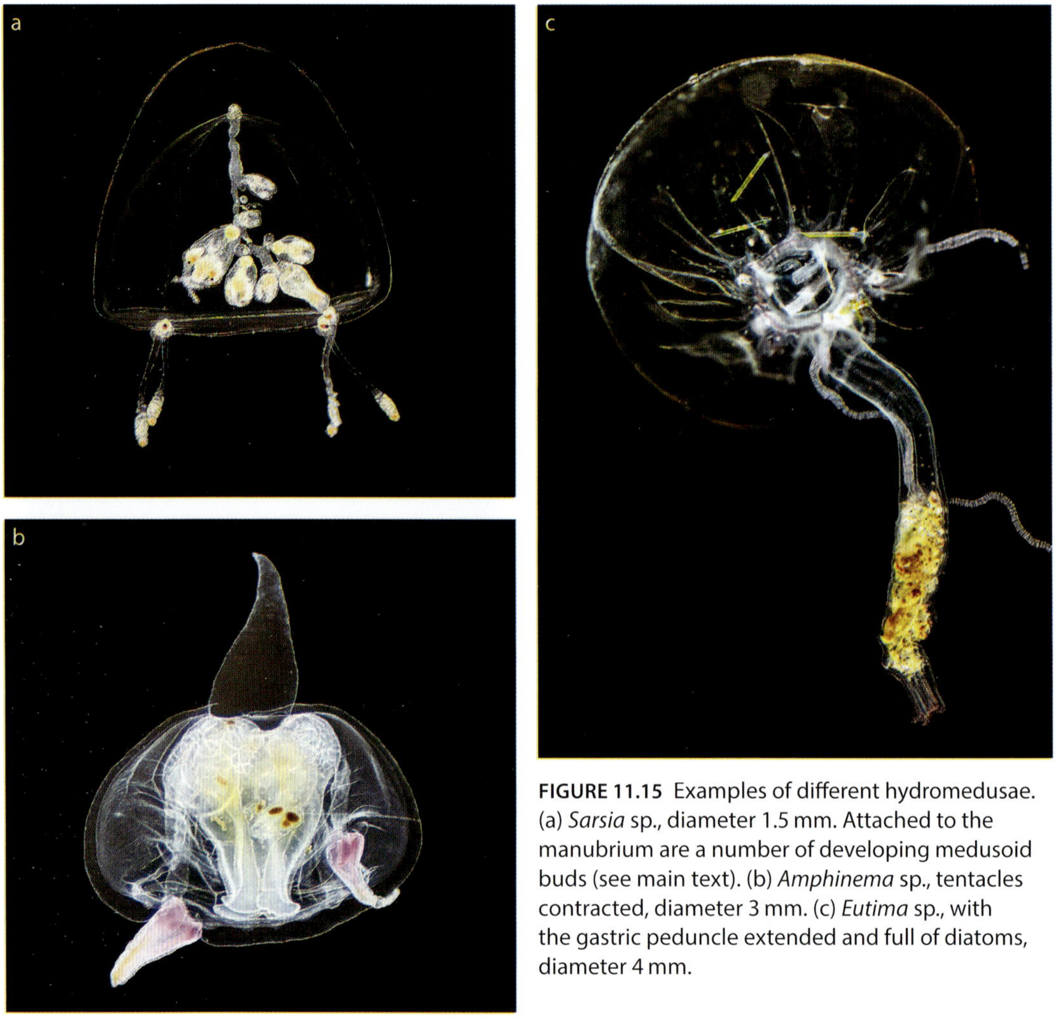

FIGURE 11.15 Examples of different hydromedusae. (a) *Sarsia* sp., diameter 1.5 mm. Attached to the manubrium are a number of developing medusoid buds (see main text). (b) *Amphinema* sp., tentacles contracted, diameter 3 mm. (c) *Eutima* sp., with the gastric peduncle extended and full of diatoms, diameter 4 mm.

There are plenty of exceptions to the *Obelia* example. Some hydroids, such as *Plumularia*, fail to release medusae, and they remain growing on the colony. Equally, some medusae live in the plankton with no known polyp phase – *Sarsia gemmifera* for one. Instead of producing gametes, buds are formed on a stalk growing from the manubrium which can be seen trailing out of the bell. The released offspring are small medusae that swim off into the plankton to grow and enlarge to form new adult medusae. Some colonial hydroids skip a phase. For example, in *Tubularia mesembryanthemum*, a quite different dispersal stage is produced, the actinula larva. This is a sac with a small number of tentacles that tends to crawl rather than swim but does occur in the plankton.

Hydroid siphonophores

There is a large hydroid taxon that has some spectacular worldwide species living their entire lives floating in the plankton, the order Siphonophora. The unique feature of the group is that they have combined the two phases of polyp and medusa into one complex formation. The most notorious of these siphonophores is the large and dangerous Portuguese Man o' War *Physalia physalis*. As colonies, they retain the three main polyp forms of dactylozooids, gonozooids and gastrozooids.

Living their entire lives in the sea, they have evolved either swimming bells (nectophores) or a flotation bag (pneumatophore) to support the polyps. There are several subgroups based on these two devices and the two we look at here are the Calycophorae, which have a swimming bell but no float, and the Cystonectae, which have a large float and no bell.

August is a time for visitors to the coastal beaches. Holidaymakers arrive in their thousands. Predatory jellies arrive in millions. So transparent are the gelatinous bells of *Muggiaea atlantica* that, as humans splash in the shallow waters, they are unaware that these siphonophores, up to a centimetre-long, are cruising around their bodies. Even in a sample dish they are not immediately obvious. As the neritic water warms in summer, so *Muggiaea* multiplies just off European coasts. Its nectophore is the medusa part of the colony and the thickened mesoglea provides buoyancy; there is no mouth, tentacles or sense organs. There is a long tube or stem (the siphosome) that runs from the top of the bell down to where the various zooids attach in a chain of groups called cormidia. Each cormidium is a colony of polyps, a colony within a colony. At the very top of the tube is a distinct drop of oil and – under the microscope – material, like lipids, can be seen moving back and forth along it. Cormidia regular break off from the colony to lead a separate life within the plankton and these are called eudoxids. The reproductive zooids now develop and the sexual phase begins to produce gametes.

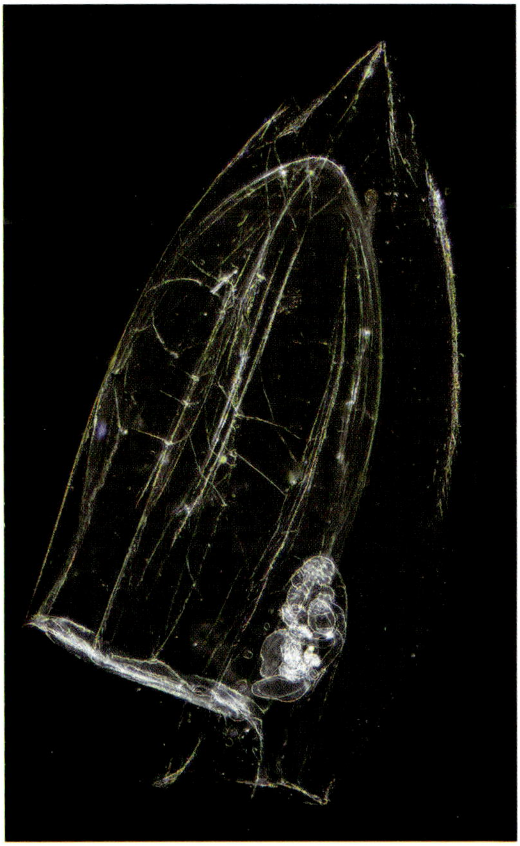

FIGURE 11.16 The siphonophore *Muggiaea atlantica* with the cormidia visible at the bottom of the bell or nectophore. Height 5 mm.

FIGURE 11.17 The siphosome of *M. atlantica*, with the oil droplets at the top running down to the cormidium and the head of a captured copepod at the base. Length 1.5 mm.

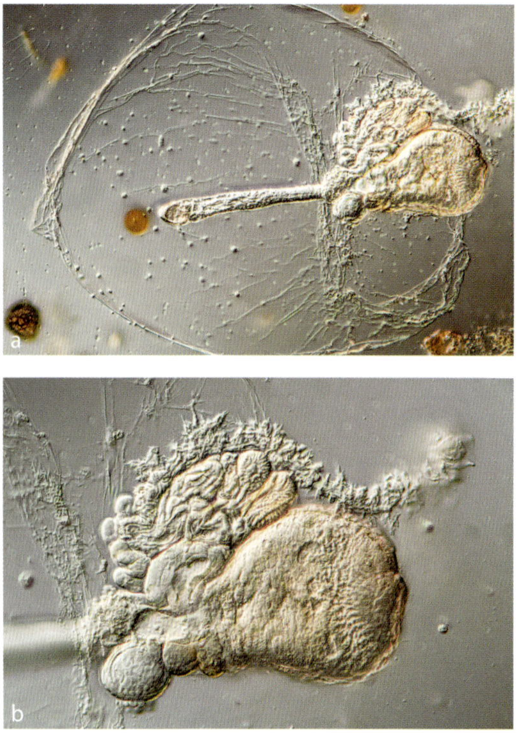

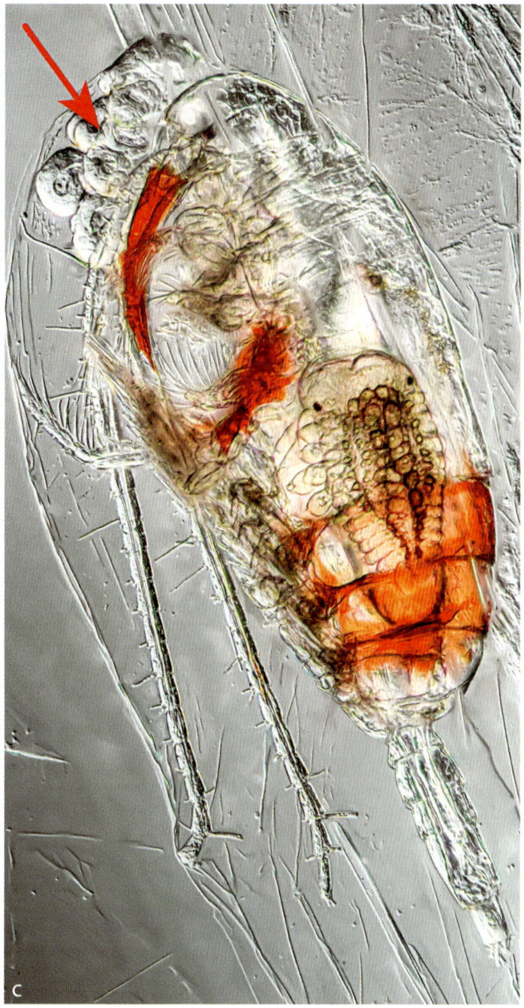

FIGURE 11.18 The siphonophore *M. atlantica*. (a) An entire eudoxid released from the main colony (width of photograph 2 mm), with an enlarged view (b) to show the zooids and tentacle (width of photograph 700 µm), (c) the cormidium (indicated by a red arrow) within the nectophore where a calanoid copepod has been caught. Note the ectoparasitic isopod, a bopyrid, attached to the back of the copepod.

You may find when viewing *Muggiaea* in a sample that many of them appear to have a copepod attached to the cormidia. This is no accident, as during the trip back from collecting the sample the siphonophore has been feeding on crustaceans. It is no surprise that the peak in *Muggiaea* occurs at the peak copepod density.

It is worth enlarging the various zooids, particularly of the eudoxid, under the microscope to view their detail and of the single tentacle. The latter may have a thin branch, the tentillum, with large numbers of cnidae which may discharge as you watch although this can be a surprisingly rapid action. In late summer and early autumn the eudoxids, in varying states of sexual maturity, can be common in seashore plankton.

Thankfully, despite the swarms that cover the beaches in summer and into the autumn, *Muggiaea* goes unknown in the water, as there is no sting to affect humans. However, specimens of the Cystonectae will be noticed when occasional rare swarms hit the beaches. The Portuguese Man o' War *Physalia physalis* is a tropical and warm-water species that breeds into large numbers out in the ocean. Strong winds and currents eventually strand them on European shores. Above the water surface it is the large pneumatophore float that is most obvious. Probably developed from the polyp, it is modified to be a double-walled chamber lined with chitin. Within it is a chamber containing a gas gland rich in mitochondria that secrete gas into the pneumatophore, the composition of which is high in carbon dioxide (8%), while concentrations of other gases are similar to atmospheric levels. The gland also regulates the volume and consistency of the released gas, so that a degree of sinking is possible. The pneumatophore has a left- or right-handed twist, so that when the wind blows it will determine the direction of travel. Under the float is a budding zone that produces long flowing tentacles, sometimes many metres long, that have multiple blue gastrozooids and large numbers of stinging cnidae for catching fish and other prey. The sexes are separate and gonophores hang in bunches off other tentacles, with fertilisation taking place externally. This results in a planula larva.

The beautiful blue By-the-wind-sailor *Velella velella* is a conundrum. Its classification has moved around as new studies bring more information to light. For years it was recorded as a floating

FIGURE 11.19 Portuguese Man o' War *Physalia physalis* stranded in a rock pool.

FIGURE 11.20 By-the-wind-sailor *Velella velella*. Note the mantle edge is a deeper blue than the rest. (a) Top view. The main disc, or pneumatophore, is divided up into flotation chambers and these are the visible rings (oldest in the middle) underneath the sail. (b) Lower surface. In from the mantle (deep blue) are large numbers of tentacles (dactylozooids) and a white gastrozooid tube in the centre which opens to the mouth. Length 9 cm.

mass of polyps, a colonial hydroid, but more recently scientists have tended to think that it is not a colony. It is ovoid, with a wide mantle edge. The main disc is the pneumatophore, a series of round flotation chambers, with the central, oldest, area forming a raised mound on which is the upright sail. This can be right- or left-handed, like the pneumatophore of *P. physalis*, and so catches the wind direction differentially. This upper surface repels water and so keeps the float upright. If it does get caught or flipped over, the mantle will assist in the recovery. During spring huge numbers can appear out in the warm, subtropical Atlantic. Wind and currents bring them on to the shore, where mass strandings can turn entire beaches blue in May and June. As they die and dry out the colour is lost, but the clear sail and disc can remain and are sometimes mistaken for fine pieces of plastic.

Underneath there is a single central gastrozooid terminating in the mouth. That is surrounded by zooid buds, the location for the reproductive gonozooids. At the outer edge are tentacles, or dactylozooids. There is debate about how functional the mouth and gut are, as the latter is quite vestigial. However, the top of the float chambers is transparent, allowing light to pass through to the tissue below, and inside there are large numbers of zooxanthellae, consisting of a symbiotic microalga *Scrippsiella velellae*. So much of this symbiont is present that it is possible nutrition is derived primarily from the microalga's photosynthesis. While cnidae appear to be present on the tentacles there is conflicting evidence on their effectiveness, but one report suggests that small fish can be trapped. Gonozooids produce medusae that slowly sink away from the surface, later developing gonads and releasing gametes for external fertilisation, creating planulae. The medusae quickly die thereafter, and the planulae form into new floating By-the-wind-sailors.

Out in the ocean *V. velella* is eaten by another ocean drifter, the Violet Snail *Janthina*. There are a number of species, but *J. janthina* is the most abundant (see Chapter 12).

Ctenophores: the comb jellies

Another group of organisms, once considered to be cnidarians, now have their own phylum, Ctenophora, the comb jellies or sea gooseberries, which makes up a substantial biomass in the gelatinous plankton. Although there are some similarities, such as the primitive nerve network and gelatinous mesenchyme (a more advanced form of mesoglea), they have far more differences.

During spring they begin to appear along the coast. *Pleurobrachia pileus*, a good example of the order Cydippida, is the species most likely to be encountered in Europe and the only one to appear in a basic plankton net, as it is a centimetre at most; a second species appearing in midsummer, *Beroe cucumis*, is much larger, at around 10 cm. If you do find one of the latter, never put the two species together, as the large one feeds on *P. pileus*. A sudden appearance of *Beroe* on the coast is a sure sign that the *Pleurobrachia* population will be wiped out within a few weeks.

The name of the phylum is derived from its most distinctive feature: rows of ciliary plates, the ctenes or combs. The body is divided into four quadrants and each has a pair of flashing ctene rows. While they might appear like flaps, each flattened piece is an amalgam of cilia that have fused together. Each ctene beats in a coordinated way with the others in the row, powering away from the oral end and driving the gelatinous ball forward. Two rows are linked by a furrow between them that makes up part of the nerve network, helping to synchronise them; any problem with this part can stop the two rows from working. The coordination of all eight rows is done by the apical sense organ around the arboral end (posterior). This contains a calcareous statolith balanced on four tufts of cilia. Any shift in the statolith will stimulate the cilia and trigger action to correct the organism's orientation and control of the locomotion. Ctenophores are the largest animal known to be moved through the power of cilia.

More complex structurally than the cnidarians, these animals have both a mouth and an anal pore, enabling food to pass right through and waste to be ejected. Feeding habits vary. In some species, including *Pleurobrachia*, prey is collected on two long, trailing tentacles with small

FIGURE 11.21
The Sea Gooseberry
Pleurobrachia pileus.

FIGURE 11.22 Sea Gooseberry *Pleurobrachia pileus*. A close-up view, with the mouth, on the right, feeding into the wider pharynx and gut. The ctenes and rows are distinct. Where the tentacles touch the body they pass into a clear tube, the tentacle sheath, where they can be withdrawn. Body length 2 cm.

FIGURE 11.23 *Beroe cucumis* is a large ctenophore, 5 cm or more in length.

branches, the tentillae, which have swollen sections containing colloblasts, cells that release a strong adhesive. When triggered they shoot glue out and the prey items stick to the tentacle. The tentacles are then swept past the mouth and beating cilia around the gut entrance trap the material in mucus and move it into the oral canal. The tentacles emerge from two ciliated sheaths inside the rear of the body, into which they can easily be retracted for safety. *Beroe*, lacking the tentacles, collects microplankton over its surface with streams of mucus that move by ciliary action to the mouth. However, *Beroe* also feeds on quite substantial prey, including other, smaller ctenophores, which it grabs with its slit-like mouth. The area between the gut and the epidermis is filled with mesenchyme that has developed several important tissues, including some muscles.

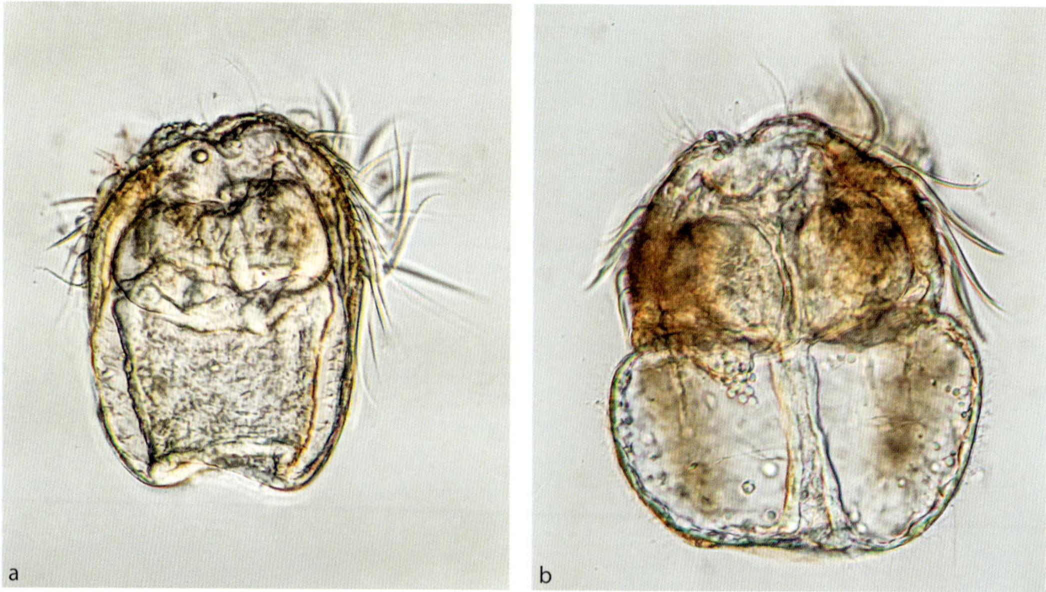

FIGURE 11.24 A young cydippid larvae (a) and one that is older (b), showing the development of the gut. 100 and 180 μm, respectively. In both cases the apical sense organ is at the top and the mouth at the bottom.

Ctenophores around Britain

Beroe cucumis and *Pleurobrachia pileus* are the species most likely to occur on European shores. Several other species make occasional appearances, especially in the North Sea and Baltic. These include species in the order Lobata, which is quite different from those already discussed, having two large lobes or folds used in a flapping motion for swimming. Further species include *Bolinopsis infundibulum*, another favourite prey item for *Beroe*, which is steadily increasing, and an invasive species, *Mnemiopsis leidyi*, which has turned up in the last few decades. Originally from the North and South American coast of the Atlantic, it has appeared in various places, including the Mediterranean and Black Sea, since the 1980s, helped by its particularly high tolerance of low salinity and low oxygen.

FIGURE 11.25 *Bolinopsis infundibulum*. Approximately 3 cm across. (Blair Morris)

Bands of latitudinal muscles are located around the edge, under the epidermis; stretching from these to the gut are radial muscles. None of these play a role in locomotion, but help the gut to expand and contract, so that it can swallow food and move the tentacles. In the genus *Beroe*, giant smooth muscle cells have been found.

The body has a low specific gravity that produces a neutral buoyancy, making it stable in the water, so that it does not sink or rise. Ctenophores have the amazing ability to regenerate themselves – not just components like tentacles, but an entire quadrant or half of the body. It is thought that mitosis, which may help to form new individuals asexually, plays a part, but quite how this might work is unknown. Mature adults are hermaphrodite and shed gametes into the water via the mouth. Once in the water they can self- or cross-fertilise to produce a zygote that quickly forms a larval stage. (There is no larval stage with *Beroe*.) This is the cydippid larva, which looks like a miniature individual of the ctenophore order it is named after. As the cydippid larva develops it starts to produce small eggs, although in a mature adult the eggs produced are much larger.

The importance of planktonic jelly

The gelatinous plankton has a huge significance in the ecosystem as the members are major predators and consumers of marine plankton. Additionally, they are an important link with the next level of consumers. The Sun Fish *Mola mola*, the heaviest bony fish species, will follow the jellyfish in the Gulf Stream eastwards, as we saw earlier. Another great jelly consumer, the Leatherback Turtle *Dermochelys coriacea*, also does this. At up to 500 kilograms, it is the heaviest turtle in the world. All these large creatures are dependent on gelatinous plankton.

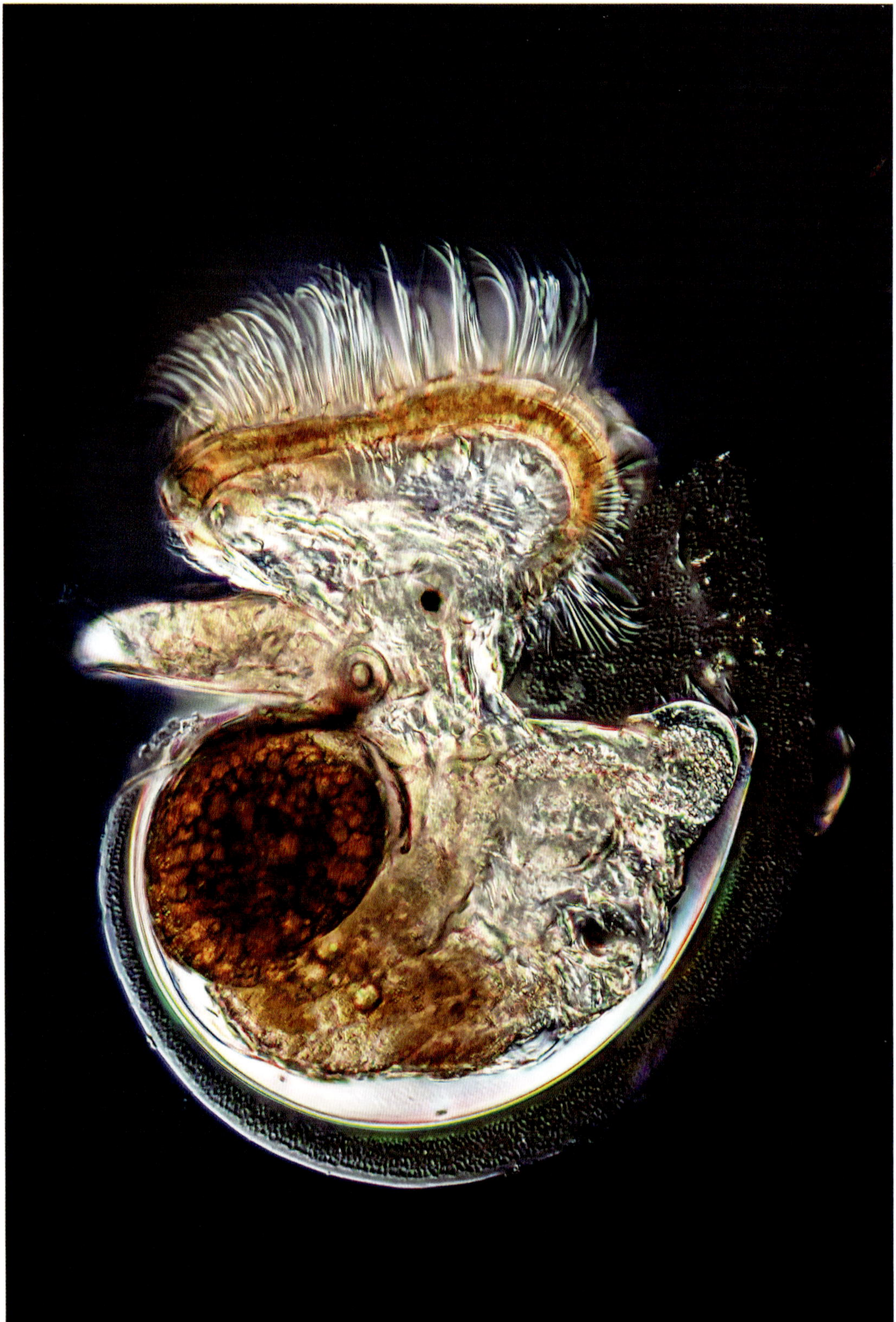

FIGURE 12.1 A veliger larva of a gastropod, probably Edible Periwinkle *Littorina littorea*. They are beautiful to watch, like microscopic whirling dervishes. Height 400 µm.

CHAPTER 12

Molluscs: the whirling dervishes

Imagining snails living in the plankton may be difficult, but they do exist and the world's oceans have such prolific numbers that the seabed can be composed of thick layers of empty shells, known as the pteropod ooze. But this specialised living is for a few holoplankton species only. Near-shore waters are used by molluscs as a nursery for their eggs and larvae, which develop and disperse in the zooplankton. Many marine molluscs, unlike freshwater species, have a lifecycle that starts with an egg hatching into a larval form called a veliger, which then lives in the plankton. This applies to the two main classes: the gastropods and the bivalves. First of all, we need to take a quick look at the permanent plankton dwellers.

Molluscan holoplankters

An initial caveat is that molluscan holoplankters are oceanic drifters and thus most unlikely to be present in plankton samples. Any found are typically small and stranded on the tide line. Molluscs living permanently planktonic lives are limited to a group of specialists totalling fewer than 150 species worldwide. Only a handful look like the snails we are familiar with on land: the janthinids, also known as the purple storm snails, are an example. *Janthina janthina*, a snail of classic appearance, has a very thin, fragile shell with a strong violet tinge to the base. In the plankton it floats at the surface, buoyed up by a bubbly mucus raft that it secretes. Here it lives among its prey, *Velella*, which we saw in the last chapter, and when they are cast up on beaches so too is the occasional violet shell.

The remaining holoplankters are also predators and are made up of the pteropods (the sea butterflies) and the heteropods. The latter can reach many centimetres across, although the only genus that may appear on North Atlantic coasts is *Atlanta*, just a few millimetres long. To maintain buoyancy the shell is very thin and the body tissue is less dense than seawater. The whole body is thin, flattened and with a keel to provide maximum surface area for drag, reducing sinking and enhancing stability. The foot is enlarged for swimming and the animal has a long head and proboscis for feeding.

The pteropods include the Sea Angel *Clione*, which has no shell or mantle but a pair of wings for swimming. Living in the pelagic zone of the Arctic Ocean, it can descend to great depths and is amazing for its ability to suddenly grab shelled prey and suck their entire contents out in one go. Another sea butterfly, *Cavolinia tridentata*, is common in the Atlantic and Mediterranean. It too is spectacular, with a straight shell approximately 1–2 cm in length that is covered by a mantle that has many lobes for swimming. None of these, however, are likely to appear in your plankton samples.

Gastropod molluscs

One of my prime plankton collecting sites is a short Victorian jetty constructed from stone and concrete. Its plentiful crevices hold a large population of the Small Periwinkle *Melarhaphe neritoides*, near the top, and Edible Periwinkle *Littorina littorea*, at the bottom. Both produce copious numbers of lecithotrophic eggs that are released into the plankton. As November approaches so eggs begin

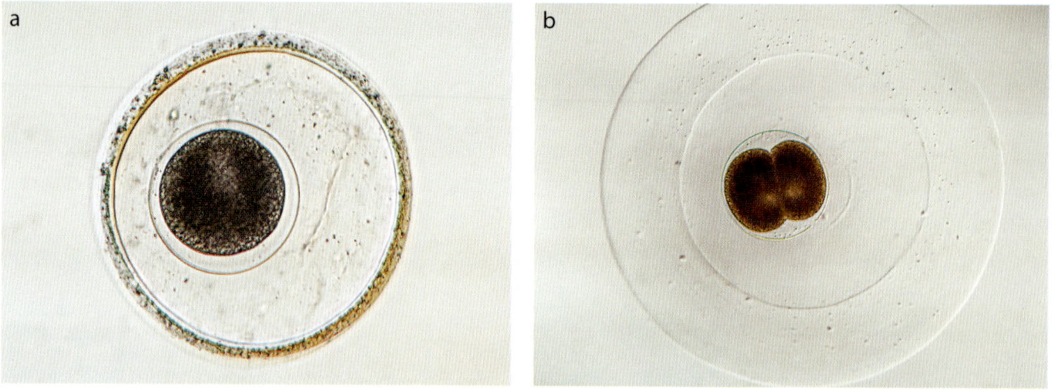

FIGURE 12.2 Gastropod eggs. (a) Small Periwinkle *Melarhaphe neritoides* and (b) Edible Periwinkle *Littorina littorea*. Diameters 200 μm and 700 μm, respectively. Note that the latter is a capsule and usually has more than one egg.

to appear, with *M. neritoides* reaching a peak in late December and into January. As I collect at high tide the net will inevitably be passing close to the adults releasing eggs. Within these eggs can be seen embryos at different stages of development. It will be a fully formed veliger larva on release and often retains a small amount of yolk to complete development when it descends to the shore (see Figure 12.3), where it will, with luck, find the required microhabitat to complete development to the adult. Egg capsules of *L. littorea* are only slightly denser than seawater, and so they can easily move with the tides. If they enter an estuary the reduced salinity causes them to slowly sink and become entangled in the salt-marsh grasses, such as *Spartina* spp., where edible periwinkle populations can be quite high.

Both these periwinkle species produce similar eggs, although *L. littorea* usually has more than one embryo inside, while *M. neritoides* has only one. The eggs look spherical but are actually disc-like, with *L. littorea* having a bulge where the embryos are located. *M. neritoides* eggs, possibly

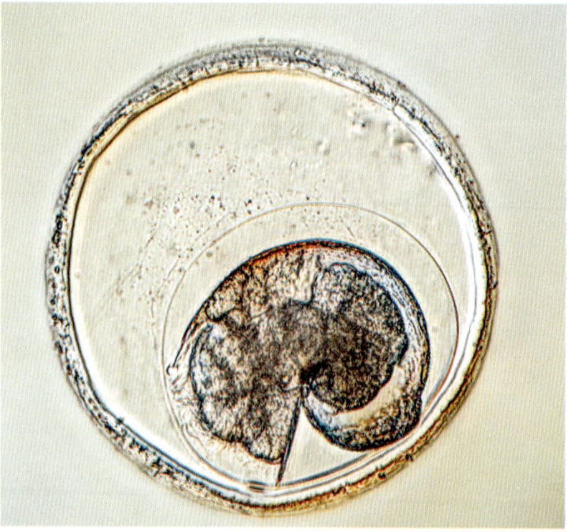

FIGURE 12.3 Developing egg of *M. neritoides*, with a well-developed veliger larva inside. Diameter 260 μm.

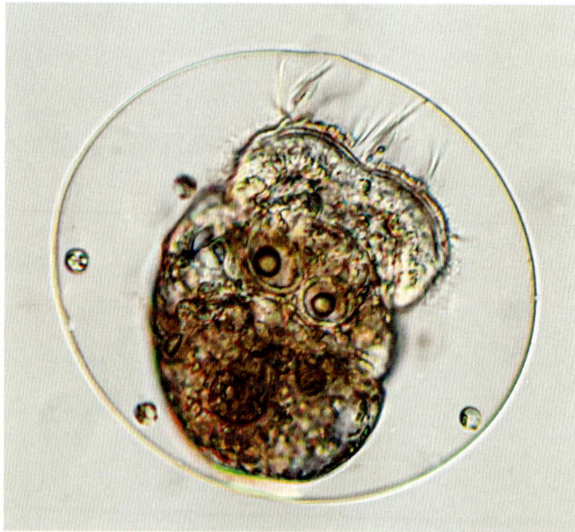

FIGURE 12.4 Developing egg of *M. neritoides*. A different view to Figure 12.3, and showing some remaining yolk cells near the veliger. Egg diameter 260 μm.

FIGURE 12.5 Veliger, possibly a periwinkle. Length 300 µm.

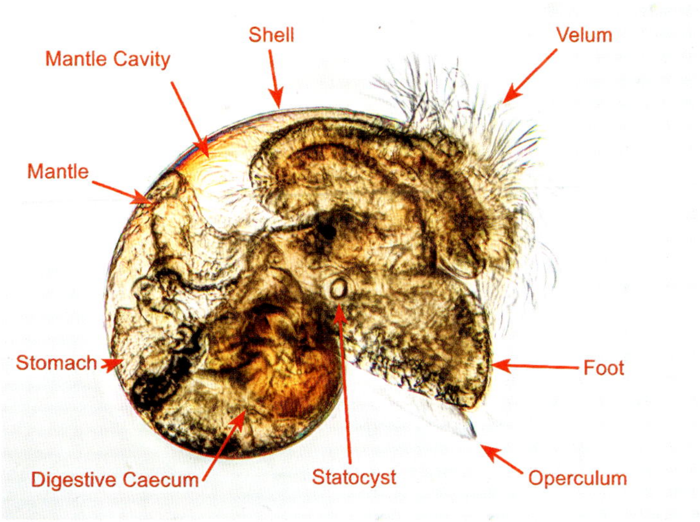

FIGURE 12.6 Veliger with the key structures labelled.

because of the amount of stored yolk, are much thicker and uniform in shape, with the veliger returning to the shore in only 24 hours or so.

While most gastropod eggs hatch into veligers, there is often a short stage prior, the trochophore. These are found in other phyla, such as polychaete worms, where they are retained for much longer than in molluscs. The trochophore is like a spinning top with a mass of cilia in a ciliary ring around the middle. A rounded top is above this, crowned with a tuft of cilia. Superficially, from above it appears radially symmetrical but on one side, just below the ring, is the mouth, which becomes a deeper groove with more cilia. Below that and near the base is the anus. In molluscan development the region between the mouth and anus becomes the foot of the veliger and, ultimately, the muscular foot of the snail.

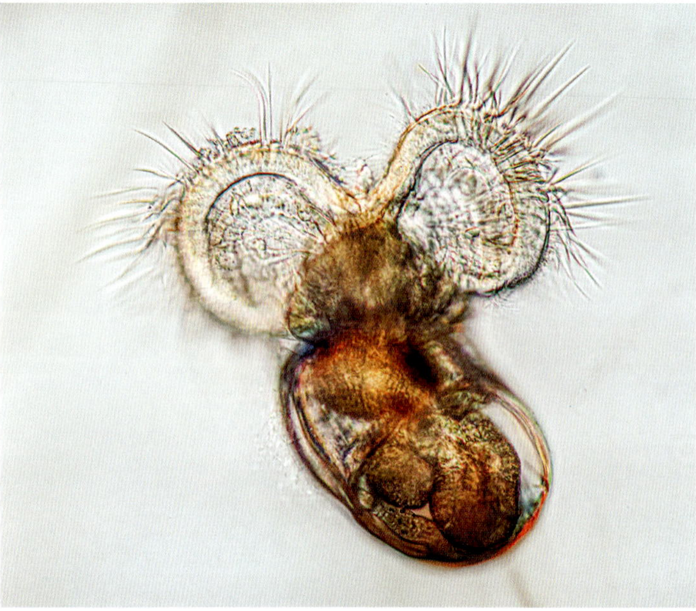

FIGURE 12.7 Veliger with a fully open velum. Height 320 µm.

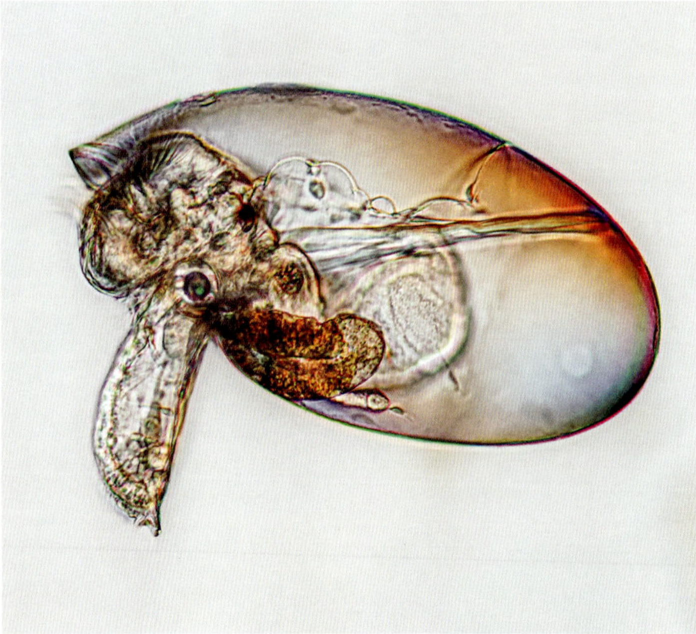

FIGURE 12.8 Side view of a nudibranch veliger. Note the retractor muscle from the shell across to the larva. Length 300 µm.

In Figure 12.3 the ciliary ring on the larva is beginning to form two lobes, which, in the newly hatched veliger, can be seen more clearly. These lobes form the velum, an organ that will provide power for locomotion through the water – the moment the velum stops beating, the veliger quickly sinks. The current produced by the velum brings particulate matter in towards the mouth, situated at the velum's base. In some cases, additional cilia guide the food into the gut. Phytoplankton is a typical food source, although only small organisms can be taken, including bacteria. Around the larva, beneath the velum, is the mantle, which secretes the shell, that steadily expands to accommodate the growing gut and other organs. A space near the anus called the mantle cavity

Mud snails

Hydrobia is a genus of small gastropods that are extremely abundant in coastal mud, feeding on bacteria and fine algae on the surface. It has a non-feeding veliger that persists for just a day or two in the mud. Dispersal is mainly by the adult, which floats on a secreted mucus raft (somewhat like *Janthina janthina*), drifting with the tide. Taking samples around large estuaries can often pick up the odd *Hydrobia* snail, along with those in other mud snail genera such as *Peringia*.

will later contain gills. Beneath the foot a small plate, called the operculum, begins to grow that will close off the entrance to the shell when the animal withdraws inside.

Quite early in the veliger development a process called torsion occurs. In some organisms, such as limpets, this happens within minutes of hatching, while in others it may take hours or days. Torsion is where the mantle cavity, with anus and attached organs, rotates 180 degrees so that the cavity, gut and anus are above the head. This is achieved using the retractor muscle, which can be seen clearly in Figure 12.8. Torsion moves everything into the correct position to allow the operculum to close snugly over the shell opening when the animal is withdrawn, in both the larva and the adult. The retractor muscle is employed in swimming and feeding to adjust the head, foot and especially the velum.

Although several organs can be seen under the microscope the pair of statocysts really stand out, looking like a pair of eyes. Other sensory receptors include ocelli (appearing as black dots), which detect light and are positively phototactic. When the matured veliger needs to search out a substrate on which to metamorphose into the adult, it becomes negatively phototactic and the velum reduces its action so that the organism sinks to the substrate, where there is a distinct change in behaviour, as it creeps rather than swims. Violent contraction of the velar lobes causes the velum to be lost and marks the point at which metamorphosis has occurred, and food collection can take place. The first meal is the old velum, which is sucked into the mouth and consumed. The formation of a radula and its muscles is necessary and starts some hours before feeding can begin.

Trochophores, if present in the lifecycle, are a very uniform larval form, whereas the veliger varies considerably depending on their time spent in the plankton. Identifying different species can be difficult and requires specialist knowledge. The molluscan development described here has been considerably simplified, as this phylum shows huge adaptive radiation and variation. Nudibranch gastropods, the sea slugs, are characterised by not having a shell in the adult, although the veliger does possess one.

Bivalve molluscs

The veligers of bivalves are like miniature clams with two shell valves. Although some bivalves retain gametes to brood the eggs for a short while, others liberate the sperm and eggs into the water for fertilisation. Like gastropods, the eggs may be lecithotrophic, for a quick life in the water before settling, or planktotrophic, hatching into a veliger that feeds for weeks and ensures a greater dispersal, albeit at a greater risk of predation. If your plankton collection is near a sheltered sandy shore you may find that your samples for a period in spring lasting several weeks are completely

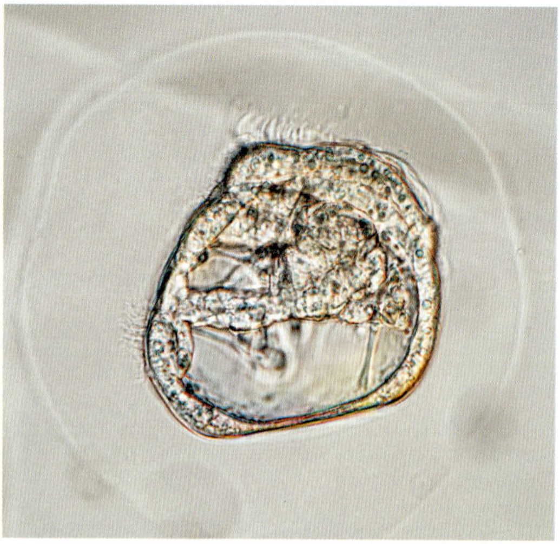

FIGURE 12.9 Developing egg of the Common Cockle *Cerastoderma edule*. Egg width 240 µm.

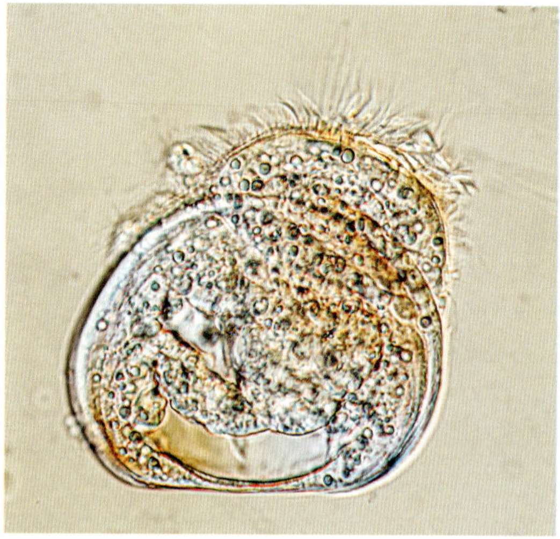

FIGURE 12.10 Recently hatched veliger of *C. edule*. Width 140 µm.

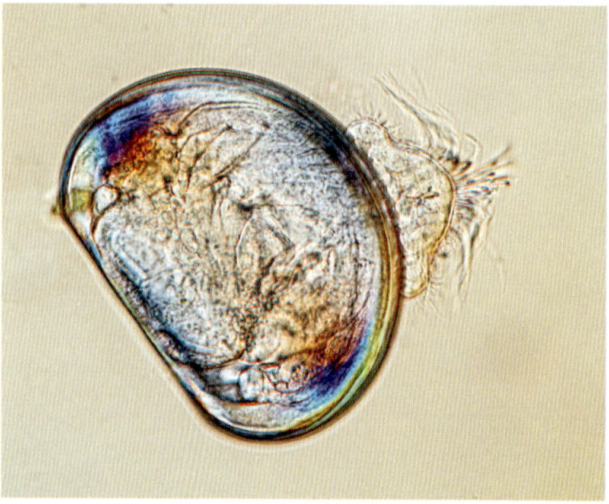

FIGURE 12.11 Maturing *C. edule* veliger. Width 240 µm.

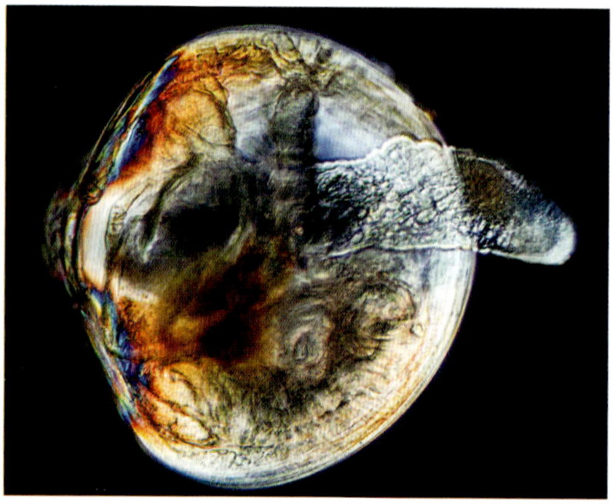

FIGURE 12.12 An advanced bivalve veliger with extended foot. Veliger width 300 µm.

dominated by cockle veligers. An adductor muscle, which pulls the two valves shut, is just visible when the shells are open. At this time the velum protrudes and occasionally the foot, which is longer and more flexible than that in the gastropod veliger. The velum is usually only one highly ciliated lobe, which is used for swimming and making a feeding current.

Cephalopods

This phylum comprises highly specialised individuals with good locomotion as well as excellent sensory and nervous systems. The adults living in the pelagic zone, therefore, are generally nektonic, like fish, and unlikely to appear in plankton samples. The Common Cuttlefish *Sepia officinalis* lives around sedimentary shores and estuaries and lays large yolky eggs, blackening them with ink so they look like a bunch of grapes. On hatching, the larvae remain in the benthos, where they live and develop.

FIGURE 12.13 A pair of Common Squid *Loligo vulgaris*, stranded on a beach.

In contrast, the Common or European Squid *Loligo vulgaris* lives above the benthos along the coast and produces very small eggs with little yolk. They hatch into what are termed paralarvae and remain dispersing and feeding in the plankton for up to three months, depending on the temperature. After this phase, they becomes subadults. The population size depends on the survival rate of the larvae and so the warmer the conditions the quicker the larva can pass from egg to subadult. *L. vulgaris* is particularly abundant in southern Europe and the mediterranean, and the UK is at the northern limit of its range; therefore, it is mainly found on the south-west coast of the UK, where young individuals quite regularly become stranded on the shore, particularly after gales in the summer. Paralarvae may be found on occasions.

The planktonic paralarval stage of cephalopods is an important dispersal phase, particularly for benthic species. During their time in the plankton the paralarvae change their vertical position by active migration up and down in the water to find microcurrents bringing nutrients up from the benthos. Unlike veligers, the paralarva, which does not undergo metamorphosis, looks rather like a miniature adult, but with very short tentacles.

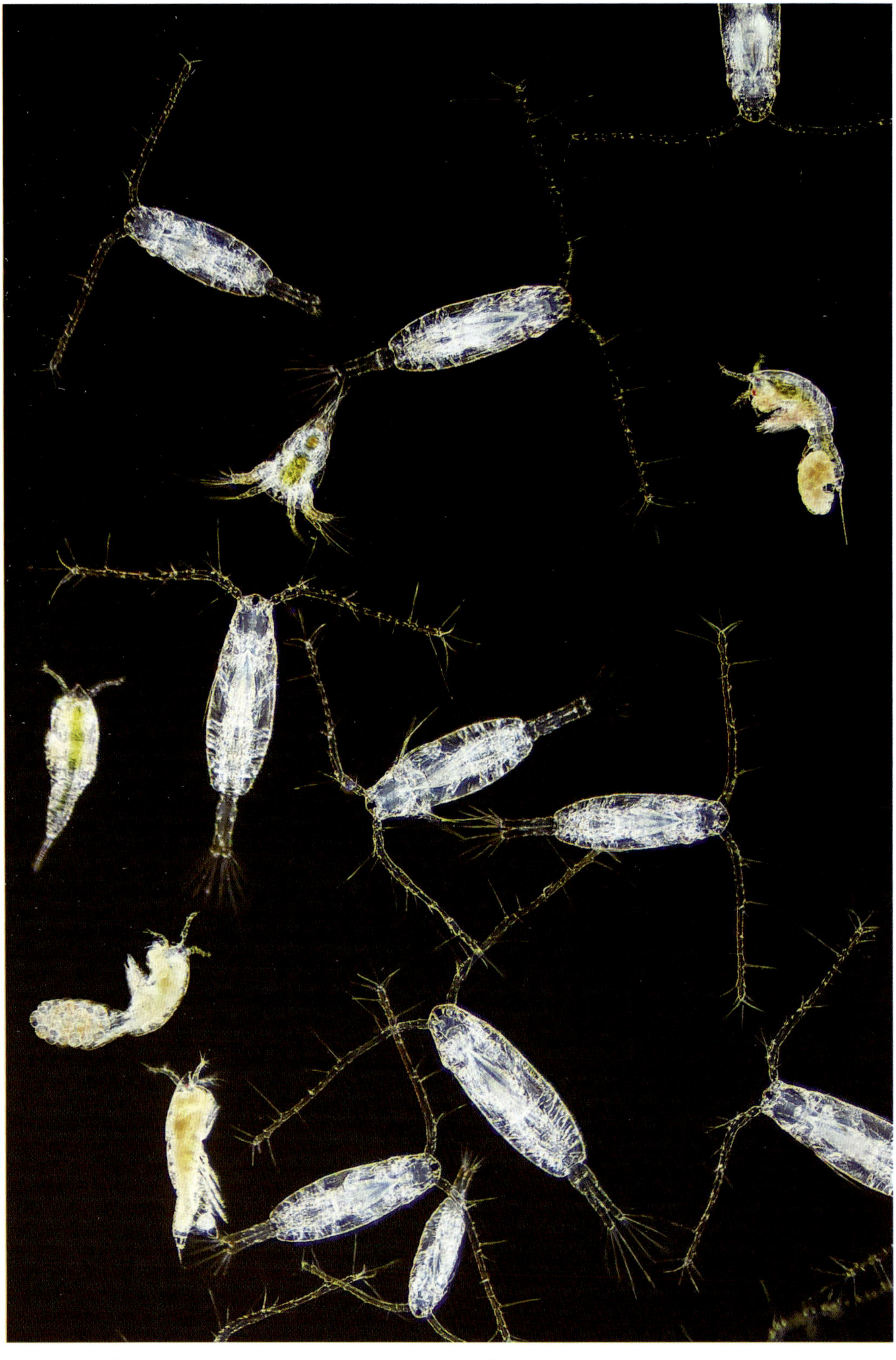

FIGURE 13.1 A mix of calanoid and harpacticoid copepods. Note that the calanoids have extended antennules at the front, used for sensing the water. Near the top and left is a solitary barnacle nauplius.

CHAPTER 13

Insects of the sea: crustaceans and other arthropods

A plankton sample without a crustacean is rare. At times they are so profuse and manically active, bumping into every other organism, that it can be almost impossible to see anything else under the microscope. Nauplii larvae are usually the most prolific over the shore and are found in every month of the year. When first discovered at the end of the eighteenth century they were initially thought to be a separate species and given the name *Nauplius*, from marine heroes in Greek mythology. It would be years before these creatures would be recognised for what they are: larvae of two very common groups that are very different when mature – copepods and barnacles. The former are permanent plankton dwellers and said to be the fastest animals on the planet (relative to size), while the latter glues itself to a hard surface to live a sedentary life in adult form. Copepods have periods of extreme abundance, both on the coast and as dominant organisms in the oceans. It is this dominance that has led to the phrase 'insects of the sea'.

True insects are very rare in the plankton and many will be accidental, such as the midge larva *Clunio marinus* that are washed out of tidal pools. Live ones are rare and specimens are usually dead and mostly consist of the head capsule, which is tough enough to resist decay.

Cumaceans, commonly called Hooded Shrimps, are a strange-looking crustacean. With a bulbous head end and only simple eyes, they dwell in coastal sediment, sifting through the grains and feeding on organic debris. A few millimetres or more in length, they are an example of tychoplankton, appearing suddenly in the plankton to live for a while before returning to the sediment. It is probable that they swim into the plankton for breeding and dispersal along the shore. Like many common creatures, they are poorly understood.

Beside copepods, water fleas (Cladocera) can live as holoplankters. They are very common in fresh water but the marine forms are easily overlooked, particularly as there are only a few genera

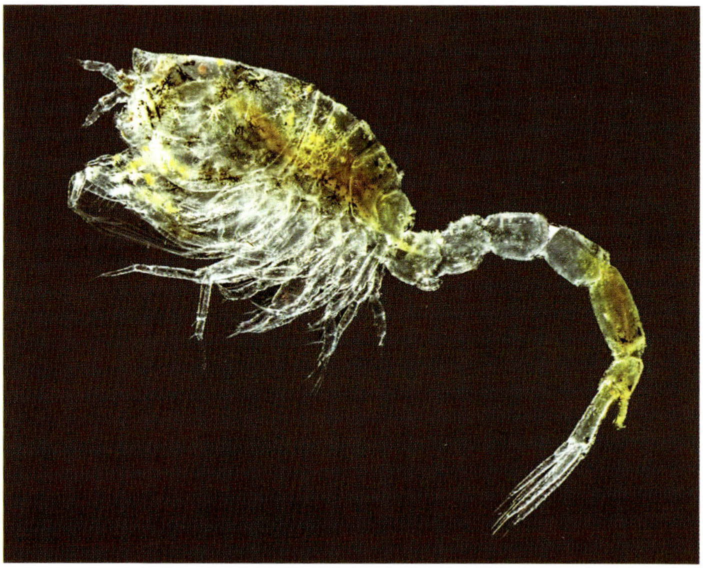

FIGURE 13.2 A cumacean, an occasional tychoplankter. Length 3 mm.

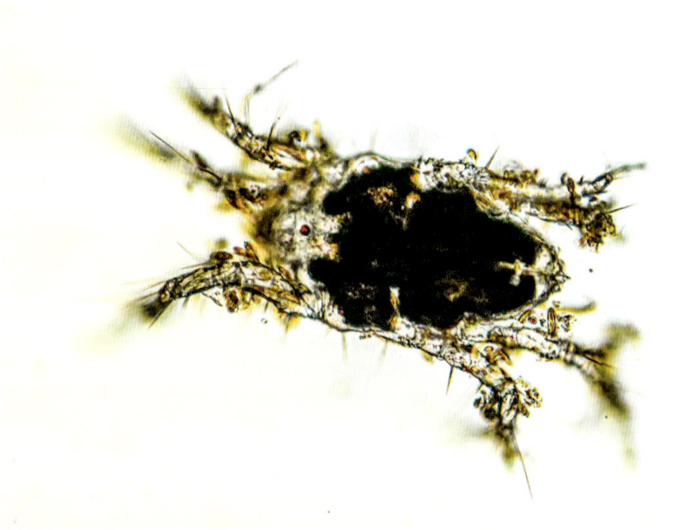

FIGURE 13.3 Halacarid mite with benthic diatoms attached. Length 1.5 mm.

and they are very different from their cousins in ponds. This also applies to the ostracods, which are not common in shore plankton but a few species can occur.

While nauplii, the principal crustacean larva, are abundant, there are plenty of other larval forms to be found, especially those belonging to the Decapoda (crabs and shrimps), which can be fascinating to watch. As Crustacea is such a large and diverse taxon this chapter will be split into the main subgroups, but first, another subphylum of the Arthropoda: Chelicerata.

Chelicerata

Acarina (mites) are found just about anywhere, and the marine environment is no different. The principal family is Halacaridae. Look closely on a rocky shore and dark mites will be crawling over rocks and seaweed. They are regularly washed into the water column, where they seem to thrive and can be found way offshore into deep water. Being related to spiders, they have four pairs of walking legs, but there is no obvious division of the body into segments. They can easily be identified as mites by their stocky, oval shape and a minimum of setae covering the body. Although common, rarely will there be more than the odd one or two present in a sample.

FIGURE 13.4 A female Sea Spider *Anoplodactylus* with maturing eggs inside the femurs. Height of photograph 4 mm.

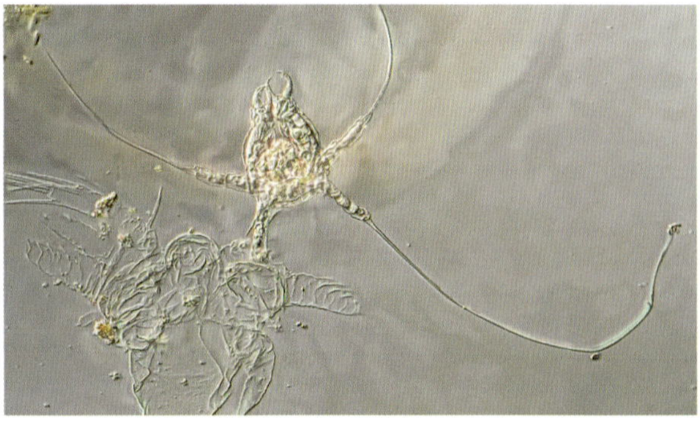

FIGURE 13.5 Protonymphon larva of the Sea Spider *Anoplodactylus*. Bottom left is an exuvia of a nauplius. Body width 35 µm.

Closely related is another group of arachnids, the sea spiders or pycnogonids. Tiny creatures, they live among hydroids and fringe weeds in rock pools and can be tricky to find. The body is so narrow that the female has ovaries inside the femurs. Upon hatching, a small larva is produced, called a protonymphon. Characteristically, they have just three pairs of appendages: the first form the pincers, or chelae, and the others are stumpy legs. The exception is the genus *Anoplodactylus*, of which at least one species has a planktonic larva possessing long whip-like extensions to the four legs (Figure 13.5). These may be to assist stability in the plankton, although it is mostly thought that they help the larva catch hold of a hydroid colony such as *Obelia*. Once there, it burrows inside to become a parasite within the polyps and grows to become an adult. Larvae of this genus have also been known to hitch a ride inside medusae.

Copepoda

Of all the creatures in marine plankton, the copepods have probably attracted the most research and study. Johan Ernst Gunnerus discovered and described the first copepod in 1767, calling it *Monoculus finmarchicus* after Finnmark, in northern Norway, where he found it, and its single central eye. After several false starts, the genus was renamed *Calanus* in 1819 by Leach and remains so today. The genus gave rise to the key order of Calanoida, the single most important group in the zooplankton. Their form contrasts, as we will see, with members of another order, Harpacticoida, which live primarily in sediment but that often appear in the plankton. As early as 1878, G. Stewardson Brady wrote a three-volume monograph *Copepoda of the British Islands*, a comprehensive, well-illustrated survey of the group. Around this time, realisation was growing about the importance of the large copepods in the food chain; these abundant crustaceans eat phytoplankton and in turn are the principal food of fish and even whales. Their importance to the fisheries industry was the key prompt in the formation of the UK's Marine Biological Association.

Within the Crustacea, copepods form a unique group that does not entirely stick to the Arthropod plan of head, thorax and abdomen. The body is divided into 16 somites, or segments, but many may be fused. The first six are usually fused to create the cephalosome. Behind this

FIGURE 13.6 Calanoid copepod swimming with the front side appendages: second antennae, mandibles and maxillae. Note what looks like a parasitic dinoflagellate attached to the right side of the body. Length 2 mm.

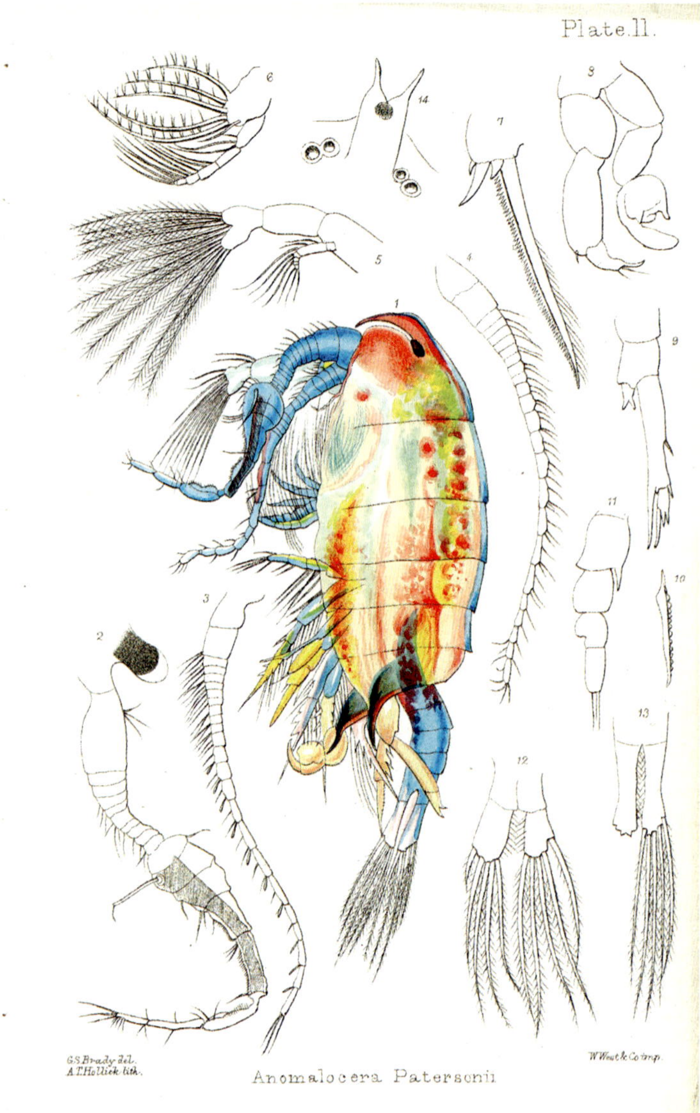

FIGURE 13.7 One of the few hand-painted illustrations in G. Stewardson Brady's *Copepoda of the British Islands*.

is the metasome, of five somites. Together these two sections are called the prosome, the main copepod body, with a tail-like urosome behind. Segments are added to the latter as the creature matures. The first of these is the genital somite, with a genital pore. At the very end is the anal somite and a pair of feathery caudal furcae assisting locomotion. Each somite of the prosome has appendages. In order from the front, the cephalosome has pairs of antennules, antennae, mandible, maxillules, maxillae and maxillipeds. The first of these is sensory, the antennae for locomotion and the rest for feeding. Each somite of the metasome has a pair of legs, although the fifth pair can be missing in females or modified in males. Copepods can move slowly in circles, sometimes spiralling up or down, with the water swirling like whirlpools on each side of the prosome. This is down to the pair of antennae behind the antennules, which remain out in front. Sudden darting actions are caused primarily by the legs on the metasome.

Copepods pick up vibrations in the water using the first pair of antennae (also called antennules), which are held out in front at 90 degrees to the body. Sensitive setae or hairs along these can pick up the slightest change in pressure along the length and comparisons between the two will help to accurately determine the position of the cause. Rapid transmission of impulses along neurons activate muscles believed to have the greatest power output of any animal muscle, gram for gram. When the organism stops moving for at least a few seconds or more they go into a resting posture, in which the body adopts a vertical, hanging position with the antenna stretched out above them.

The link these organisms forge between phytoplankton and fish protein is of huge importance, but their second claim to fame is speed. The swiftness they can call upon as a means of escaping predation is hugely impressive, as anyone who has attempted to catch *Calanus* with a pipette in a dish will attest. *Calanus* has a top speed of 80 cm per second, which will take it well clear of a predatory fish. Members of the genus *Acartia*, which average around 1 mm in length, have been recorded moving at 500 body-lengths per second. These organisms' acceleration is brought to a sudden stop by the viscosity and drag of the water, but is often sufficient to allow them to escape from danger.

In addition to their ability to flee danger, most copepods, especially the more mature and adult calanoids, have areas of red pigment, a carotenoid called astaxanthin produced by diatoms, in sections of the antenna or at the end of the body. It is thought that this colouring disrupts the light passing through the animal so that predators below do not perceive their prey. Along the back of the animal, above the gut, is the oil sac (a food store) running the length of the body. When resting, as noted, copepods hang vertically in the water and the oil sac acts like a perfect lens, channelling the light down to the end so that it stands out as a bright pinprick of light, ideal for any predator below to see. The pigment at the end of the oil sac absorbs and prevents this burst of light (Miller 2023). If this is the case, however, why would pigment in the middle of the copepod's anterior be advantageous? Look closely, and you will see it is typically arranged around the mouth. When the calanoids are feeding messily on hundreds if not thousands of diatoms a day, their prey diatoms release chemicals to warn others of their kind that there are grazers about. Crucially, the diatoms release sparks of cold light (bioluminescence) in and around the mouth. Again, the red pigment around the mouth will minimise this light display to potential predators. It also seems likely that randomly spaced splashes of pigment will disrupt the copepod's body outline in a manner akin to camouflage.

FIGURE 13.8 Adult calanoid, side view, showing the large setae towards the end of the antennules that detect water vibration. Above the gut, with pale green debris, is the oil sac, with red pigment at the lower tip. Below the main body, the prosome, are the various appendages, with the forward-facing legs at the rear.

FIGURE 13.9 Juvenile calanoid, ventral surface, showing the single naupliar eye at the front and the red pigment at the tip of the oil sac. Between is the pigment around the mouth, with the antennae and some mouthparts just visible to the side.

Despite these adaptations, copepods are still a major food source for a wide and diverse array of predators, from fish and whales down to protists and even feeders such as barnacles. To avoid being eaten they feed in the surface layer under the cover of dark and by dawn are migrating into deeper water, where light levels are dimmer, to wait out the day before migrating higher again as the night closes in. This movement is called diel vertical migration (Miller 2023). Zooplankton other than copepods migrate vertically, although not necessarily to the same extent. I am regularly given zooplankton samples collected offshore around midday. Instead of collecting near the surface, a sampling depth of between 30–40 m has to be used, as by this time the copepods will have moved to deeper water.

The number of diatoms consumed by an individual copepod in a day is extremely variable and depends on the size of diatom available, but research suggests that a maximum of a third of a million individuals may be consumed per day. Copepods are messy feeders, especially if they are eating large diatoms. *Calanus* might use their mouthparts to break down almost 2,000 individuals of *Odontella sinensis* (one of the largest diatoms) daily, although much of the material will be lost as debris. This process releases huge amounts of dissolved organic nutrient into the water column, which benefits the overall ecosystem. Most of what is eaten is voided as faecal pellets, which sink

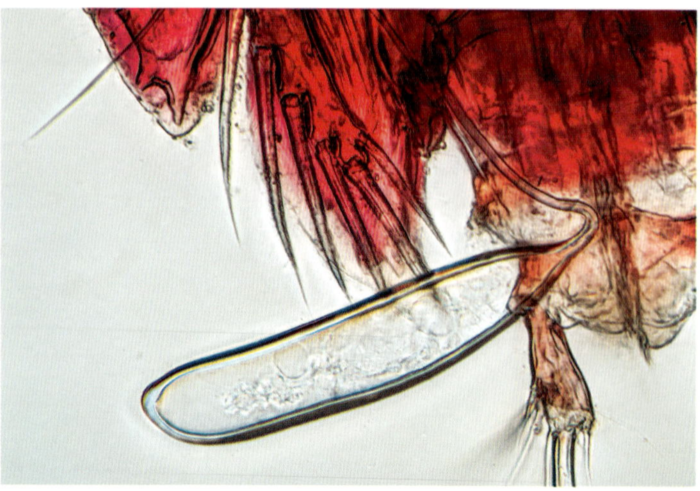

FIGURE 13.10 A spermatophore attached to the female genital pore. Length 150 µm.

Copepods at rest

Copepods may be absent from samples not just during the winter. Typically, in June a decline in seashore phytoplankton occurs before a slow recovery in August. Possibly due to the lack of food, copepods can disappear at this time, sometimes completely. The rise and fall and rise again can be followed and predicted from their forms that appear under the microscope. A fall can be anticipated when a strange egg-like structure turns up in samples. Looking rather like a marine mine from the Second World War, it is round and covered in spikes. These do vary in form and indeed are resting eggs, produced as a way to survive environmentally challenging periods. They sink slowly, remaining in the plankton for some time, before reaching the benthos and lying dormant there.

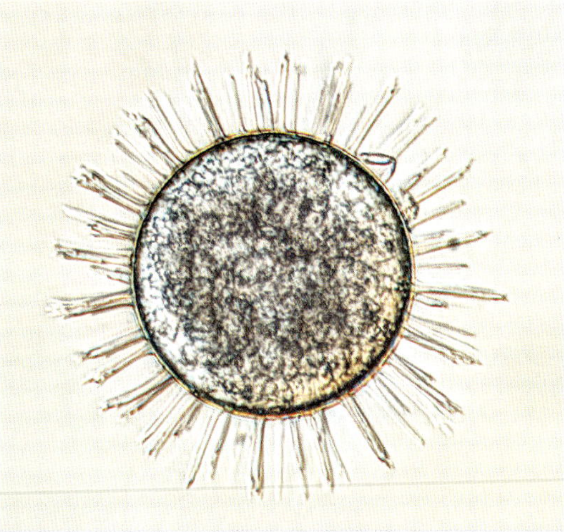

FIGURE 13.11 A copepod resting egg. Diameter 100 µm, including spikes.

to the bottom or are consumed by other organisms within the plankton. Smaller diatoms such as *Thalassiosira* spp. will be dealt with more efficiently, creating less debris but more faecal pellets.

During the autumn species such as *Calanus* begin to disappear from the seashore plankton, moving into deeper, colder water. They prepare for this well in advance by building up an energy store, the oil sac, over the summer. It is filled with a waxy lipid and the sac can expand to occupy much of the body as the contents need to see the organism through diapause, a state of physiological dormancy. This phenomenon is also associated with insects that overwinter. In this state the copepods hang suspended in the deeper water, but they are still able to be active and, if disturbed, will jump away from vibrations picked up from the surroundings. Over the winter there may be less food available and during diapause they do not feed until January or February, when the spring bloom of diatoms begins.

In the vast expanse of ocean water, finding another copepod with which to mate requires pheromones. The females release these, which disperse within the water column. The excited males begin to zig-zag back and forth, tracking the change in pheromone concentration, until they reach her. The antennules possess chemosensory sensilli as well as mechanoreceptors, and it is these that detect the pheromones. Males produce and then externally carry sperm via a spermatophore, which he attaches into the genital pore of the female. When she is ready the

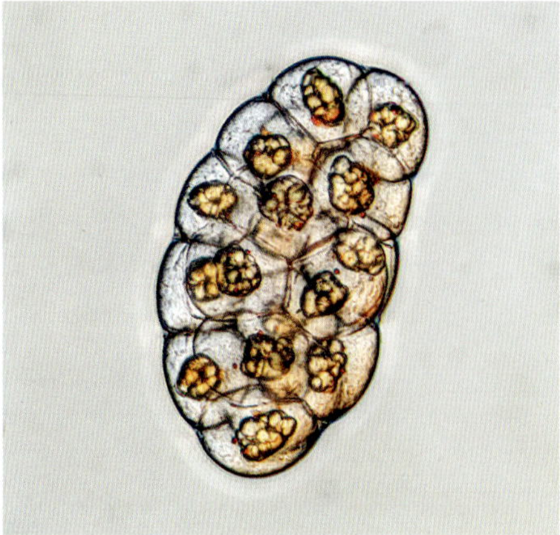

FIGURE 13.12 Copepod ovisac. Length 200 μm.

FIGURE 13.13 Eggs being released by a calanoid copepod.

spermatophore discharges and the contained sperm fertilises her eggs. Some species, such as *Calanus*, broadcast the eggs, while most retain them in an ovisac that hangs from the female urosome. Eggs can take around ten days to develop and hatch, releasing an oval-shaped nauplius larva. As it feeds and grows so it sheds the exoskeleton and passes through six naupliar stages, before metamorphosing into a juvenile called a copepodite. Looking like a miniature adult, it is small and lacks some of the adult segments, particularly in the urosome. It continues growth and passes through five copepodite stages before becoming an adult. Each time ecdysis (the shedding of the old cuticle) occurs, extra segments form and there is an increase in body size. Many different stages can be present within the same sample, adding to the confusion of identification.

When viewing samples across the year adults can often be in low numbers but nauplii will be the most common sign of copepod presence and, several times a year, will be abundant. Barnacles also have a nauplius (discussed below in this chapter), but that tends to be broader and with a pair of horns on either side of the head. Copepod larvae are typically ovoid with three pairs of

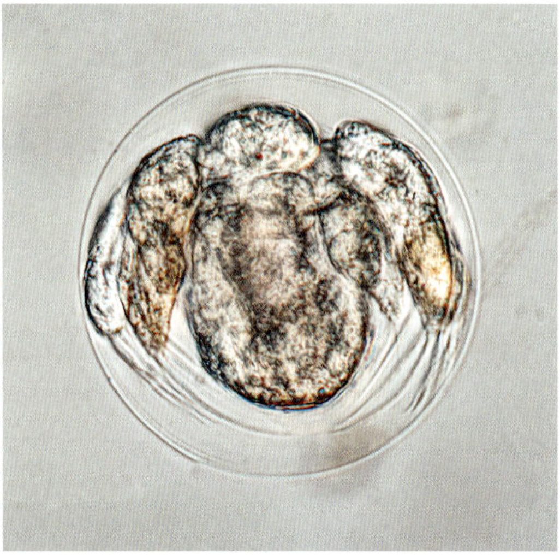

FIGURE 13.14 Calanoid egg development after 10 days with a nauplius inside. Diameter 160 μm.

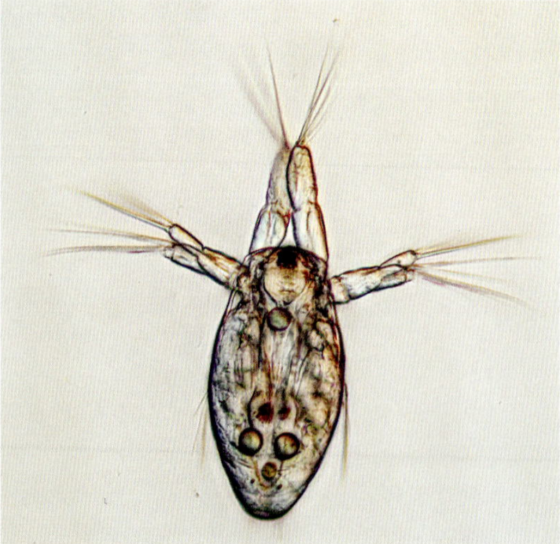

FIGURE 13.15 An early copepod nauplius, top view, with lipid droplets. 200 μm.

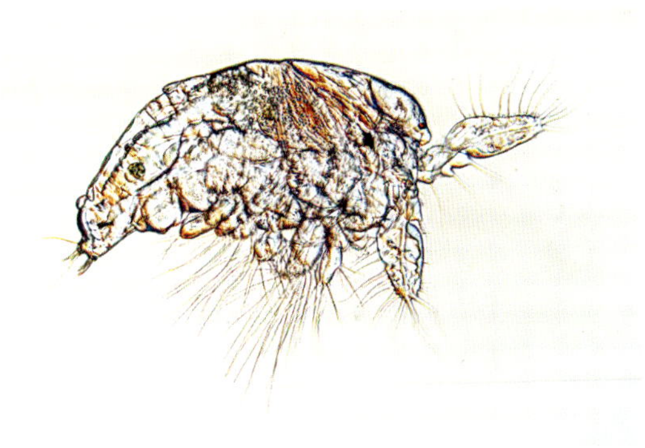

FIGURE 13.16 Late nauplius stage, lateral view. 340 µm.

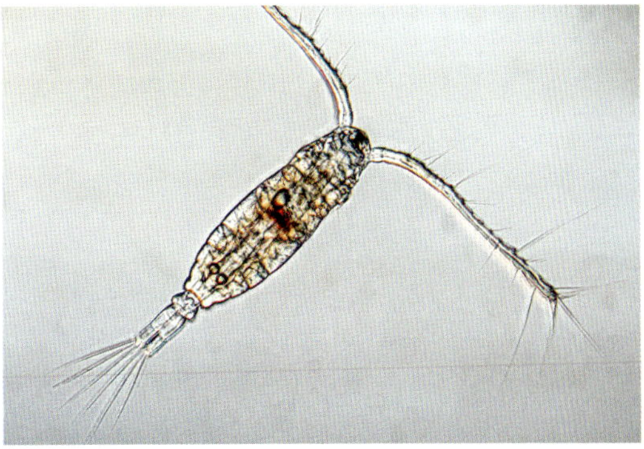

FIGURE 13.17 Young copepodite, stage 4. Length 500 µm.

appendages, all used in creating jerky movements through the water. The antennules have the most important role, while the other two pairs are used for feeding. The setae help in movement and collect fine particulate matter to pass to the mouth. A single eyespot, called the naupliar eye, is usually present. In the most typical form it consists of three fused ocellar cups, each made up of three parts – a retinal sphere, a tapetal layer and a surrounding pigment cup – although under the microscope it appears as a red blob.

Although copepods are considered important consumers of diatoms they can have quite a wide variety of phytoplankton in their diet and may be omnivorous. What is clear is that, as noted, they feed at night and by day move into deeper water to avoid predators. Eyesight is therefore not important and typically they have just a single eyespot, which is the same as that of the nauplius and may even be referred to as a naupliar eye. The larvae also use chemical or mechanical means to find food; the only requirement is that the food item must be smaller than them. First- and second-stage nauplii will feed on tiny dinoflagellates, which they will detect by tasting the water.

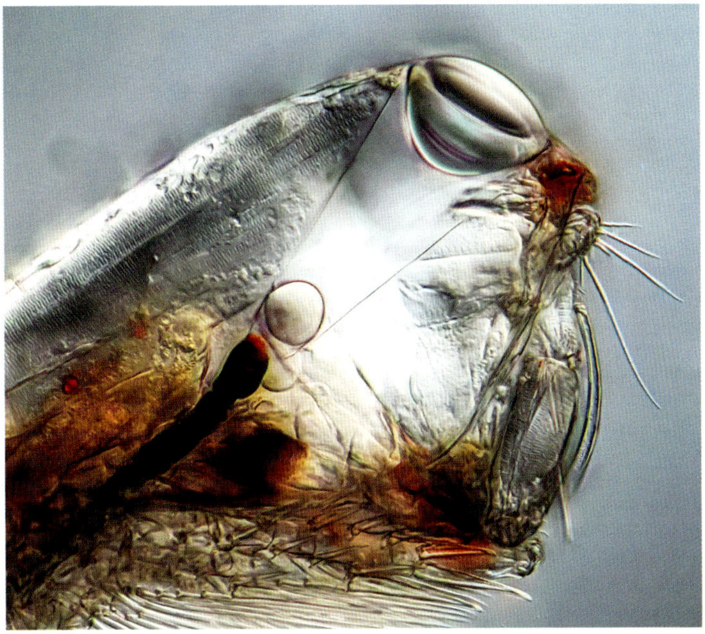

FIGURE 13.18 Lateral view of a Corycaeid copepod, probably *Ditrichocorycaeus*, showing one of two large lenses that refract light back through a crystalline cone to a secondary lens and the dark rhabdome deep in the body. Width of photo 550 µm.

FIGURE 13.19 The parasitic copepod *Caligus*. Length 5 mm.

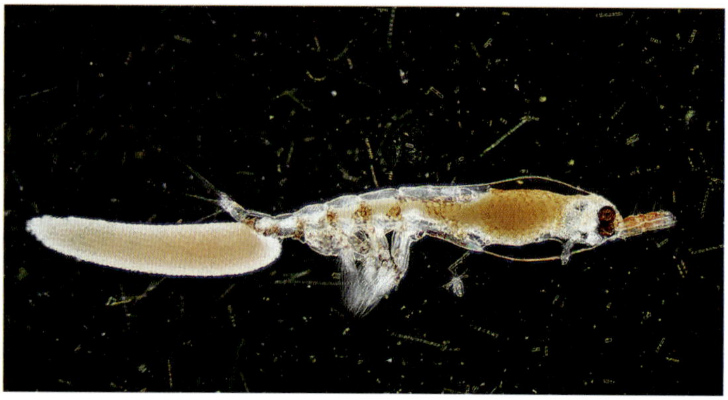

FIGURE 13.20 A Monstrilloida female copepod with eggs, *Cymbasoma rigidum*. It is non-feeding but substantial amounts of stored lipid can be seen inside. Length including eggs 5 mm.

Predatory copepods have a very different eye structure. Members of the family Corycaeidae, which appear from late summer to autumn, possess two giant lenses dominating the front of the head, like headlights. Each lens refracts light back through the body, passing through a crystalline cone which focuses it on to a rhabdome, behind which is the eye pigment that creates the image. The angle of view is extremely narrow, around 3–5 degrees. Near the mouth is a pair of claw-like pincers for grabbing prey.

Roughly half of all copepods are parasitic on fish. The family Sapphirinidae is a large group of quite sizeable parasitic and predaceous copepods notable for their iridescent colours, but are rare close to the shore. The group Siphonostomatid are exclusively parasitic and the genus most likely to occur is *Caligus*, of which there are 200 species worldwide and 12 in Europe. At 5–6 mm in size, with a chunky body, its rapid swimming quickly makes it known in a sample. Summer is the usual time for them to appear around the shore, although they are never common; usually just the odd one or two will be present. Living a parasitic way of life necessitates significant adaptation and *Caligus* species are no exception. Endoparasitic copepods tend to be small but, being ectoparastic, *Caligus* is very large and with modified limbs for grasping. The pair of compound eyes is distinct and probably important for finding a host fish, where *Caligus* can attach to the area around the gills. All caligids have the same number of stages in the lifecycle: two naupliar stages, an infective copepodid, and then four pre-adult stages (called a chalimus) before becoming adult. In the same family can be found the Salmon Louse *Lepeophtheirus salmonis* which can be the bane of fish farmers' lives.

Similarly, the Monstrilloida, a group of endoparasitic copepods, live a highly modified lifecycle. Adults appearing in the plankton live for only a short time as they cannot feed, having no gut. Their purpose is simply reproduction. Consequently, they can be a rare encounter. The female carries eggs attached to spines, rather than in an ovisac, and the nauplii that hatch drop to the benthos looking for a host, either a polychaete or gastropod. Once inside the host the parasitic nauplius feeds and grows, eventually developing into a single copepodite stage before emerging as an adult.

The first time I found a member of *Cymbasoma* with eggs it stood out, as it rapidly flipped back and forth in a manner unusual for a copepod. Initially I used a series of flashes to freeze the movement and achieve a sharp photograph. The result was an individual that appeared to have a pair of brilliantly bright headlights instead of eyes. These are known as mirror eyes, because they reflect the light. Figure 13.20 was taken with LEDs behind the animal so that the light did not enter the eye and the outline of the body stood out. Just behind the eye is an oral tube passing through to the outside. As this animal cannot feed it may not seem important to observe this part of the anatomy, but it is useful in identifying the species as members of Monstrilloida are quite different.

There is one last subdivision of copepods that requires discussion: the order Harpacticoida. They tend to receive minimal attention with regard to plankton, as they are chiefly benthic organisms living in sediments. However, they can be abundant in the shore plankton, particularly where rough weather over muddy shores has washed them up into the water column. If you take your samples from a bridge or walkway close to a saltmarsh or mud or in an estuary, expect the copepod species to be dominated by harpacticoids. Their main characteristic is that the body flexes between the fifth and sixth somite, while other copepods flex at the base of the urosome. The body is vermiform (long and narrow), giving them a worm-like ability to manoeuvre through

FIGURE 13.21 An estuarine harpacticoid copepod. Length 1.5 mm.

FIGURE 13.22 *Stenhelia palustris* nauplius, an estuarine harpacticoid. Incapable of swimming, it is broader than it is long. Width 150 μm.

FIGURE 13.23 The harpacticoid copepod *Porcellidium*. Length 600 µm.

and around the detritus on which they feed, along with the microorganisms within. The antennules and antennae are short and often quite stubby. *Stenhelia* (*Delavalia*) *palustris* has a distinctive nauplius that is often washed into samples from estuarine mud, despite being incapable of swimming. Its characteristic feature is that it is approximately round, but slightly wider than long. Abundant in spring, it easily becomes flushed into the seawater as it moves with a rolling action through the substrate. The tiny harpacticoid *Porcellidium* also bucks the vermiform trend in being dorso-ventrally flattened with a very broad, shield-like prosome and small urosome. Members of this genus typically live on seaweed, but do become detached by wave action.

True planktonic harpacticoids

Of the few species that permanently exist in coastal plankton, *Euterpina acutifrons* is most likely to be found. It is long and narrow, and the prosome is quite large compared to other harpacticoids. The legs are long and spread wide to the side and the entire body is hairy, all of which are adaptations to slow the rate of sinking. They will survive quite high up estuaries in low salinity, but can be found washed out to sea. The nauplii feed on particulate debris both in the plankton or from weed or other substrates when they settle.

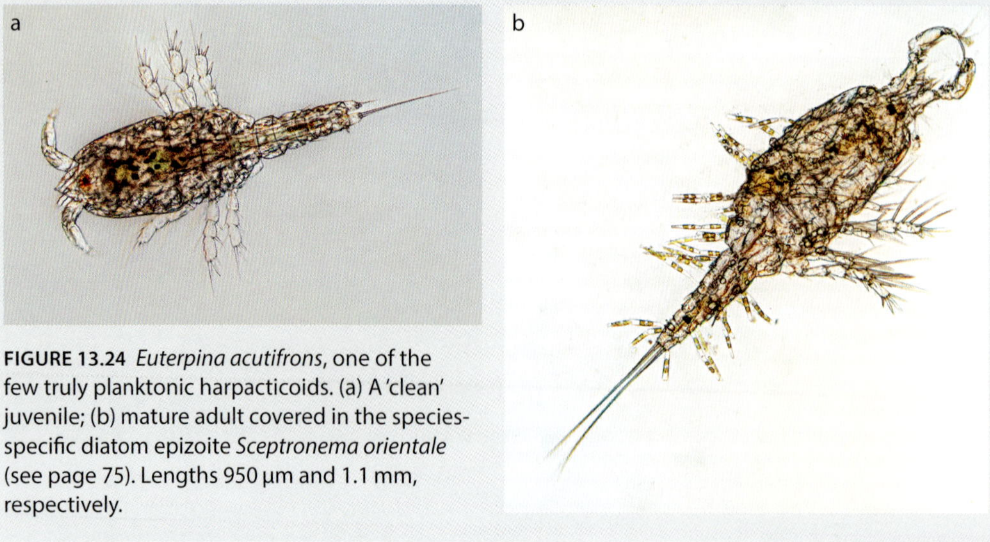

FIGURE 13.24 *Euterpina acutifrons*, one of the few truly planktonic harpacticoids. (a) A 'clean' juvenile; (b) mature adult covered in the species-specific diatom epizoite *Sceptronema orientale* (see page 75). Lengths 950 µm and 1.1 mm, respectively.

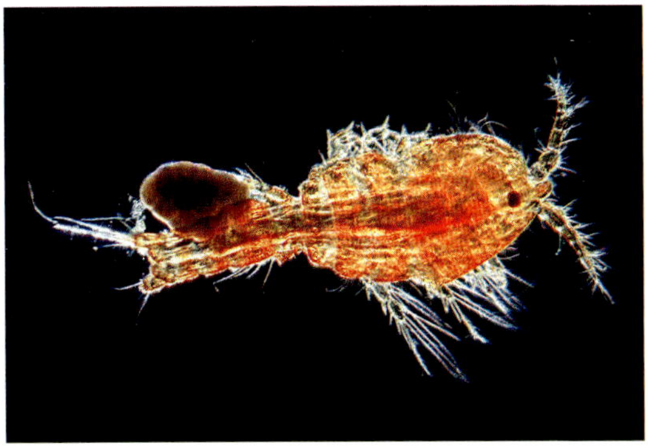

FIGURE 13.25 *Tigriopus* sp. from a rock pool. Length 1.5 mm.

Our discussion of harpacticoids cannot be complete without mention of a copepod that has adapted to survive in a particularly precarious and extreme marine environment. *Tigriopus* thrives in rock pools close to the splash zone, where salinity varies widely as a consequence of rain or summer heat. The darting action of these small crustaceans is visible in rock pools and a fine tea strainer, sieve or large pipette can be used to collect them. Consuming bacteria, fine algae and particulate matter, they remain in these pools to complete growth and reproduction. From egg to the first copepodite stage is just eight days at 15°C; a lifecycle is completed in 31 days and total longevity is around 55 days. The nauplii rarely move in open water, but drop to the bottom or sides of the pool to feed.

Harpacticoid copepods are a huge group occupying very diverse habitats, as the preceding paragraphs have shown. These include even damp areas on land. If you collect a sample of bog-moss (*Sphagnum* spp.) and squeeze out the water into a tube a surprising number of harpacticoids can usually be seen wriggling around under the microscope, particularly examples of the genus *Bryocamptus*. For more information, Hingley 2021 is very useful.

Cirripedia: the barnacles

It is not just copepods that produce a nauplius larva, as we have seen above. Barnacles, too, have a nauplius stage, which are likely to occur in any plankton collected throughout the year. However, at some points in the year, such as the spring, they can be like a blizzard filling a sample. A light placed to one side of the container will soon draw them away to amass on that side, helping to reduce the distraction.

Most people are familiar with acorn barnacles, the types commonly found on rocky shores, boats and other hard substrates, but these are not the only type relevant to our discussion. Stalked barnacles, or goose barnacles, are warm-water species that require temperatures around 19°C for breeding but may be washed ashore in the UK with driftwood and debris brought in by the Gulf Stream from the tropics and subtropics. The Pelagic Goose Barnacle *Lepas anatifera* washes up attached to hard material, such as wooden planks and plastic floats. Not so common is *Dosima fascicularis*, which secretes its own spongy white float from its cement gland. This float usually adheres to a piece of seaweed, contributing to the barnacle's appearance at the coast when the seaweed is washed up as tidewrack. While these goose barnacles are most unlikely to appear in your plankton samples, they are still planktonic and feed in the plankton.

Within the Cirripedia there are many examples of parasitic forms that infect hosts living on the shore, especially other crustaceans. They belong to the group Rhizocephala, among which

FIGURE 13.27 *Dosima fascicularis*. The float is visible at the base of the stalk. Length 6 cm.

FIGURE 13.26 A group of Goose Barnacle *Lepas anatifera*.

FIGURE 13.28 Acorn barnacle feeding with the legs extended to collect suspended particulate matter. They regularly shed the exoskeleton, so exuviae are common in the plankton.

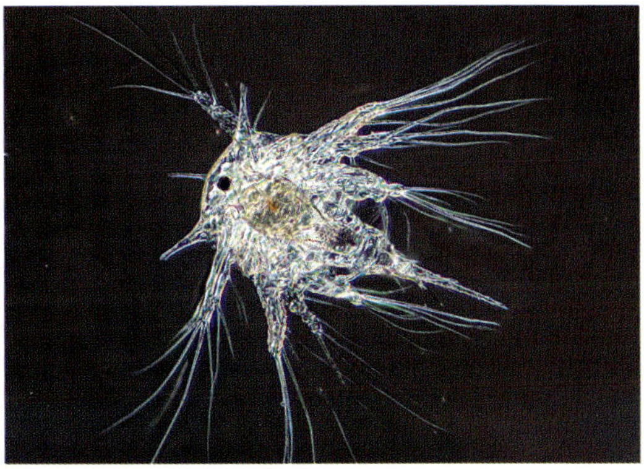

FIGURE 13.29 Barnacle nauplius, ventral view. Note the horns. Length 600 µm.

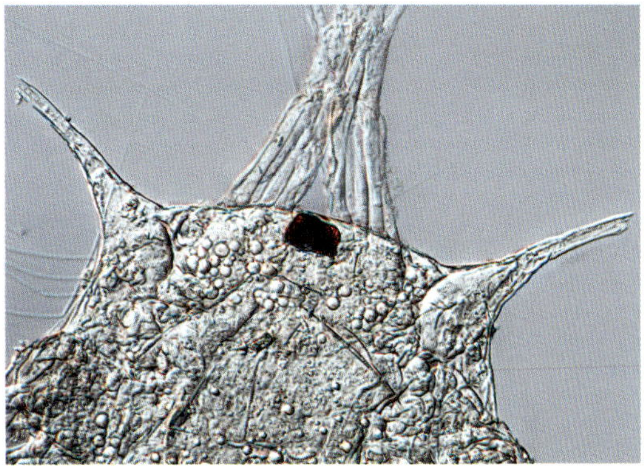

FIGURE 13.30 The head of the nauplius, showing the open ends of the perforated horns.

Barnacle discovery

We met the important figure of John V. Thompson in Chapter 2. It was he, in the late 1820s, that finally determined that barnacles were crustaceans. Previously they had been considered molluscan because of the calcareous plates surrounding their bodies. For some years after this, however, nauplii were still thought by many scientists to be a distinct species. Charles Darwin also carried out extensive work on barnacles from the 1840s which was crucial in our understanding of the species and their ecology.

Acorn barnacles are permanently glued to the rock and so mating necessitates fertilisation of immediate neighbours. The eggs are retained so that first-stage nauplii are released into the water at high tide. Despite the very different appearances of the adults, the larvae of copepods and barnacles are remarkably similar, with the latter simply being more shield-like with – most importantly – a pair of horns on either side of the head. There are also spines emerging from the caudal end of the body, with the middle one being especially pronounced. Once released, the nauplius passes through six naupliar stages of growth and development. Stage one initially has a membrane over the two horns, but this is soon shed as it enters the second stage. These horns are hollow and have ventral perforations. Glands inside the horn base produce a secretion rich in tryptophan and tyrosine, possibly as an anti-wetting agent, to prevent the larva from being caught in the surface tension. It may also deter predators. When looking at plankton in a small dish one should always check the surface of the water as so many organisms become caught here, but barnacle nauplii will be absent. As the nauplius grows and moults so the horns increase in size.

Once stage two is reached feeding begins, principally on diatoms. As the nauplius passes through successive moults, so the amount of lipid built up within the body can be seen as droplets of oil. After stage six a completely different, non-feeding larva is produced, called a cyprid or cypris larva. The stored lipid enables the non-feeding cyprid to concentrate on settling out of the plankton and searching for a suitable substrate on which to permanently settle. The cyprid structure is very different to that of the nauplius, consisting of a bivalve-like carapace with adductor muscles holding it together. The body inside can be seen, including the unique compound eyes, which are made from three crystalline cones. Six pairs of legs

FIGURE 13.31 A barnacle cyprid. Note food storage at the front, the cupped ends of the antennae, right, and the legs to the left. Length 1 mm.

FIGURE 13.32 A limpet, a few adult barnacles and a huge number of freshly attached young barnacles on a rock. The tiny dark yellow dots are cyprids that have just arrived from the plankton to settle on the substrate.

work in parallel for swimming and also help the cyprid to crawl over the rock surface. It was this locomotion that convinced Thompson that they were arthropods, and not of molluscan origin. There is only one set of appendages at the anterior end, derived from the first pair of antennae, with cup-shaped ends and setae. There are lattice-organs (chemoreceptors) on the carapace edge and along with the antennae taste and feel the substrate. Once the cyprid finds the ideal spot to settle, several glands release the necessary glue for attachment and within a few hours the larva has moulted into a juvenile without the carapace or eyes. After just another five hours, the young adult barnacle can be clearly seen. The cyprid can delay metamorphosis by a number of days if necessary as it continues searching for its final location, being limited only by its lipid resources.

Acorn barnacles continue growing and expanding, carrying out ecdysis as they develop. This shedding of the exoskeleton does mean that huge amounts of barnacle debris, mainly of the limbs, are released from rocky shores. This debris occurs in all samples and is quite distinctive. The limbs of a barnacle form a net to collect plankton and, once shed, the material continues to act as a net trapping a whole host of different planktonic organisms.

Sacculina carcini is the one most often encountered. The host is the Common Shore Crab *Carcinus maenas*; on picking one of these up, occasionally you may find that underneath there is a yellow-brown sac (the externa) protruding from under the abdomen, where a crab would normally hold its eggs (see Figure 13.42 for a female with eggs). This is part of the female parasitic barnacle, within which she has her ovaries and a cavity where her developing nauplii are retained until their release into the plankton. Released nauplii may go through five moults, during which time they live off stored yolk. Then a female planktonic cyprid larva lands on the crab's surface, searches out a suitable hair and, rather than undergoing the normal barnacle metamorphosis she becomes a kentrogon larva: basically, a hypodermic needle. This injects an infectious stage, the vermigon, into the host's blood (the haemocoel) which develops into an expanded series of threads (the

FIGURE 13.33 The Shore Crab *Carcinus maenas* with the yellow externa of *Sacculina carcini* visible under the abdomen.

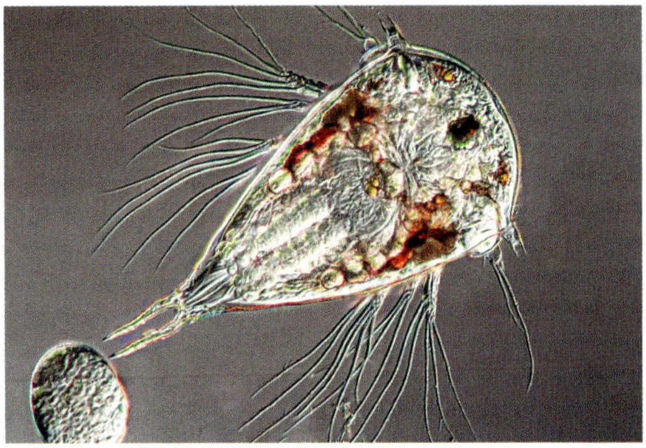

FIGURE 13.34 Nauplius of *Sacculina carcini*. Length 280 µm.

interna) that reach all parts of the crab's body. Here she remains from six months to three years. The externa is the only part to grow outside the crab and is where the male nauplius enters. Male cyprids, after they enter the crab, become trichogon larvae. One or two live a parasitic existence in a space near her ovaries, operating as functional testes to produce the sperm to fertilise the eggs. The interna and externa form the adult stage. Up to two months can elapse before the eggs hatch into first-stage nauplii within the externa, which operates as a brood pouch. The interna has a profound effect upon the host, both behavioural and morphological. For example, crabs stroke the externa as if it were eggs, keeping them clean, and parasitised crabs move into deeper water, which is likely to be a better place for the release of *Sacculina*'s nauplii.

Ostracods

Although uncommon in the seashore plankton, ostracods are one of the most abundant and successful organisms on earth. Much of this is down to their dominance in deep oceans, where the largest, *Gigantocypris*, reaches 32 mm. But they are found in almost all habitats, including fresh and brackish water. Most are benthic crawlers, feeding on organic matter, although some are suspension feeders in the plankton. It is surprising that so few occur on the shore. Those I have seen are Podocopids, some of the smallest ostracods, at just 110 microns. Interestingly, ostracods are often confused with the cyprid larvae of barnacles and there is a genus of ostracods called *Cypris*.

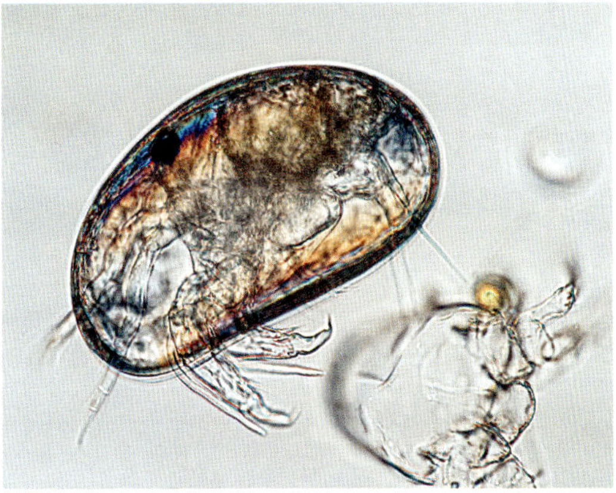

FIGURE 13.35 A Podocopid ostracod. The antennules and antennae are visible at the front, but there are no legs. Length 110 µm.

The few species that may turn up in a plankton sample are usually much smaller than barnacle cyprids, however, which are more than 600 mm in length and just visible to the naked eye. The shape of the ostracod bivalve carapace is more bean-shaped than that of the cyprids, and they have the fewest number of limbs of any crustacean. The antennules and antennae do not have the cup-shaped sensory tip of barnacle cyprids and the six pairs of legs are absent.

Cladocera: the water fleas

The Cladocera also have a bivalved carapace which is fused to two or more of the thoracic segments, although the head remains outside. A single compound eye is dominant in the head, and there are four pairs of legs on the thorax. The abdomen is small, unsegmented and ends in a furca that may protrude out between the carapace valves. The majority of these crustaceans live in static or slow-moving fresh water, where they might be planktonic in the open water or benthic, feeding on deposited detritus, bacteria and microorganisms. Some species have adapted to brackish water and, at some point, a limited number of species invaded the marine environment. Marine water fleas number only around five species, with two genera in northern Europe: *Podon* and *Evadne*. They both look quite different to their fresh-water cousins and are predators, catching protists, dinoflagellates and other suitable-sized prey. *Podon* is quite distinctive, with a depression between the head and thorax, while *Evadne* has no notch or depression but is smooth from the head along the back. The second antennae are large, separating into two extensions (biramous) with long setae. These are for swimming. The thoracic limbs are not completely covered by the carapace, projecting beyond it.

Populations along the coast can be very transient and both genera are possible early-summer visitors to the shore. Although they thrive offshore in deeper water from April through to the autumn, their appearance close to the shore varies according to weather conditions. A gale in August will almost guarantee individuals turning up a week later, abundant above the seashore. With climate change producing warmer sea temperatures in spring, they can suddenly appear in April within sheltered bays. These individuals breed and a temporary population may remain for a month or two. The first to arrive are females producing young by parthenogenesis when

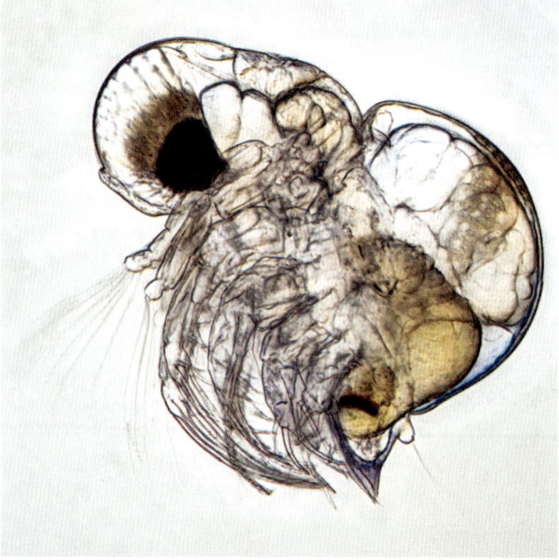

FIGURE 13.36 *Podon leuckartii*. Length 650 µm.

FIGURE 13.37 *Evadne nordmanni*. Length 400 µm.

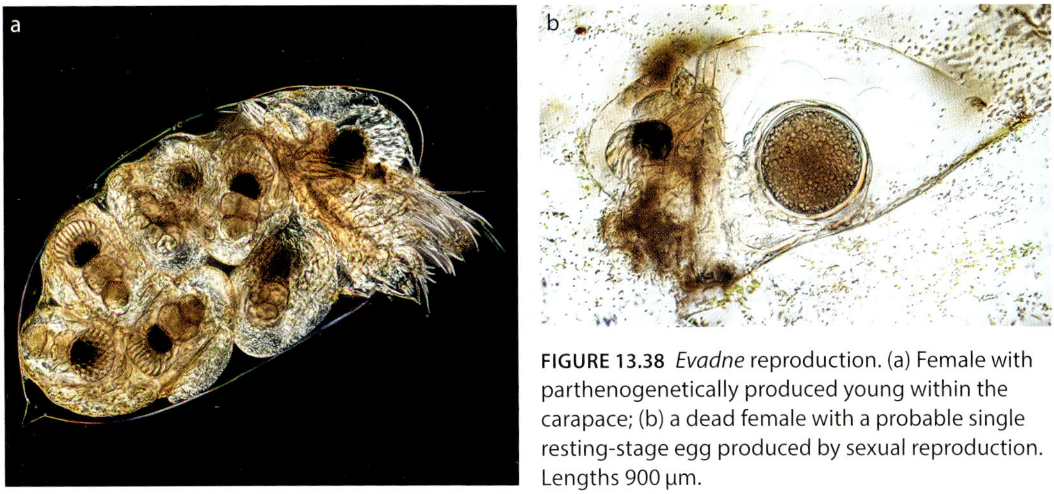

FIGURE 13.38 *Evadne* reproduction. (a) Female with parthenogenetically produced young within the carapace; (b) a dead female with a probable single resting-stage egg produced by sexual reproduction. Lengths 900 μm.

conditions are ideal, but once stressed by changing environmental factors the occurrence of males in the population takes place. These are much smaller than females, particularly in *Evadne*, and on mating a single resting-stage egg (occasionally two) forms inside the female. Shortly afterwards she dies and drops to the benthos, where the egg remains until improved conditions enables hatching to a new female.

Mysids: opossum shrimps

The opossum shrimps, which are believed to be quite primitive crustaceans, are not part of the Decapoda but have their own discrete group, the Mysida. They can be found in huge numbers within estuaries and open, offshore areas, including substantially deep water. They occur in large rock pools but remain in the depths and are not common in the shore plankton. Wave action is damaging to them and so they appear in plankton samples taken in sheltered places in or very near to estuaries. Most are transparent, blending with the sample, but the antennae are quite conspicuous and the second pair have an outer area that forms a flat plate (called the antennal

FIGURE 13.39 An adult mysid or opossum shrimp. Length 2 cm.

FIGURE 13.40 A mysid zoea larva. Note the statocysts just visible in the uropods. Mysid zoea are more abundant in seashore plankton than the adults and may occur in small numbers. Length 7 mm.

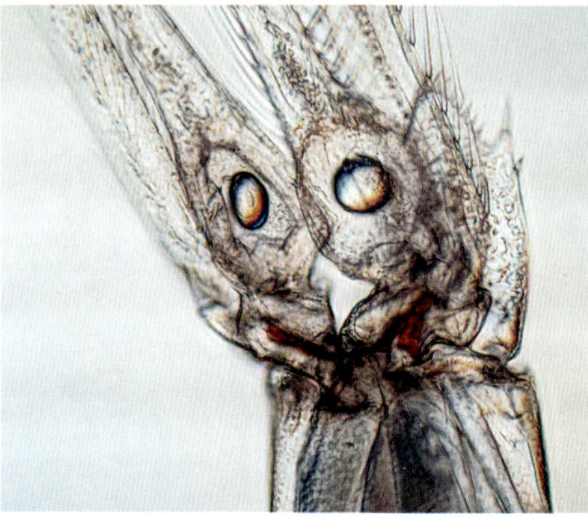

FIGURE 13.41 A pair of uropods each with a statocyst present, characteristic of mysids.

FIGURE 13.42 Female *Carcinus maenas* with eggs attached to pleopods beneath the abdomen.

scale), a useful feature to help identify this group. The eyes are large and on stalks. Unlike prawns and shrimps, and another characteristic feature of mysids, the biramous legs lack pincers or chelae at the ends. The end of the abdomen has a pair of uropods that are in two parts, with the inner one of each having a round statocyst that looks like an eye at the end of the body.

At the base of some thoracic legs are plates for a brood pouch (hence, opossum shrimp) within which are placed the sperm and eggs together for fertilisation. After a few weeks larvae are released, called zoea, which feed on detritus that they collect from a water current created by the legs; that is, they are suspension feeders. As they grow the diet extends to diatoms and small crustaceans. The young mysids live initially in the plankton but some become benthic upon reaching adulthood.

Decapoda: crabs, shrimps and prawns

Decapods are an economically important group of crustaceans that undergo distinct metamorphosis with all larval stages present in the plankton, sinking to the benthos when adult. With seven major taxon divisions, only three are really pertinent to our interest in shore plankton. A possible fourth, lobsters, are a rare occurrence. The three important groups are: Brachyura (the true crabs),

Anomura (squat lobsters, hermit and porcelain crabs) and Carida (the true shrimps and prawns). In all cases the larval stage is called the zoea. Using the scientific terms correctly is useful to try to reduce the confusion that can arise with the use of common names. For example, the mysids that we have already seen are called shrimps, but are not true carids. True crabs such as *Carcinus maenas* have short stubby antennae while the porcelain crabs have long whip-like antennae.

The feisty nature of zoeae, especially those of the brachyurans, soon makes their presence known in samples. They flip, hover and maniacally move at speed, pushing aside everything in their path. They can survive for several months in the plankton, plenty of time for the tide to sweep them along the coast and disperse them offshore, although many are predated. They are common in samples taken on the seashore, but never abundant. By contrast, a summer collection a mile or so offshore can be significantly high in numbers and much more diverse. This group is typically part of the meroplankton: the larvae feed and develop before settling in the benthos to become adults. Typically, the female carries the fertilised eggs under the abdomen, attached by a sticky glue to the pleopods. They remain here for up to two months, slowly changing colour as the embryos develop from bright red through yellow to a darker brown. During this time they pass through what appears in most cases to be a nauplius phase, before metamorphosing into the zoea. This is remarkably well formed, with limbs and body segmentation, when it is finally

FIGURE 13.43 Zoea larva of the brachyuran Shore Crab *Carcinus maenas*. 1.2 mm. The dark markings are chromatophores embedded in the epidermis.

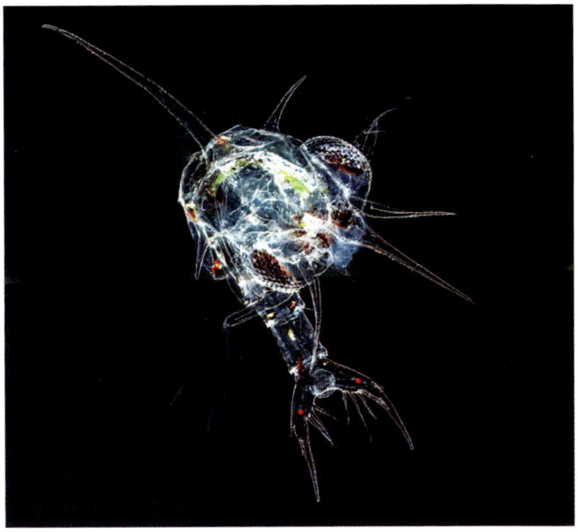

FIGURE 13.44 Zoea larva of a *Necora* crab. The frontal view shows the long dorsal spine and lateral spines. 2 mm.

FIGURE 13.45 A megalopa crab larva, 2 mm across.

hatched and released. Crabs may have 1,000 or more eggs, prawns and shrimps usually rather fewer, at 800 or so. As well as limbs and a pair of compound eyes, spines are common, especially in the crabs. The Shore Crab *Carcinus maenas* has a dorsal spine on the back of the carapace; the swimming crabs *Necora* have even longer spines and smaller lateral ones. Spines help in stability, providing both resistance to sinking too fast and help to remain upright when actively swimming. As zoeae develop and grow in the plankton they pass through a number of instars, a figure that varies with species and environmental conditions but is usually around four to seven stages. For example, low salinity will increase both the number of instars and the duration of life in the plankton.

Slabber and the Zoea

The first description of a zoea was by Martinus Slabber in 1775, and probably related to *Carcinas maenas*, although Slabber linked it with another crustacean. The term zoea was actually coined in 1802 by L.A.G. Bosc, but as a name for a crab species; like nauplii, the identity of larval decapods was a matter of uncertainty and argument for decades. Indeed, in 1840 a prize was offered by the Royal Holland Society of Sciences and Humanities for research that would determine whether crabs exhibit metamorphosis. To date the prize has not been claimed. Our old friend the citizen scientist John Thompson had actually produced evidence in his 1828 report but, as we have already seen, influential scientists of the day came out against him. By the middle of the century the controversy faded, as knowledge improved and the facts were brought together, proving Thompson's work that crabs exhibit metamorphosis.

Once the final crab zoea stage is complete, ecdysis occurs to produce a non-feeding form called a megalopa, sometimes referred to as a post-larva. Looking like a miniature crab, a few millimetres in length, the megalopa continues to swim for several days in the plankton before gradually sinking to the bottom. In the right conditions, crab megalopa finally moult to become tiny juvenile crabs. Members of Anomura, such as the porcelain crabs, also have a zoea and megalopa, and are very distinctive. The Broad-clawed Porcelain Crab *Porcellana platycheles*, a species common under rocks of the lower shore, has quite spectacular zoeae. With a rostrum spine extending out in front of the carapace and two more that stretch behind, when confined in a dish of plankton they seem very unwieldy. On rocky coastlines they can be quite common in summer plankton, with the megalopa appearing in autumn.

FIGURE 13.46 Anomuran zoea larva of the Broad-clawed Porcelain Crab *Porcellana platycheles*. 2 mm including spines.

Isopoda

Very occasionally *Idotea*, a common isopod living on seaweed, swims for a short period at high tide and can thus be captured on occasion in plankton samples. This genus is a good example of this form of crustacean, flattened dorso-ventrally like woodlice but with gills under the telson (tail). *Idotea* (see Figure 1.6) is an incidental drifter and as an omnivore it is not like true planktonic isopods that have large numbers of parasitic species, second only to the copepods. Isopods most likely to be found are one of the larval stages of the large group Epicaridea. The family Bopyridae is common in shore plankton. All epicarideans are parasitic, both internal and external, and constitute

Carids have zoeae that look more like small juvenile adult shrimps. Some have small spines and those in deeper water can develop red carotenoid pigments, rather like the copepods, and perhaps for similar reasons – to prevent the larva being seen from below by a predator, standing out against a blue sky. Metamorphosis lacks a megalopa, with the zoea gaining extended spines, hairs and appendages as it develops.

FIGURE 13.47 Megalopa larva of a porcelain crab, 3 mm across.

FIGURE 13.48 A young carid zoea. Note the red pigment providing disruptive camouflage. 900 μm.

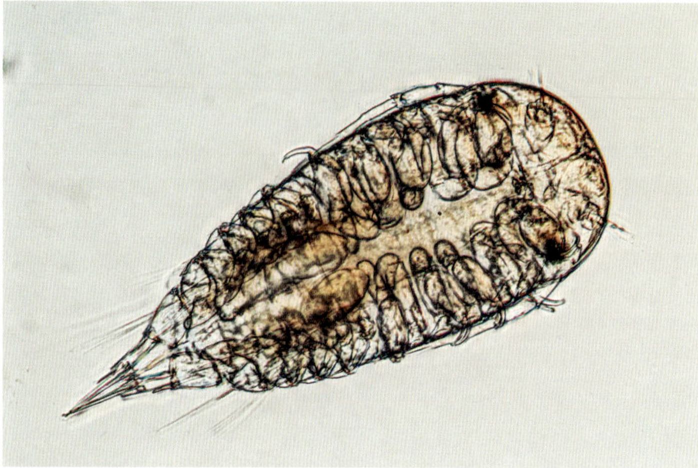

FIGURE 13.49 A parasitic bopyrid cryptoniscus larva (Epicaridea), underside view. 150 µm. See also Figure 11.18c.

around 8% of all Isopoda. Juveniles are pelagic and the first stage, an epicardium, initiates the parasitism by finding a copepod on which to feed, before growing into a larger cryptoniscus phase. Ultimately, they parasitise other crustaceans such as barnacles or decapods.

Gnathia maxillaris is a particularly unusual isopod, looking very like an insect, that can appear in the UK a few times a year. Typically, *Gnathia* is up to half a centimetre long and highly active. The female is very different from the male and from juveniles. Adults can be found in kelp holdfasts, crevices on rocky shores and old barnacle shells. Surprisingly little is understood about the adult phase, although it is almost certain that they do not feed. The female also occurs in the plankton, probably when she releases the larvae, called zupheae. Up to 100 of these are produced at a time and swim in the plankton to find a fish host such as the Common Blenny *Lipophrys pholis* or Corkwing Wrasse *Crenilabrus melops*. Attaching themselves temporarily to the fish, they gorge on blood before dropping to the ocean bottom, where they become a first-stage praniza larva, characterised by a developed thorax. After a break of a day or so the praniza swims back into the plankton, again to find a host for blood before dropping to the bottom once more to become a second-stage praniza. This is repeated a third time before metamorphosing into an adult. The blood meals are believed to sustain the individual through adulthood.

FIGURE 13.50 A female adult *Gnathia maxillaris*. 2.5 mm.

FIGURE 13.51 A praniza stage of *Gnathia maxillaris*. 1 mm.

Amphipoda

Unlike isopods, the Amphipoda are crustacea that are flattened laterally. They tend to be a centimetre or less in length and are also called sandhoppers, after their behaviour in the strandline at the top of the shore. However, they may appear in plankton samples, having been washed off the shore by the rising tide. One subgroup, Hyperiidae, has examples that live among the jellyfish that we encountered in Chapter 11. Some are commensal, while others are carnivores. The species that occurs occasionally in the plankton net is *Hyperia galba*, which is a parasite on medusae and jellyfish. Searching when these jellyfish are stranded on the shore may reveal these inside, attached to the gut. Here they feed on the food the jellyfish has gathered or, if this is not on offer, they start eating the jellyfish's gonads. Individuals living in jellyfish are very pale in colour, while those caught in a plankton net are a much darker brown and have a squatter body with distinct green compound eyes. The ability to change body colour is down to the chromatophores dispersed over the exoskeleton.

FIGURE 13.52 *Hyperia galba* taken from the plankton. The dots across the surface are chromatophores. 1.2 mm.

FIGURE 14.1 The larva of *Magelona* sp. Just visible are the long fine hairs, or chaetae. On each side of the head is a coiled palp that extends when swimming. 2 mm.

CHAPTER 14

Worms and more worms

There are true marine worms, and there are many other invertebrates called 'worms' that appear worm-like but are not. The former are the Polychaeta, a class of the phylum Annelida. All possess the characteristic bristles, or chaetae. Non-worms do not have chaetae, and some of the important planktonic worms-that-aren't-worms are included here, as well as Polychaeta, including the nemerteans, nematodes and flatworms (Platyhelminthes). The arrow worms (Chaetognatha), a particularly important group, have their own chapter below.

The polychaetes, or bristle worms

As well as the chaetae already mentioned, these worms have a bilaterally symmetrical body pattern showing metameric segmentation: that is, repeating segments containing the same organs. The exceptions are the first and last segments. The head may have a number of palps, tentacles and cirri for sensitivity and feeding. The gut is well developed, with a mouth and anus. The fore-gut, in

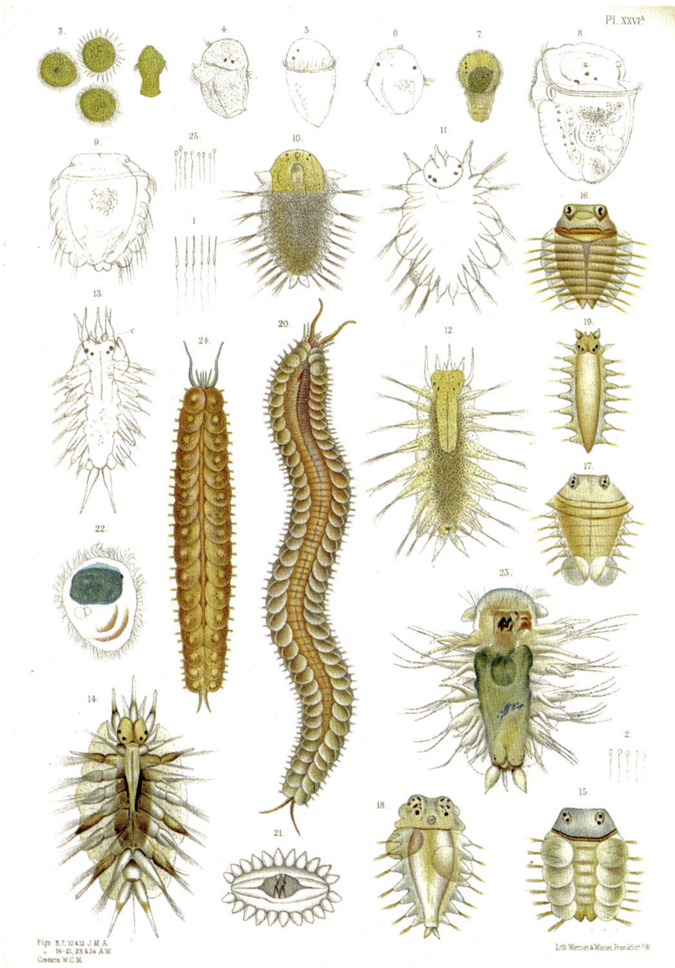

FIGURE 14.2 A page from W.C. McIntosh's *British Marine Annelids*, showing adults, larvae and spermatozoa of some scale worms.

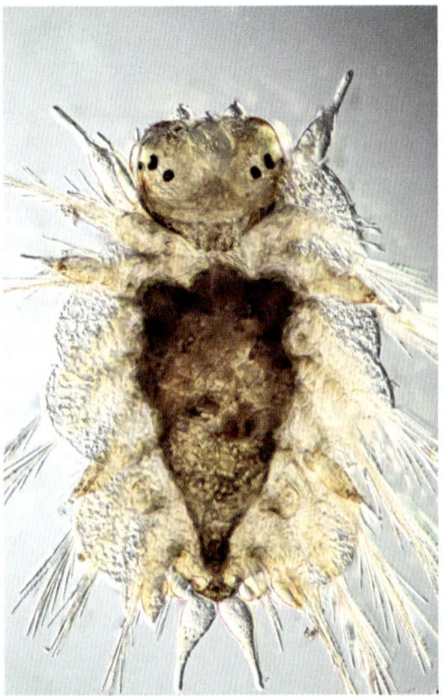

FIGURE 14.3 Scale worm *Harmothoe* nectophore larva. 800 µm.

FIGURE 14.4 Paddle worm *Phyllodoce* nectophore larva. 2 mm.

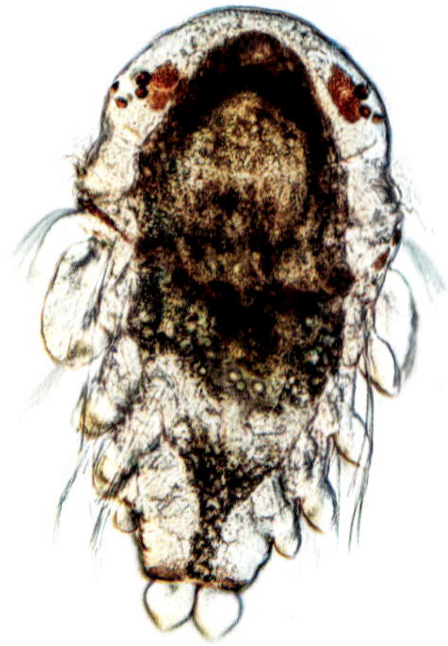

FIGURE 14.5 Cat worm *Nephthys* young larva. 600 µm.

the predatory species, has a modified pharynx which everts out of the body to grab prey. These worms tend to be dioecious, meaning that the individuals possess either male or female gonads. This class, which exhibits a huge variety of size and form and may be found in every habitat of the marine environment, from sediments to rocky shores, can be simply categorised into two types: the errant worms and those that are sedentary.

Sedentary worms have minimal locomotion and remain largely where they settle, living within a tube that they construct. They are reliant on a dispersal phase for the next generation. Most will be suspension or deposit feeders, which means that they eat what arrives in the water that they can filter or collect. They have vestigial to no parapods – extensions on either side of each body segment – but can have well-developed palps, tentacles and cirri that are numerous and long, as seen in the order Terebellida. Terebellids live in tubes that they build out of proteins, to which they attach sand grains and shell fragments. The tentacles have cilia that use streams of mucus to bring particulate matter to the mouth. Others have tentacles bright red with blood, which are modified to act as gills. Worms that burrow in the sediment, like lug worm *Arenicola* species, release and disperse larvae into a shifting tide before coming to rest again. Other than this, they keep out of the plankton. The family Capitellidae, also sediment dwellers, have adults that look more like earthworms. *Capitella* species can produce a planktotrophic larva which is common in coastal plankton outside the winter period.

The errant species are often predatory, seeking prey. Many of these, such as the Common Ragworm *Hediste diversicolor*, do not have a dispersal stage and larvae remain in the benthos. Errant worms have well-developed parapodia; the paddle worms (Phyllodocidae), for example, have enlarged parapods for swimming. A closely related family are the scale worms (Polynoidae). These are shorter and stubbier predators and, as the name suggests, have scales or elytra covering the dorsal surface. Again, the parapods are well formed.

Some polychaetes, often errant worms, produce large, yolky egg capsules that they glue to a substrate on the shore, while sedentary species will tend to broadcast large numbers of eggs

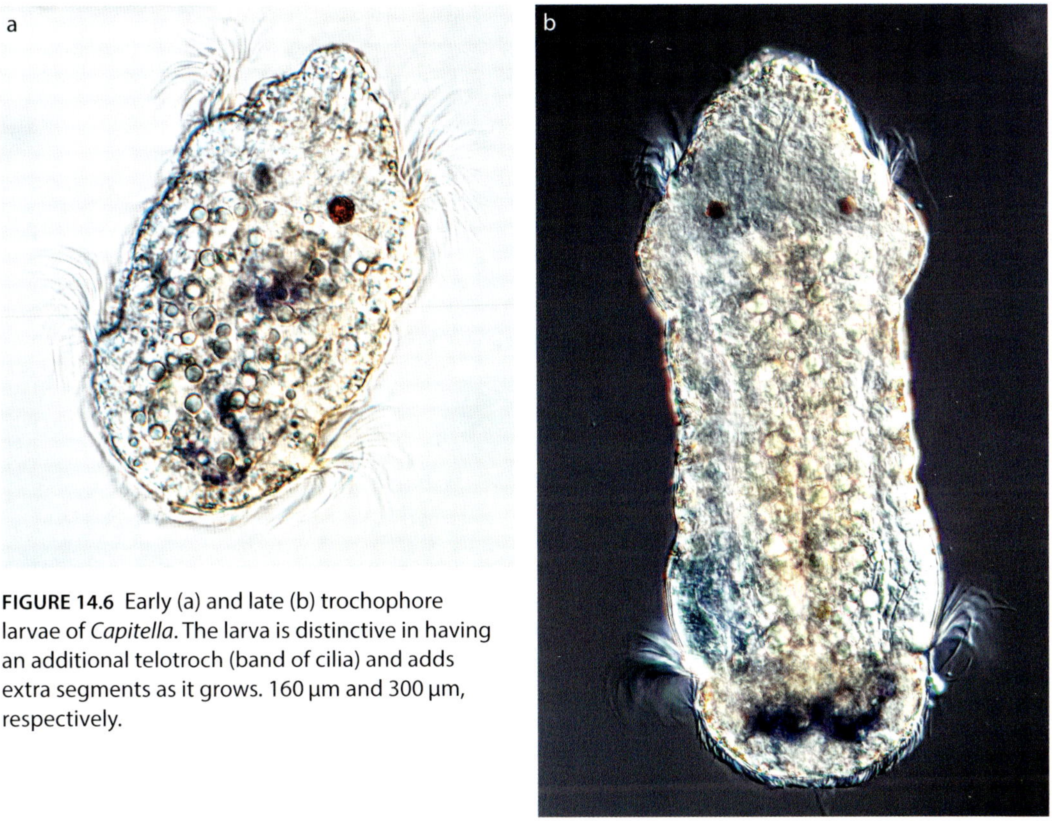

FIGURE 14.6 Early (a) and late (b) trochophore larvae of *Capitella*. The larva is distinctive in having an additional telotroch (band of cilia) and adds extra segments as it grows. 160 µm and 300 µm, respectively.

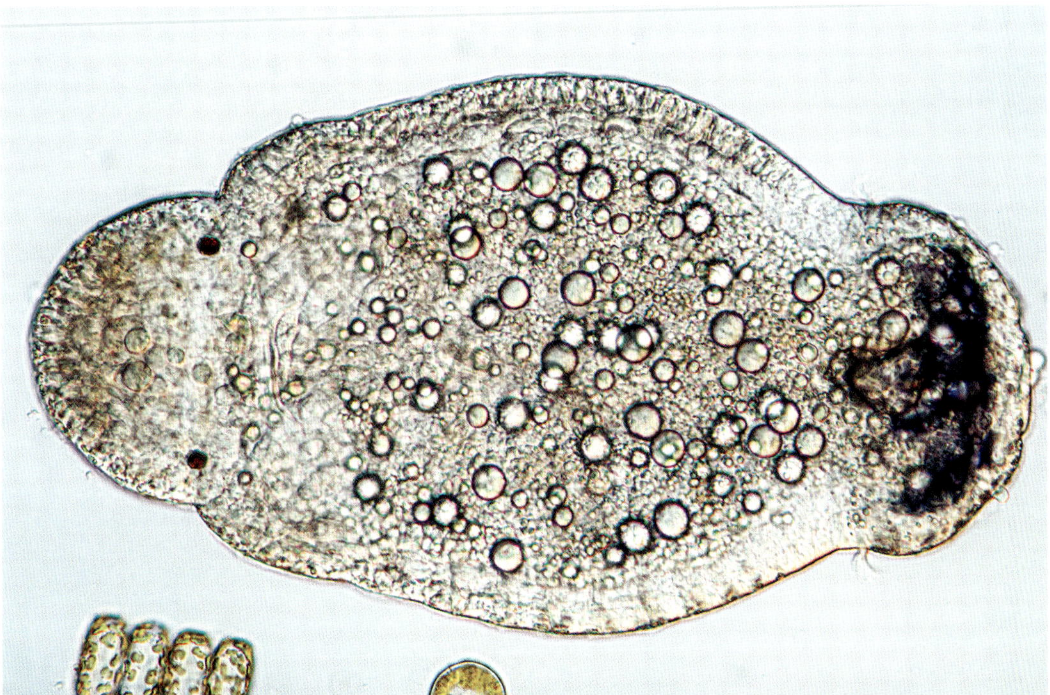

FIGURE 14.7 Late larval stage of *Capitella*. Note inconspicuous chaetae, typical of adults. 900 µm.

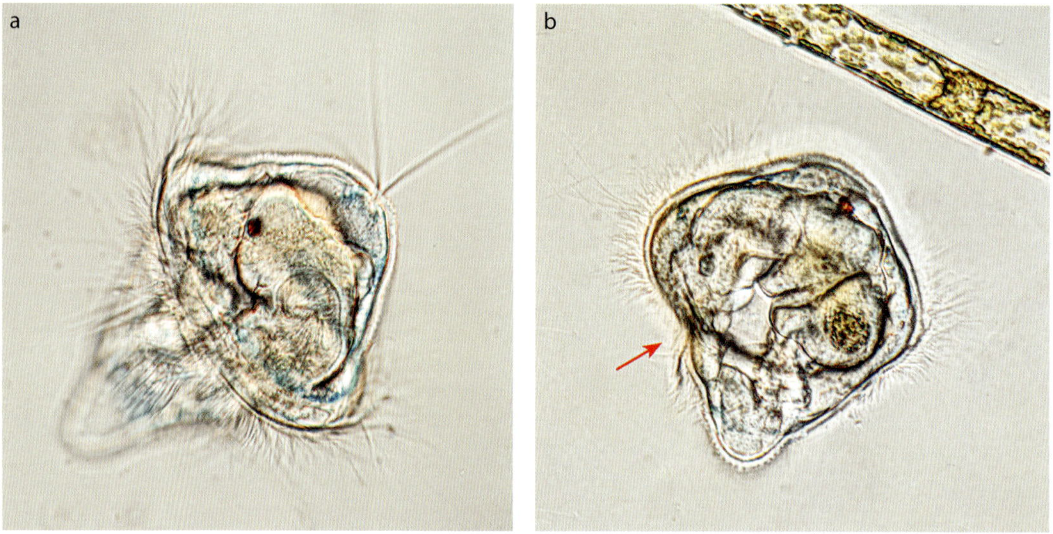

FIGURE 14.8 Early trochophores from different views. Note in (a) the apical tuft; in (b) the arrow indicates the area of mouth development. 60 µm diameter.

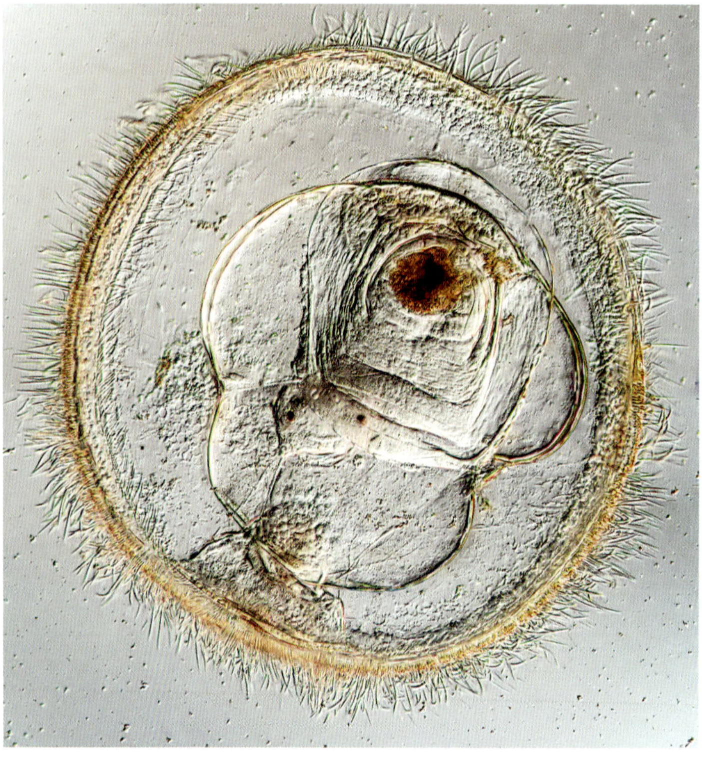

FIGURE 14.9 Larva of a Knot Worm *Polygordius*, one of the largest trochophores, viewed from the top. 1.1 mm.

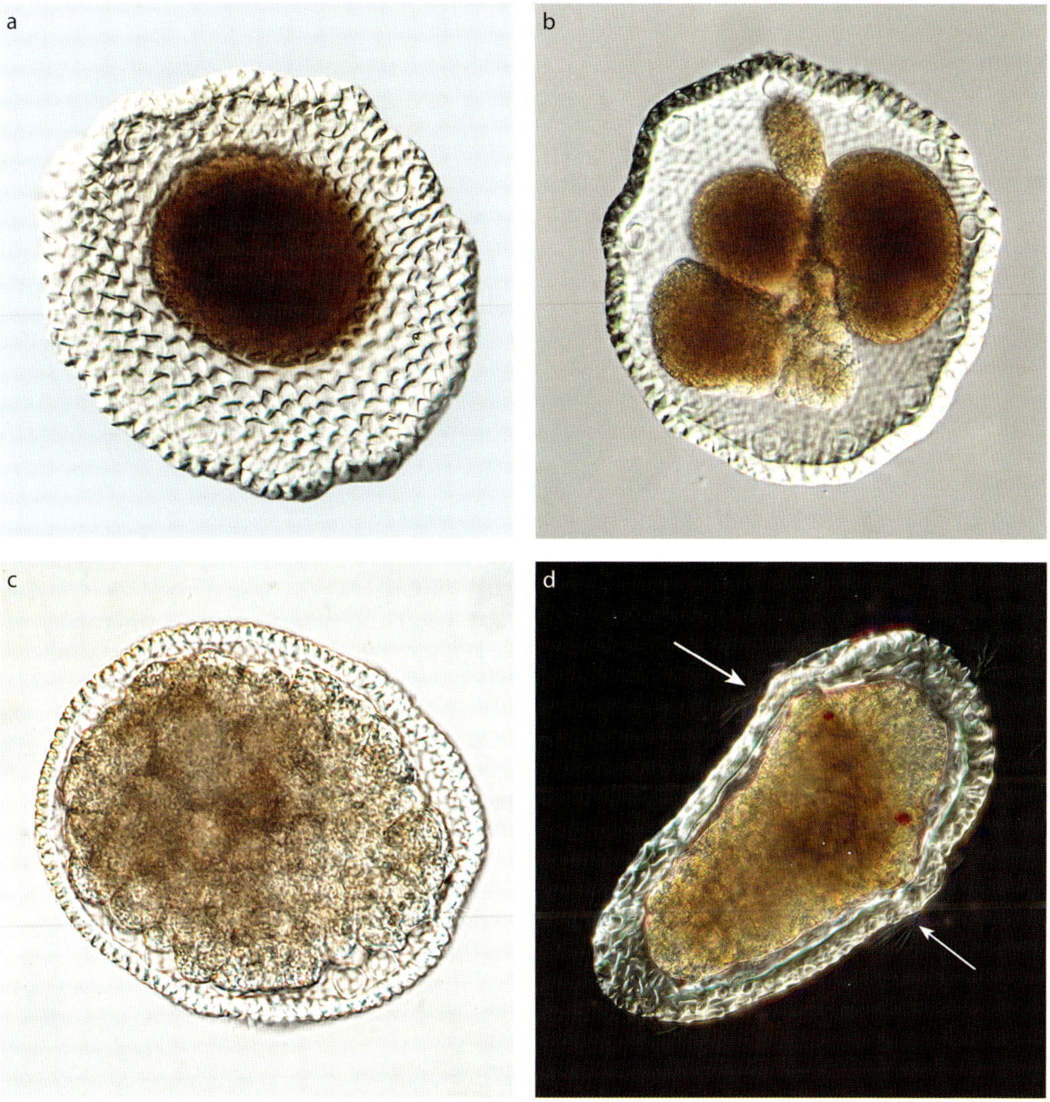

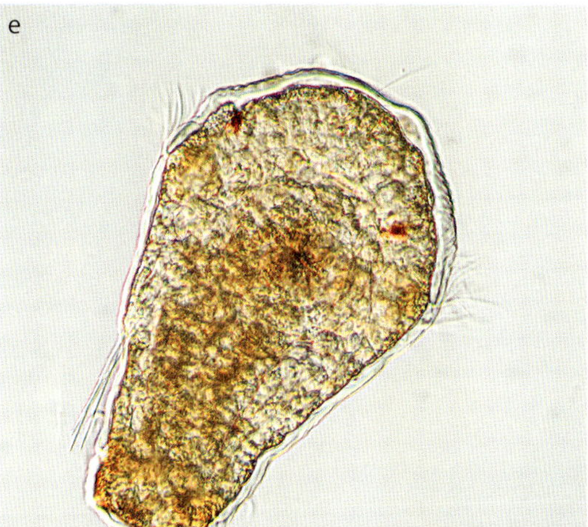

FIGURE 14.10 Egg to trochophore development in *Malacoceros fuliginosus*, a spionid polychaete common on the middle shore, where it lives in a mucus tube. (a) Planktonic egg with tough honeycomb covering and cortical vesicles; (b) early egg development called cleavage; (c) developing embryo; (d) the young trochophore with two red eyespots and apical cilia; arrows point to the now developed ciliary band, the prototroch; (e) late trochophore with chaetae beginning to form. Egg diameter 180 µm.

Larval stages

Terminology for polychaete larvae can be confusing, especially as each slight change during development has a different name. The sequence that applies to most planktonic polychaetes is: Trochophore – Metatrochophore – Nectochaete (or Nectotrochophore) – juvenile or Erpochaete (varies with species). The addition of '-chaete' refers to the appearance of the bristles characteristic of the class.

We have come across trochophores before, in the molluscs, where the duration of the stage is short. It is within the polychaetes that we see the trochophore as a more significant phase. The unsegmented trochophore larva hatches from the fertilised egg in around six hours. The early trochophore is a hollow ball of cells moving apparently at random through the water. Within a day or two it becomes pear-shaped and the division between an upper region, with an apical tuft of cilia attached to the apical sensory organ, and a lower region becomes more obvious. Between the two a ring of cilia called the prototroch encircles the trochophore, providing the means of locomotion. On one side, below the prototroch, is a mouth passing into the stomodeum and broad stomach, with an anus at the lower apex. At first there is a tuft of cilia here that may extend to form another ring, the telotroch. The groove passing into the mouth is lined with cilia for collecting particulate food known as the neurotroch. In the upper dome are usually two eyespots, or ocelli, that in time may disappear or increase in number.

It is about now that the larva could be referred to as a metatrochophore, as a second ring of cilia, the metatroch, has formed beneath the mouth and become a separate segment. It is

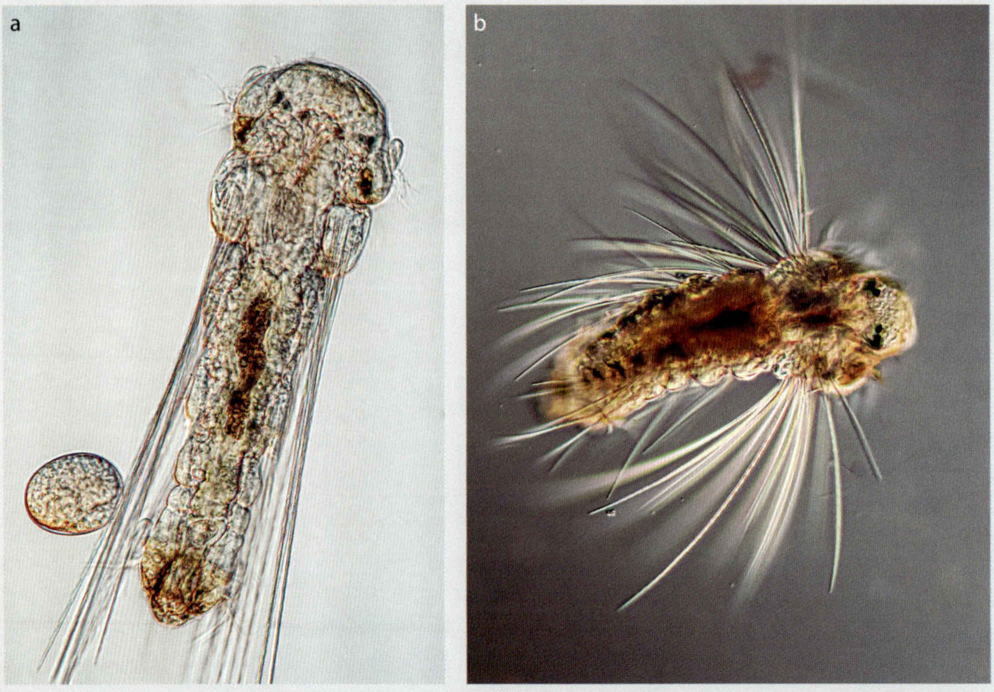

FIGURE 14.11 Young nectochaete of a spionid, *Polydora* sp. (a) Swimming with chaetae at the side; (b) defence mode with chaetae spread wide. 400 μm.

FIGURE 14.12 Late nectochaete of a spionid, *Polydora* sp. 900 µm.

from here that more segments will be added over time. Each segment bears a cilia ring, or metatrochal band, which are important for adding extra motive power as the body becomes larger. The bristles (chaetae) start growing on the segments, with some being especially long and well developed. This will increase the surface area and help with buoyancy.

After a week or so of increased segmentation, the free-swimming larva can be called a nectochaete. By now the apical sense organ has become the cerebral ganglion, which is eventually joined to the ventral nerve cord as that develops. All the while the worm continues to elongate as more segments are added, eventually becoming a juvenile, when parapodia may form, along with eyes, jaws and palps. At this time it begins to sink and settles on the bottom.

Trochophores vary in size from around 100 microns upwards to some sizeable ones at 1 mm. *Polygordius*, known as the Knot Worm, is particularly large. A long thin worm with two tentacles, it lives in sediment along the coast among coarse sand grains. It is superbly adapted with no obvious signs of bristles or segmentation, despite having over 200 segments, and is highly fecund, producing many planktotrophic larvae that can travel great distances.

With the development of the nectochaete, variations between the different species become more apparent and identification is a little easier. Members of the family Spionidae have particularly distinct larvae, like *Polydora*, as they have small pouch-like palps on either side of the head, becoming more elongated and pigmented with chromatophores the more mature the nectochaete becomes. Adults live in sediment. Samples taken near muddy shores in May can be dominated by larval stages, although these can be found at most times of

FIGURE 14.13 *Sabellaria* trochophores. (a) Resting; (b) alarm response with chaetae held wide. Body width 90 µm.

the year. Around the coast they are the polychaete most likely to be encountered, not only because of the huge numbers of adult worms but also as a result of their long survival in the plankton. Unlike the weaker yolky-egged larvae, which barely live for a few days in the plankton before settling, these stronger planktotrophic types can be present for weeks.

Magelona is another spionid species (Figure 14.1) associated with sediment and the larva is quite spectacular in the plankton. With long, fine chaetae flaring back from the front two segments, they have a pair of very long lateral palps on the head used in feeding. The prostomium, a flattened extension to the front of the body, gives the group its common name: shovel-headed worms. *Magelona* larvae feed exclusively on bivalve veligers and so to find them in the plankton it is best to sample near sandy shores. Spawning can occur over long periods and is probably synchronised with the release of eggs by bivalves; early summer through to autumn is typical in northern Europe, and winter in the Mediterranean.

Some of the sedentary tube worms have quite characteristic larvae. *Sabellaria* adults build tubes of sand, both singly and in colonies on rocky shores where there is a plentiful supply of nearby sand. The trochophore have long, serrated chaetae, characteristic of the group. While swimming these trail behind them, but if disturbed the trochophore will stop and erect the bristles in a complete defensive circle around their body. Another builder is the Sand Mason *Lanice conchilega*, a member of the Terebellidae, commonly known as the spaghetti worms as, out of the burrow, they look like a pile of pasta strands. Adults are found

and sperm into the water for external fertilisation. On hatching, a larva called a trochophore emerges.

There are several polychaete families with unusual methods of reproduction, but Syllidae is particularly astonishing. This is a large group of predatory marine worms found in a variety of microhabitats, such as *Laminaria* holdfasts and among colonies of hydroids. Hiding themselves away and living in fine, mucus tubes, they use their eversible pharynx for grabbing prey. Like most worms, they have good powers of regeneration. However, while most worms can replace at low tide on open sandy beaches within the long tube they produce, encrusted with sand, shells and foraminiferan shells (see Figure 9.3) and protruding from the sand surface. An unusual feature of the larvae is that they enter the plankton twice. During the first phase the trochophore develops and elongates before settling on the sand. The larva, now called an aulophora, returns to the plankton, secreting a clear hyaline tube within which it lives and which provides buoyancy. Detritus may adhere to the tube surface, but the inhabitant is quite visible inside. Movement through the water is achieved by cilia-induced jets of water out of the end of the tube. As it feeds on diatoms development proceeds and the number of tentacles increases until it eventually drops to the sand and the juvenile begins life on the shore.

FIGURE 14.14 Sand Mason *Lanice conchilega* tubes at low tide. See Figure 9.3 for a close-up of the tube surface.

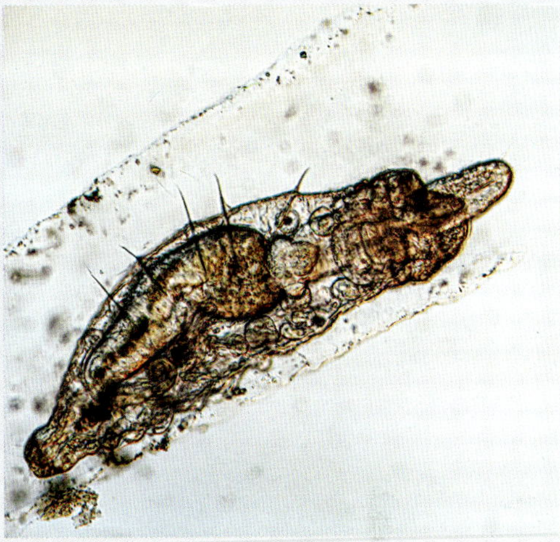

FIGURE 14.15 Sand Mason *Lanice conchilega*, early-stage aulophora larva in its hyaline tube. 750 µm.

FIGURE 14.16 Sand Mason *Lanice conchilega*, late-stage aulophora larva with multiple tentacles in its tube. 2.5 mm.

damaged posterior segments, Syllids and Sabellids are the only ones able to regenerate heads. With that ability it is just a brief step to growing whole new worms, which are then separated by the growth of a macroseptum between the two. They can also reproduce asexually; sometimes entire chains of asexual individuals may be present. Synchronised by the moon, Syllids may also, through a process called epitoky, generate sexual individuals. The process varies between species, but in essence the parent (an atoke) on the shore grows a new individual at the posterior end of its body; this is the epitoke, a sexual worm which may have either eggs or sperm – so is either

FIGURE 14.17 Syllid worm with epitoke, which is on the left. Epitoke 5 mm long.

female or male. There may be a chain of these, but I have only seen single individuals produced. When mature, the epitoke is released to swim up into the plankton. The synchronised nature of this phenomenon means that many of these non-feeding worms occur together, increasing the chances that these broadcast spawners will ensure fertilisation of the eggs. The males soon die after releasing their sperm, but the females can survive for another few weeks, looking after the eggs in the plankton (see the frontispiece photo at the start of the book). The male and female differ slightly in appearance, and the latter may initially carry the eggs on her back. After fertilisation she curves her body around a cluster of them and, once they have hatched into trochophores,

FIGURE 14.18 Female epitoke with eggs. 4 mm.

she dies. This form of epitoky is called schizogamy to produce the planktonic epitokes but other forms exist in polychaetes, all as a way of increasing the chance of finding other members of the same species for reproduction.

The elegant worm: *Tomopteris*

As described in Chapter 2, the naturalist Henry Gosse spent 1852 in Devon studying marine life. While using his net in Ilfracombe harbour he made an exciting discovery. 'I have the pleasure of announcing a new animal of much elegance, which I believe to be of a hitherto unrecognised form', he wrote in *A Naturalist's Rambles on the Devonshire Coast*. He had found several examples of the genus *Tomopteris* but, unfortunately for him, it had already been recognised. Gosse was so convinced he had discovered this worm he named it *Johnstonella catherina* after the botanical illustrator Catharine Johnston, as a tribute to her excellent engravings of marine algae. Three years later, in his two-volume *Marine Zoology*, he notes his mistake, although Johnston is often added in parentheses today. *Tomopteris* had first been recorded several decades earlier, in 1825, by the Baltic German scientist Johann Friedrich Gustav von Eschscholtz, who travelled worldwide, contributed extensively to all areas of zoology, died in his early 40s and has never received the acclaim he deserves.

The genus *Tomopteris*, often called the gossamer worm, contains around 70 known species, all of which are bizarre in some way.

Gosse and others struggled to classify their specimens, as they frequently did not seem to fit the characteristics of the group. They are truly holoplanktonic worms – unlike the other polychaeta, where any planktonic life will be a larva – and spend their entire lifecycle in a pelagic world, some going to great depths; several species, however, commonly live in neritic waters and therefore appear in shore samples.

FIGURE 14.19 *Tomopteris helgolandica*, 8 mm.

FIGURE 14.20 *Tomopteris helgolandica*, a close view of the head. Note the eversible pharynx at the top of the gut to the left of the two red eyes; the yellow spots on the parapods are rosette glands.

FIGURE 14.21 *T. helgolandica*, parapod close-up showing the two paddles and yellow rosette glands. Parapod width 750 μm.

These animals are so transparent that several may be invisible in the sample dish, making their presence known only if they begin to move rapidly, which produces great swirls of water. Even though I see them regularly, I still find them one of the most exciting animals to observe. Individuals of various sizes may be found in seashore plankton, growing up to 2 cm in length. The first and second segments of the body are fused to form two palps and on the head are two red eyes. The segment behind this has a few very long sensory hairs – the only presence of bristles on the body. The remaining segments have parapodia, paddle-shaped outgrowths on each side. These produce waves of synchronous movement to propel the worm quickly through the water, a type of propulsion called metachronal paddling. This can be maintained for long periods of active swimming, and they have the highest metabolic rate among annelids. Each parapodium has two broad and flexible paddles. The blade folds on the return stroke, ensuring an efficient stroke with a minimum of drag. Each blade has at least one yellow spot called the rosette gland, although pigments can also occur elsewhere on the body. These special bodies produce bioluminescence but, unlike that of other organisms, this light is yellow and is not derived from the luciferin–luciferase process. A most spectacular sight occurs when *Tomopteris* is surprised by a would-be predator. It not only flashes yellow but also fires off (or excretes) flares of yellow 'chaff' reminiscent of the countermeasures used on military aircraft to disrupt any attack.

The body cavity, or coelom, is filled with fluid of the same density as seawater to maintain buoyancy. The coelomic space so dominates the body that there is little room for anything else, and musculature is limited to a mesh beneath the epidermis. There is no blood or circulatory system either. The small body size and large surface area, especially with the extended parapodia, mean that oxygen and nutrients are transported by diffusion through the thin cuticle and coelom. The movement of coelomic fluid is maintained by an extensive internal network of cilia. Unlike other polychaetes, there are no internal membranes separating the segments and so this ciliary movement is sufficient. Fine, internal tubules with ciliated funnels at the coelom end (metanephridia) connect to the outside by a pore for excretion. Females produce huge gelatinous egg cases that float in the water column and the eggs are fertilised with sperm that, unusually for annelids, have two tails.

Tomopteris are important planktonic predators, catching large prey such as fish larvae, arrow worms and appendicularians. They do this with an eversible pharynx that is fired out from the mouth and can be seen inside the head. *T. helgolandica* is the species perhaps most common around Europe and can be distinguished by its characteristic long tail, which most other species do not possess.

Nematoda: round worms

Ubiquitous is the word used for nematodes. There are 28,000 species of these creatures and wherever you look they seem to be there: in just about all environments on earth they will be found if moisture is present. They are the most abundant multicelled animal living in marine sediment; they also occur on rocky shores, but appear as tychoplankton only on seemingly random occasions. Those in the plankton are small, less than a few millimetres in length, although as a group the nematodes vary widely in size. The body structure across the group is very uniform, and they are protected from hostile habitats by a thick, poorly permeable cuticle made from collagen. In the marine environment they are essential in helping to recycle organic matter and releasing nutrients. They are also an important food source, as they are eaten by a wide variety of species. Populations in inhospitable places tend to have short lifecycles. In estuaries and muddy shore sediments densities of 28,000 individuals in an area of 100 square centimetres have been recorded, with sparser numbers in sand, where there is less organic matter.

Identification is difficult, as they are so similar in appearance, but the mouth can vary and is an important feature in classification. As well as the detrital feeders, there are species adapted to consume diatoms and others that are predatory. Some of the latter may even enter the plankton to feed when phytoplankton are blooming, as this is also when zooplankton numbers rise. Nematodes can survive in very low oxygen levels, but one problem they cannot seem to cope with is any degree of oil spillage.

FIGURE 14.22 An example of a nematode in the plankton. Determining species is a highly specialised task. 900 µm.

Nemertea: ribbon worms

Ribbon worms can be quite bizarre. Finding the Bootlace Worm *Lineus longissimus* under a rock on the lower shore can be a memorable experience. Initially looking like a large gelatinous, purple-black slime ball, it unrolls into a worm stretching many – sometimes very many – metres long. Specimens up to 55 m have been reported. McIntosh could be forgiven for placing them with the true worms, but they lack both chaetae and segmentation. The smooth head, lacking palps,

FIGURE 14.23 A nemertean worm from the shore that has appeared in the plankton. 2 cm.

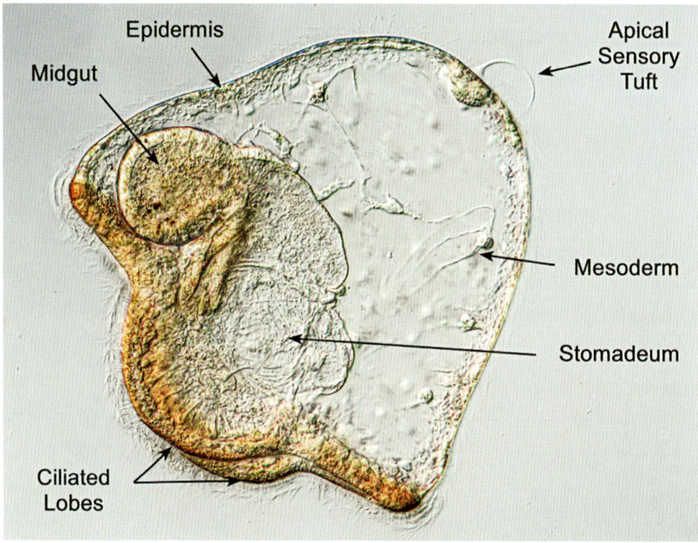

FIGURE 14.24 A pilidium larva of a heteronemertean ribbon worm, probably *Cephalothrix* sp. 400 μm.

may have a pair or two of eyes, and the bodies of some, like *Lineus*, are very long and flexible. Adults tend to be benthic, but do occasionally show in the plankton, and most European species release their eggs or young direct into their habitat, including the water column. Members of Heteronemertea, a sub-group of Nemertea, disperse by producing a planktotrophic larva called a pilidium, which is found in no other taxon or group, even those closely related.

These nemerteans release large numbers of tiny eggs into the water that hatch into the pilidia, which look like a winter hat with ear warmers at the side. These ciliated lobes, or lappets, hang down on either side of the mouth, opening into the stomodeum. The gut is quite visible inside, under a layer of epidermis crowned at the apex with an apical sensory tuft. Just under the surface of the epidermis a few mesodermic cells can be seen. The development of a juvenile nemertean is far from straightforward. Sections of specialised cells around the gut invaginate to ultimately form the next phase and utilise the pilidium's gut to form the immature adult. Development occurs inside the old larva, which is eventually shed or consumed.

Nemerteans, like many benthic organisms, have a well-developed microbiome of bacteria rich in species capable of producing neurotoxins. Adults of *Cephalothrix simula*, an invasive species of concern discussed in the final chapter, can use neurotoxins to stun prey and although the larvae are not known to use this method of predation they are still believed to carry these bacteria.

Platyhelminthes: flatworms

Where the cnidarians have two layers of cells sandwiching the mesoglea, the flatworms step up a level and have three, with the middle layer, the mesoderm, forming organs within the body. Free-living flatworms (class Turbellaria) are common on the shore and several are particularly beautiful, such as the Candy-stripe Flatworm *Prostheceraeus vittatus*. This is a member of the order Polycladida, a group that is almost entirely marine and has multibranched guts and an egg with yolk granules in the cytoplasm, called an entolecithal egg. The specialised yolk speeds up development and the egg, soon after release, hatches into a larva. Several forms occur, but the most common in the group is the Müller's larva, with eight lobes. Additional rarities include Goette's larva, which is produced in a few species and has four lobes. Müller's larva is an active swimmer covered in cilia, particularly the eight lobes, although they can be found with only seven or as many as ten. A pair of ocelli (eyespots) are present and development is quick, with the larva remaining in the plankton for only a day or two. As they mature extra ocelli may appear. At the apex of the larva there is a tuft of long sensory hairs that connect back to a glandular frontal organ. This marks the formation of a head, which, in the adult, develops further. The frontal organ is largely for chemoreception: tasting the water around the larva's head. Behind the front lobes

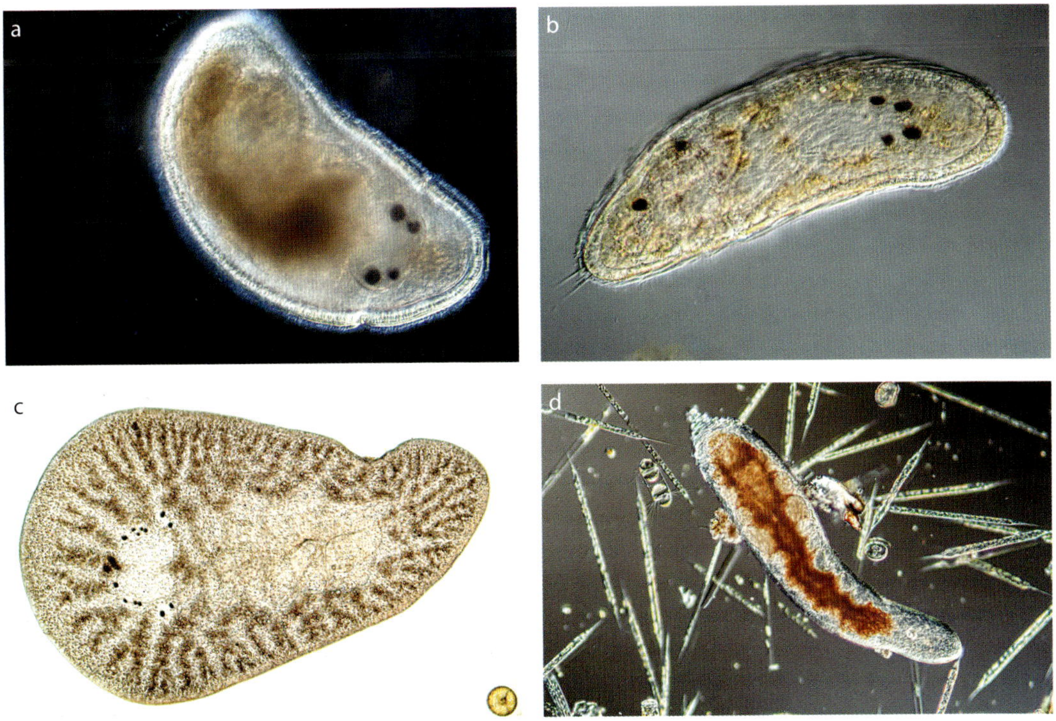

FIGURE 14.25 Planktonic flatworms. (a) and (b) Rhabdocoelan flatworms, 700 µm and 900 µm, respectively. (c) and (d) undetermined species, 800 µm.

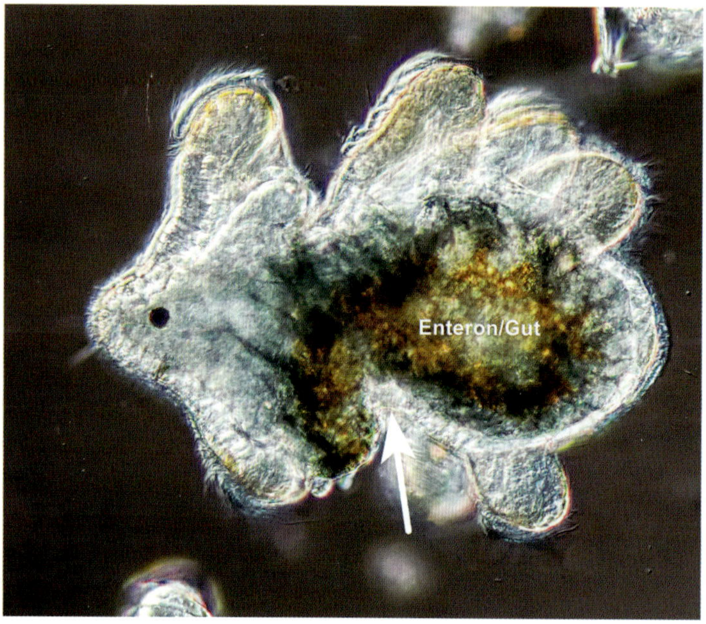

FIGURE 14.26 Müller's larva. Between the dark ocellus (eyespot) and the sensory tuft of long hair there is a frontal organ. The eight ciliated lobes provide locomotion and the arrow denotes the position of the stomodeum, or mouth. 180 µm.

on one side is the mouth, which passes through a pharynx into the intestine. When the larva first forms the mouth is much further back, moving forward during development to the adult position. The pharynx inside, when developed, is large and can be everted to collect prey.

Microscopic flatworm species are particularly common living in sediments, but they also occur in the plankton and a few are common occurrences in plankton samples. The Rhabdocoelan flatworms are the most developed, some living as permanent members of the plankton, and can be abundant in the spring. Flatworms are hermaphrodites and release eggs and sperm through separate pores in the body, the gonopores. When mating, the individuals align themselves so that cross-fertilisation can take place. A penis, or cirrus, is enlarged by hydrostatic pressure and inserted into the female pore. After internal fertilisation has occurred the eggs are usually laid in gelatinous masses. It is from here that any Müller's larvae will hatch, live in the plankton for a few days and then metamorphose back on to the shore.

The flatworm phylum has several classes of parasitic forms, including Trematoda. This is an important group in the plankton, as these flukes use their larval stages for dispersal and finding new hosts. Mature trematodes will usually be found parasitising coastal birds, such as gulls, as well as fish. The lifecycles are complex, as the flukes find a new primary host via secondary and sometimes tertiary intermediate hosts. The adult releases eggs that enter the water and soon afterwards hatch into a miracidium larva. These are not common in the plankton, as they quickly penetrate the body of a mollusc, perhaps a periwinkle or a cockle. Soon after entry it develops to become a mobile, worm-like sac full of embryos called the sporocyst. Much of this happens within the digestive gland of the mollusc. Within this first intermediate host the embryos eventually produce another larval form called a cercaria, which are then released into the plankton.

Cercaria are the forms most likely to be found in plankton samples and can appear at almost any time, although they are particularly abundant in late summer and early autumn. Identification to species level is difficult, although some of the family groups can be distinguished by the presence of suckers and spines. On the larger cercariae are rows of spines on both sides of the body to provide drag against sinking. These are all very active wrigglers, swimming rapidly with

FIGURE 14.27 Trematode cercaria, Opecoelidae, 300 µm.

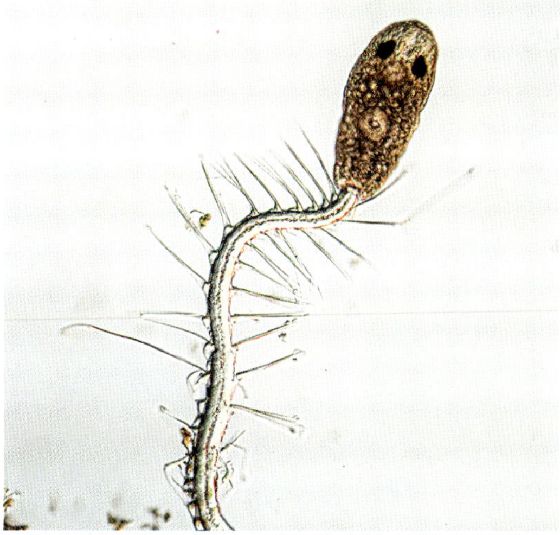

FIGURE 14.28 Trematode cercaria, Lepocreadiidae, probably *Opechona*, 450 µm.

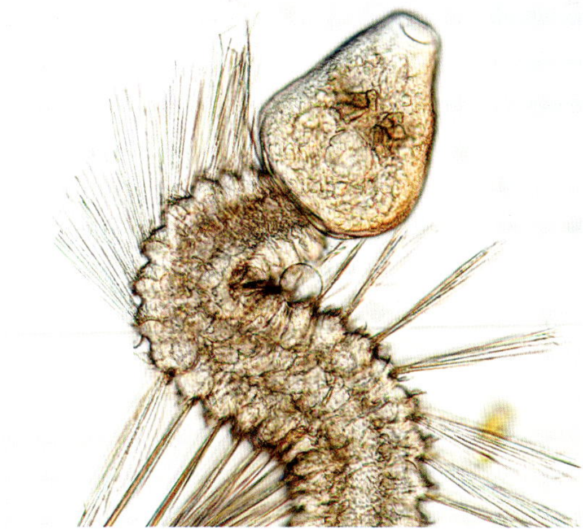

FIGURE 14.29 Trematode cercaria, undetermined species, 600 µm.

FIGURE 14.30 A large trematode cercaria, undetermined species, 1 mm.

snake-like movements through the water. Many of these belong to the family Fellodistomidae, but larvae of the Opecoelidae, which have a distinct sucker in the middle of the body, can be found swimming and crawling through detritus. These cercariae disperse, searching for another intermediate invertebrate host to enter. This host may at some point be eaten by a vertebrate – a fish or bird – that will form the new primary host, thus completing the cycle. At least 95% of gulls are thought to be carrying trematode parasites.

FIGURE 15.1 A juvenile *Parasagitta elegans*. Dark field illumination allows observation of the sensory bristles on the body edge, particularly on the tail. Length 9 mm.

CHAPTER 15

Chaetognatha: the arrow worms

Early autumn is my time to find arrow worms. Offshore they occur at almost any time, in some cases in huge numbers, but for much of the year the seashore plankton is free of these fast, transparent holoplanktonic predators. Like so many gelatinous and transparent zooplankton they appear almost invisible in samples, although back lighting will help to increase their visibility. Most arrow worms arriving in late summer are juveniles, around 1 cm in length, and few possess gonads. Adults will be between 2 and 4 cm in length. For a few months they may be regulars in shore samples, but catching them in a dish with a pipette can be a challenge.

The 100 or so worldwide species are classified within one phylum, Chaetognatha, the name of which describes the head: bristle jaws. Around the UK and Europe two species are most likely: *Parasagitta elegans* and *P. setosa*. Alister Hardy describes their distribution in *The Open Sea* (1956); the former species is found from the tip of Brittany, around Ireland, up the Irish Sea, around Scotland and down the UK's east coast to Suffolk. *P. setosa* dominates in the English Channel and the North Sea, Scandinavia and the Baltic.

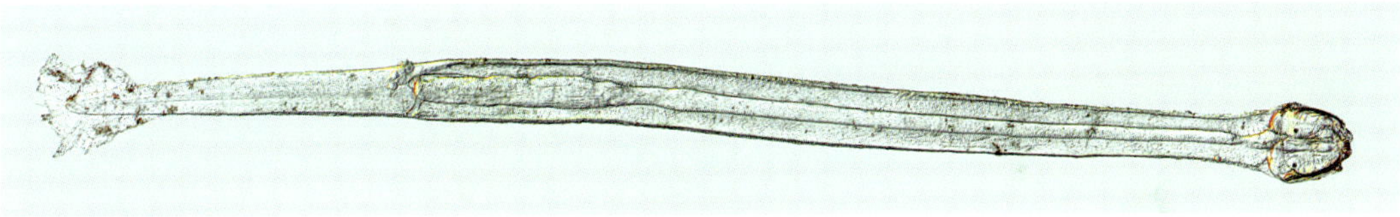

FIGURE 15.2 *P. elegans* with DIC illumination. Length 1.5 cm.

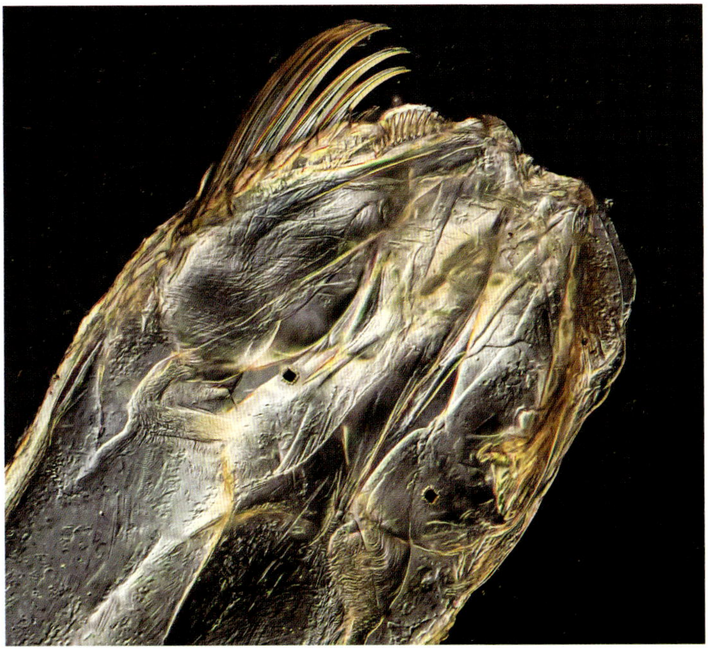

FIGURE 15.3 *P. elegans* head view, showing the grasping bristles open on the left side and the pair of eyes. Head width 650 μm.

FIGURE 15.4 *P. elegans*, close-up of bristles and teeth. Bristle length 150 µm.

Parasagitta shows well the streamlined, bilaterally symmetrical form with a slightly broader head, narrow body and lateral fins. There is a wide tail fin, which, along with the other fins, provides stability. The darting locomotion is achieved by rapid flexing of the body musculature to left and right. The 'skeleton' on which these muscles work is a hydrostatic one, the fluid-filled coelom that runs most of the body length. A very thin cuticle helps the worm maintain its body shape, while small tufts of sensory hair called ciliary fans along its sides pick up vibrations in the water emanating from both predators and prey. On either side of the head are two curved sets of bristles that grab prey, while, near the tip of the head, a number of teeth help in the grasping process before prey are pushed to the broad central mouth to be swallowed whole. The head also has other sensory cells, including chemoreceptors, statocysts and unusual eyes. Around the base of the head is an appendage that acts like a hood and can be extended to provide some cover and protection.

As predators arrow worms have amazing acceleration, which enables them to catch prey such as copepods (their primary food item) that are themselves extremely quick off the mark, jumping away when they detects vibration in the water. The arrow worm is an ambush predator, waiting in the plankton for copepods to pass near and jumping forward to grab them. They react only when they detect pressure changes along the body picked up by the ciliary fans, which act in this respect a little like the lateral line canal in fish. Differences in pressure sensed by different sensory tufts along the body enable good prey location and accurate attacks. Small fish and fish larvae, which can easily be as large as them, are also caught. Digestion is quick.

Chaetognaths are hermaphrodites. The slender gonads fit along the worm's narrow body, with a pair of ovaries in front of the anus and the testes behind. When mature, two swellings containing small sacs of sperm appear just in front of the tail. They are transferred to the outside, where they burst, allowing the sperm to swim into another individual to fertilise the eggs. No larval stage occurs; instead there is progressive growth to the adult form. Offshore arrow worms can constitute up to 10% of the total zooplankton biomass and are important food for fish.

FIGURE 16.1 A bryozoan colony with a calcareous skeleton on a seaweed frond, showing the lophophoral tentacles out feeding.

CHAPTER 16

Lophophorates and the sea mats

The Lophophorata, which include the phylum Bryozoa, are a collection of organisms with a lophophore: that is, a fan or ring of ciliated tentacles around the mouth, often horseshoe-shaped, that is used for suspension feeding. They are typically small and form extensive colonies on substrates such as rocks and seaweed, as suggested by their names: Bryozoa means 'moss animal', and they are also called 'sea mats'. The largest species, *Pectinatella magnifica*, is found in fresh water and can form colonies half a metre across. It has spread from its native environment in North American waters to become an invasive species in Europe.

The colony, called a zoarium, consists of a base skeleton that can be gelatinous, chitinous or calcareous, according to taxon. Each compartment of the skeleton holds a zooid and in many species they interconnect through a stolon, sharing nutrients. Each colony is usually the result of asexual reproduction from a single sexually produced individual, the ancestrula. Bryozoans are mainly hermaphrodites, although the sequence of gonad development varies across the species. In some, ovaries and testes develop at the same time, enabling self-fertilisation to occur.

The species most likely to be associated with shore plankton are in the order Cheilostomatida and two types of larvae can be found: cyphonautes and coronate. A third form, the vesiculariform, is also planktonic but are rarely found. The Sea Mat *Membranipora membranacea* is one of the most abundant bryozoans along the lower shore and both male and female gametes may mature at the same time, allowing self-fertilisation, although sperm released through the tentacles may be drawn in by neighbours to cross-fertilise their eggs. Numerous small eggs are liberated into the water and hatch to become planktotrophic cyphonautes larvae. The length of time spent feeding in the plankton will depend on environmental conditions such as nutrients, food and temperature, but one or two months is quite possible.

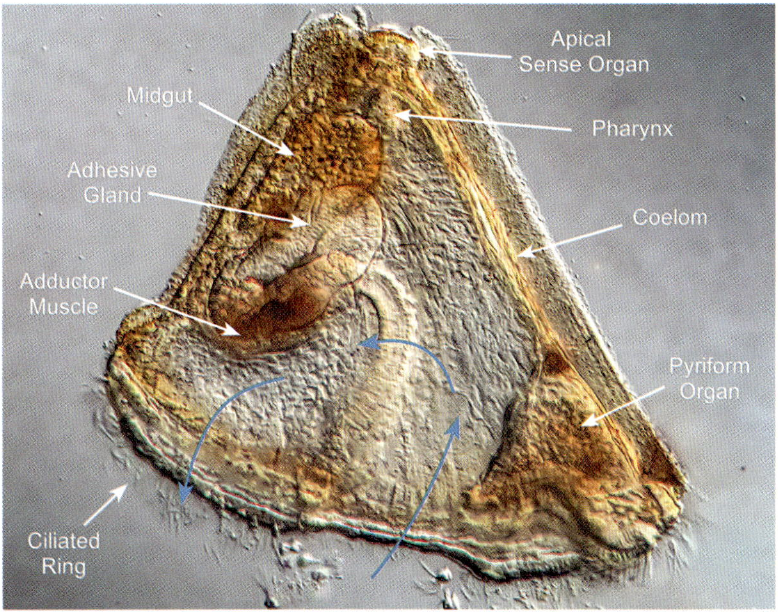

FIGURE 16.2
A *Membranipora* sp. cyphonautes larva. The blue arrows indicate water flow. Width 450 µm.

FIGURE 16.3 An early cyphonautes, *Membranipora* sp., height 90 μm.

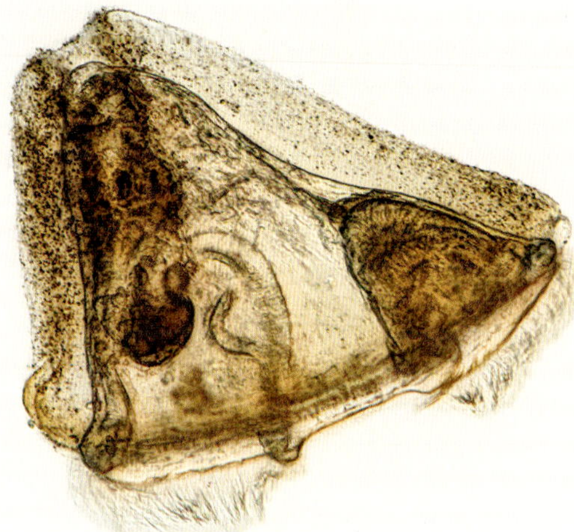

FIGURE 16.4 A late-stage cyphonautes, *Membranipora* sp., width 550 μm.

Cyphonautes are a common feature of shore plankton and can be found any time from early spring through to autumn and occasionally outside this period. They are thought to be the ancestral larval form of the Bryozoa and are typical of the family Membraniporoidea. Their triangular shape is formed by two shell valves that are flattened together, as can be clearly seen when they are swimming in a sample dish under low magnification. Once under the compound microscope the organs can be just made out between the valves. The gap at the top, between the shell valves, may have the apical sense organ protruding and around the edge of the lower shell is a corona of cilia providing locomotion. A ciliary feeding current brings water to a space within the valves, through a filter and back out again. The filtered food is passed into the mouth and pharynx, near the apex, and on round to the main gut. Just beneath the gut is a pouch-like tube, the adhesive

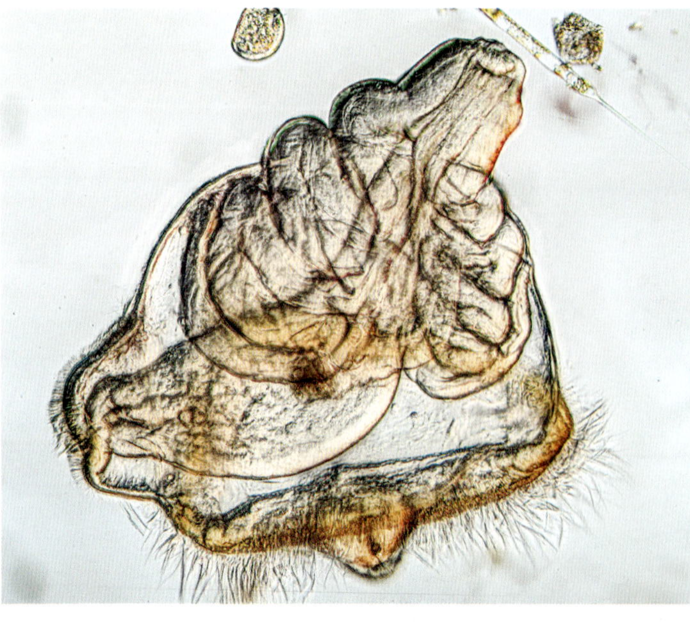

FIGURE 16.5 Mature cyphonautes, undetermined species, 800 μm.

sac. On the other side of the valves, near the entrance, is a large mass, the sensory pyriform organ. During the life of the larva it is positively phototactic, but when the individual is fully mature this response to light reverses to ensure that it finds a substrate on which to attach. The cyphonautes then uses its pyriform organ to chemically sense the substrate, seeking a good place to establish a colony before using glue from the adhesive sac to stick itself firmly to the surface. Dramatic metamorphosis of the cyphonautes creates the ancestrula, which will become the first primary zooid, which will reproduce asexually to create the zoarium.

By contrast, other lophophorates produce lecithotrophic larvae that live in the plankton for a brief period of hours before settling. These are the coronate larvae and because they are present for such a short time they are only occasionally found in shore samples. For example, *Bowerbankia imbricata* is an upright, tube-forming species common in rock pools, where it grows on Coral Weed *Corallina officinalis*. It produces an egg that is brooded inside the zooid and hatches into a non-feeding vesiculariform larva. Five coronate variants have been identified – although identifying individual specimens under the microscope can depend on which view of the larva presents itself. Those of the genus *Bugulina* (formerly *Bugula*), such as *B. turbinata*, are among the easier species to identify. The largely hollow ball of cells has an extensive corona of cilia around it (hence the name coronate), with a distinct apical cilia tuft. The coronate larva retains the pyriform organ, which is sometimes visible within the ball.

At dawn on the seashore some species of Cheilostomatid are stimulated by the appearance of light to release larvae, but as light intensity increases the process is inhibited. This ensures that larvae spend the morning in the plankton being locally dispersed on the shore before a settlement later the same day. Delays of 12 to 24 hours can occur before settlement and it has been seen in several genera, such as *Bugulina*, that longer delays reduce the resulting colony's viability. The best chance of a colony forming is when the coronate settles during the day on which it was released,

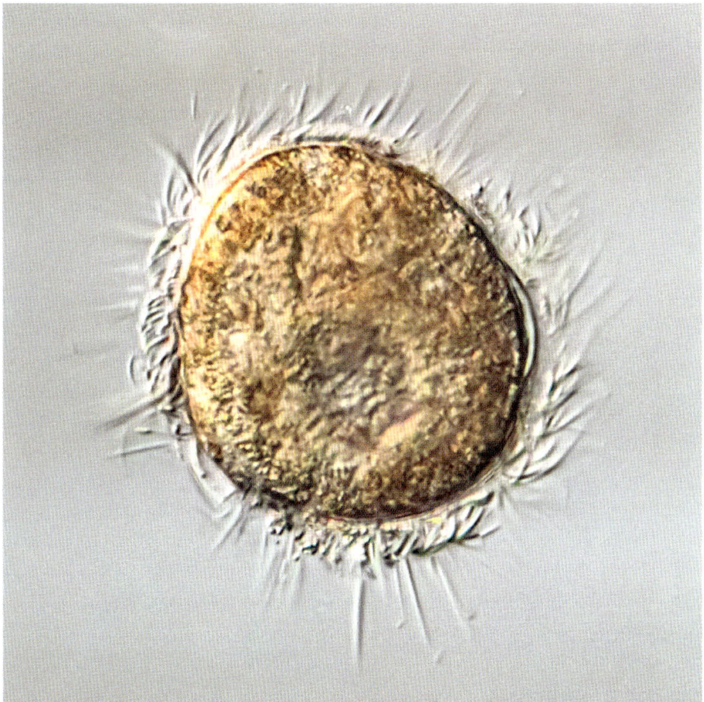

FIGURE 16.6 A coronate larva, possibly *Bugulina* sp., diameter 100 µm.

An invasive alien species

Similar to *B. turbinata* is the Red Ripple Bryozoan *Watersipora subatra* (Figure 16.7). This is an invasive species from the Pacific that probably arrived in Europe on the underside of ships during the early 2000s. Since then it has colonised rocky shores and harbour walls, where it produces striking red patches.

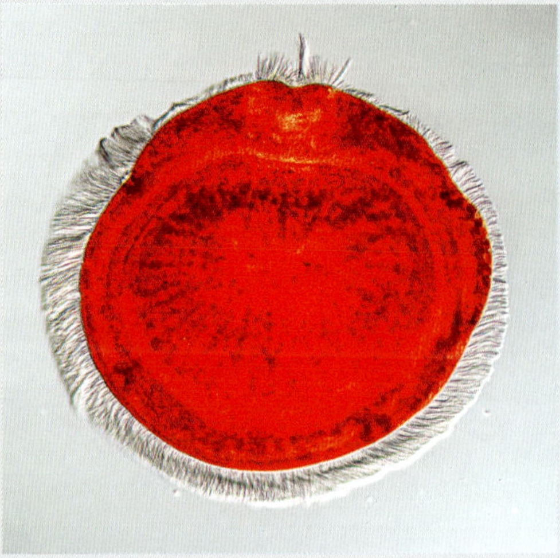

FIGURE 16.7 The large and striking coronate larva of *Watersipora subatra*, 1 mm diameter. The bright red colouration is typical of this species.

although this may vary between species. The downside of a short period in the plankton is poor dispersion and genetic similarity between colonies, reducing variation. The coronate explores the surface that it will settle on before inverting the adhesive sac inside the body that will glue the now metamorphosing larva into an ancestrula.

A second phylum, the Phoronida, are remarkably similar to and often confused with sea mats. Despite 25 or so different larvae being found worldwide, barely half that number are known as adult species and fall into two genera, so our knowledge of the group is still limited. Their planktotrophic larva is called an actinotrocha and when fully grown is around a millimetre in length and

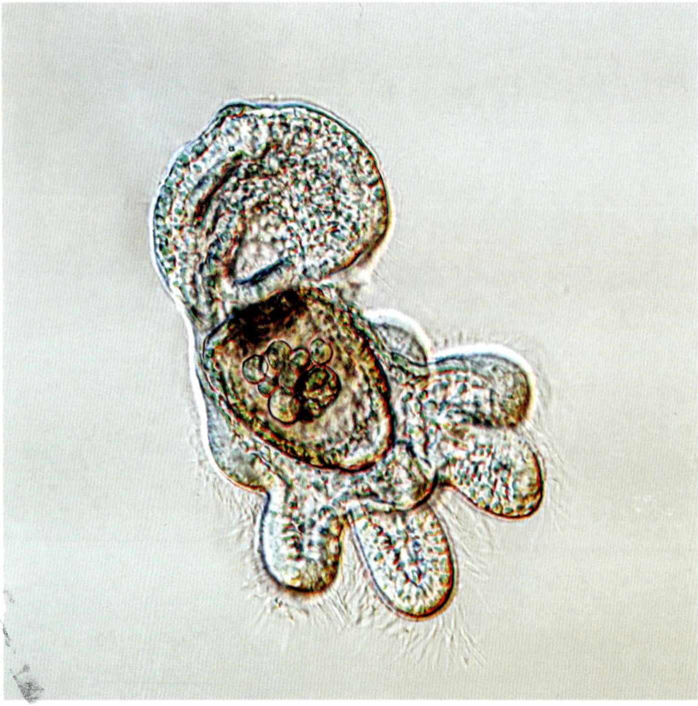

FIGURE 16.8 Developing actinotrocha larva less than 12 hours old. 350 µm.

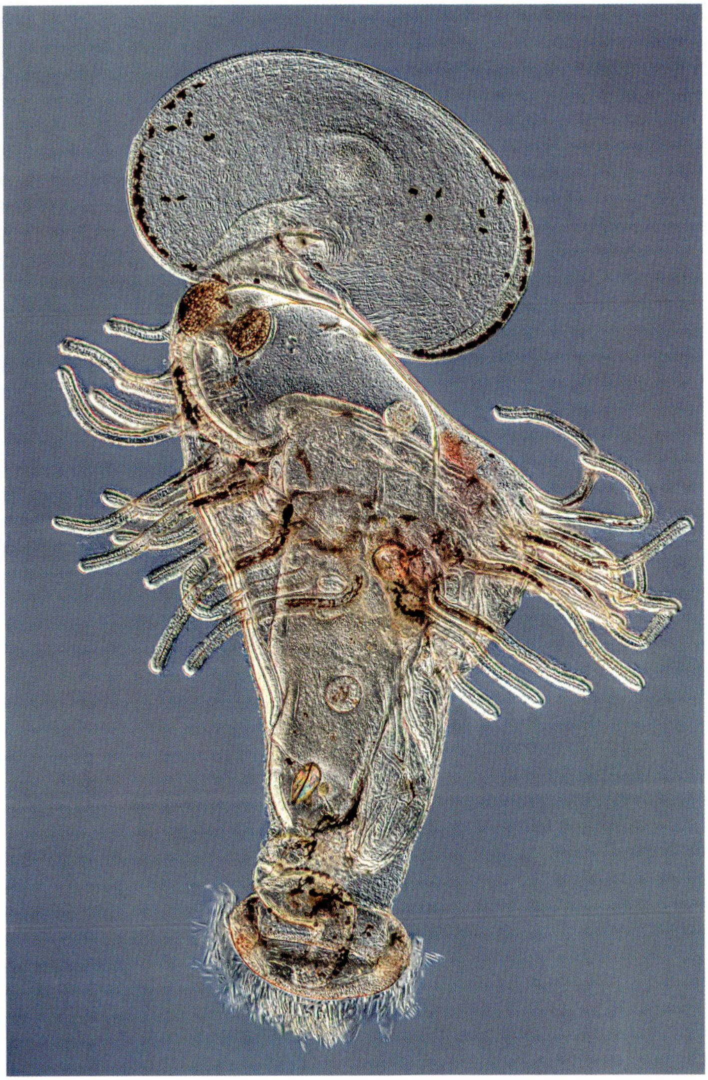

FIGURE 16.9 A late actinotrocha larva of *Phoronis* caught just off-shore. Height 1.4 mm.

quite spectacular in appearance. Although widespread across Europe, adults of the few known *Phoronis* species are not that common and it is only on rare occasions that the larvae may appear in plankton samples. The eggs are brooded on the lophophore before the actinotrochas are released and it is at this time that they are most likely to be caught in the net. The main body is sac-like and has a number of lobes that increase in number as the first 24-hour period advances; Figure 16.8 shows a specimen caught within the first 12 hours of growth. These lobes and tentacles continue to develop in size and number, but older larvae tend to be swept away from the shore and I have not been fortunate enough to find any large near-shore specimens. The lobes are covered with cilia for locomotion and the production of a feeding current.

FIGURE 17.1 Juvenile starfish. One is still metamorphosing inside the base of the brachiolaria stage larva and is quite rounded. Juvenile is 2 mm wide.

CHAPTER 17

A change in symmetry: the echinoderms

Despite the nearly 7,000 species found worldwide, the phylum Echinodermata has declined in diversity since their peak 250 million years ago, in the Palaeozoic era. They are an ancient group with some very recognisable individuals, such as the starfish, sea urchins and brittle stars, as well as the sea cucumbers and feather stars. The last of these share their group, the crinoids, with sessile sea lilies. Crinoids are unique among the echinoderms in having retained their ancestral structure of a mouth facing upwards, surrounded by cilia-laden tentacles for suspension feeding; collected particulate matter is transferred to the mouth by mucus streams. The eggs are cross-fertilised on the arms and hatch into a planktonic larva called the vitellaria, which is a fairly primitive, barrel-shaped form. It has a very short stay in the water column before dropping to the bottom and sticking to the substrate, where it grows a short stalk to become a pentacrinoid larva. During the brief stay in the plankton they are difficult to find, and I have yet to see one. The vitellaria may well be an early stage in other echinoderms, occurring during egg development or very briefly after hatching.

Members of this phylum share three main features: they possess a water-based vascular system, which operates tube feet and feeding tentacles by hydraulic pressure; underneath the epidermis is a system of calcareous plates or ossicles, often loaded with spines, which in late larval

FIGURE 17.2 Adult crinoid, the Rosy Feather Star *Antedon bifida*.

development may be seen forming; and the symmetry is primarily radial, with the body divided into five sections. Some bilateral symmetry is found in the group, particularly in the planktonic larval forms, which develop radial symmetry on maturing and metamorphosing. Echinoderms do not have a differentiated head and the sexes are separate. Except for the crinoids, the larvae all share a similar level of complexity in their development. Echinoderms are therefore a uniquely placed phylum, not just because of their strange anatomical structures but because they are the only successful complex animals to have descended from an attached sessile group. Their comparatively large size and ability to produce huge numbers of gametes (one female Common Starfish *Asterias rubens* could produce more than two million eggs) has allowed the embryonic development of echinoderms to be extensively studied in the laboratory.

Each class of the Echinodermata has a very different larva and some possess several types. A slightly controversial form should be mentioned here: the dipleurula. Upon hatching from an egg, the early echinoderm larva is very basic (see Figure 17.4a) and certainly in the past this has been called the dipleurula – books and articles can still be found with this name in use. Rather than being a term for a distinct larval stage, it is now applied as the name for the hypothetical larva of ancestral deuterostomes, a superphylum that includes echinoderms. Larval development of echinoderms can be complex and varied with quite a few having no larval phase at all. Those that do can remain in their pelagic life for many weeks or even several months, during which time they are swept into deeper water by the tides and currents. The consequence for the planktologist working on the shore is that these spectacular forms can be a rarity, if they are seen at all. For all these reasons the following sections are brief.

Asteroidea: the starfish

Like most intertidal organisms, starfish are affected by temperature on the seashore. As they do not tolerate drying out they will generally be in the lower shore, and if that still means being exposed to the sun they will avoid the shore completely. For example, in Scotland several species occur among the *Laminaria* (kelp) zone, but further south in Cornwall they would be in the sublittoral or below, rather than the intertidal area.

The Common Starfish *A. rubens* is distributed around most of northern Europe and can be abundant just below the shore. The lifecycle is not particularly typical of the group, but it is a common species so it will be described here. The separate sexes shed huge numbers of gametes and fertilisation occurs in the water, where the egg soon hatches. When the larva first emerges from the egg, it is in a basic form like those found in some other echinoderms. Within a day the gut is formed and other structures are developing; now it is a bipinnaria larva. The tide will carry the gametes and eggs away from the shore and so they are not common in the shore samples, but do occur in the summer months. This planktotrophic bipinnaria steadily develops over several weeks, becoming increasingly complex. As the larva begins to grow arms, becoming a brachiolaria, the settlement process begins. This stage has adhesive cells on the increasingly long and numerous arms that enable attachment to a substrate so that metamorphosis into a juvenile can take place.

The cushion stars *Asterina*, Bloody Henry *Henricia sanguinolenta* and the Purple Sun Star *Solaster endeca* have no pelagic larval stage at all. The sand dwellers *Astropecten* spp. and *Luidia ciliaris* have a bipinnaria but no brachiolaria. The Common Sun Star *Crossaster papposus*, which eats *Asterias*, produces a large lecithotrophic egg and a non-feeding larva that spends only a brief time in the plankton.

FIGURE 17.3 Common Starfish *Asterias rubens* can be found in the lower shore and sublittoral areas of rocky shores, where it uses its tube feet to bring its mouth to its prey. It has a wide distribution and produces a bipinnaria larva that can be picked up in suitable shore samples.

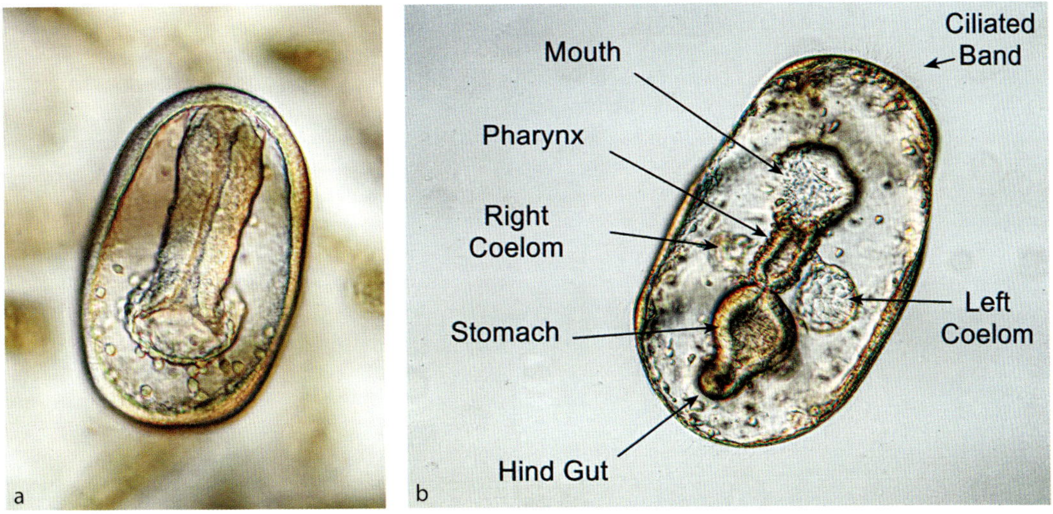

FIGURE 17.4 *Asterias rubens* larvae of different early stages. (a) an early gastrula stage in the development of a bipinnaria (b) bipinnaria larvae. Height 350 and 450 µm, respectively.

FIGURE 17.5 Brachiolaria stage of *A. rubens*. Juvenile inside is 2 mm.

The general bipinnaria larva has several ciliated lateral lobes with a mouth and fairly central anus. Rows of cilia occur around the mouth for feeding, as well as in other areas of the larva. Between the mouth and stomach is the pharynx and to each side of this are two circular cavities, the left and right coelom spaces; of the latter, the left develops into a hydropore that links the water system to the outside, while the right diminishes with time. The adult madreporite, a perforated structure which controls the entry of water into the vascular system, forms from the hydropore. The arms of the brachiolaria are extensions of the coelom and inside this transparent structure the juvenile starfish begin to develop until settlement.

Ophiuroidea: the brittle stars

The Common Brittle Star *Ophiothrix fragilis* is common throughout northern Europe on the lower shore and among seaweed and stones in rock pools. Just offshore large numbers can lie, criss-crossing each other on the bottom and suspension feeding on particulate matter and diatoms. With the ability to survive up to ten years and their sheer abundance along the coast, it is no wonder that the larva of this species is the echinoderm most likely to be found in plankton

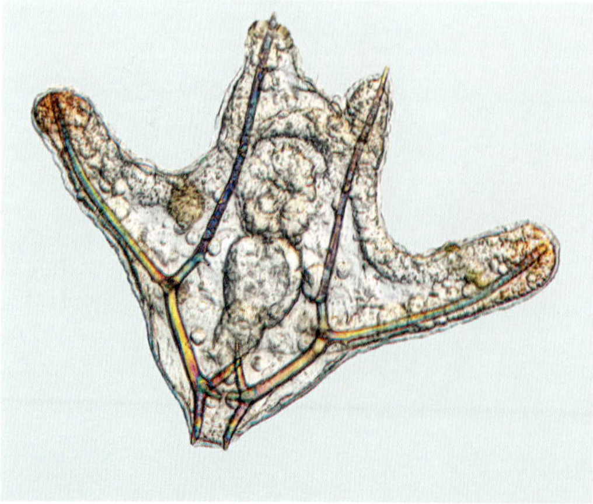

FIGURE 17.6 Early ophiopluteus of a brittle star; note calcite rods. 400 μm across.

FIGURE 17.7 Late ophiopluteus of a brittle star, 750 μm across.

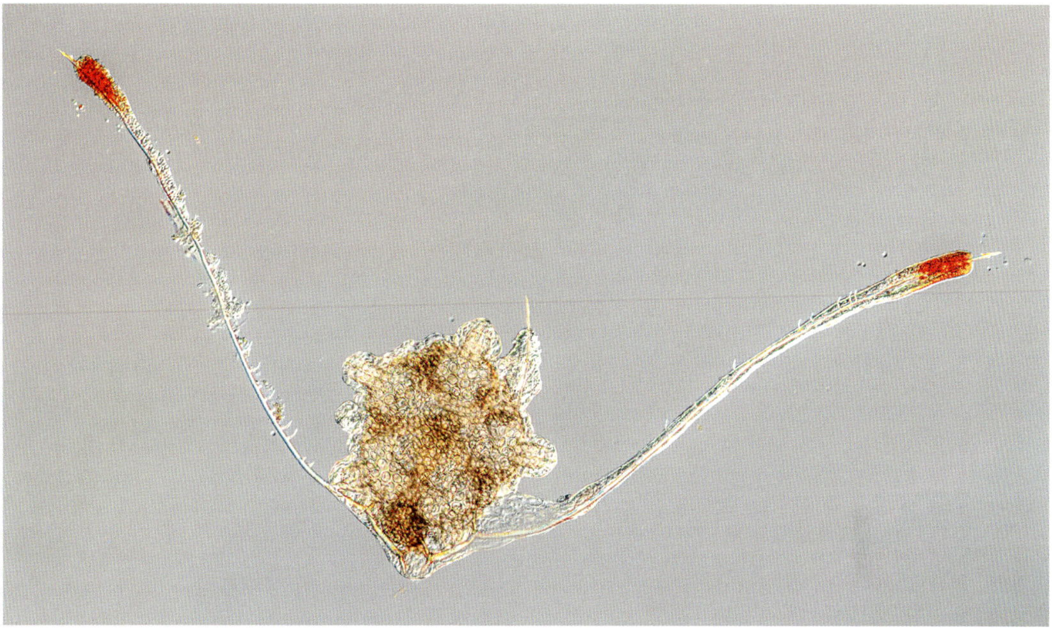

FIGURE 17.8 Ophiopluteus metamorphosis. 2.5 mm.

samples from spring through to early autumn. Complex and beautiful, they move relatively little, making them easy to observe.

Separate sexes broadcast their gametes for external fertilisation and the egg soon develops into an ophiopluteus larva. The arms point up and outwards, providing good stability in the plankton, and are supported with calcite rods, the number and pattern of which vary with age

FIGURE 17.9 Juvenile brittle star, 1 mm across.

and species. The mouth is located at the top of the main body, and cilia all over the arms generate a current collecting particulate matter and streaming it to the mouth. The anus is towards the base. Juveniles begin to form while the larvae are still in the plankton (see Figure 17.8) and then descend to the bottom, where they are known to associate initially with adult brittle stars and sponges. Here they live a semi-parasitic existence for some months, using some of the suspended material being brought in by their adult or sponge hosts for food.

Echinoidea: the sea urchins

Like the starfish above, sea urchins are also affected by temperature and water loss. In northern Scotland and further north, they can be encountered in lowest part of the intertidal rocky shore, but generally will be in the sublittoral or further into deeper water. One of the smaller and most widespread species is *Psammechinus miliaris*, which can be found among the *Laminaria* fronds in the lowest parts of the rocky shore. The Sea Potato *Echinocardium cordatum* is common on sheltered sandy shores, where it burrows deep below the surface. Both species shed gametes into the water for fertilisation, the result of which is a pluteus larva. Called the echinopluteus to distinguish it from that of the brittle stars, it has many similarities. The arms point upwards and the body and arms are well ciliated for movement. To bring food into the mouth the ciliary beat reverses. The pluteus can remain feeding and growing in the plankton for around a month. Sea urchins may take two or three years to reach sexual maturity and can live for up to ten years.

FIGURE 17.10 *Psammechinus miliaris* is a common sea urchin in the lower shore. The pelagic larva is at sea for four weeks.

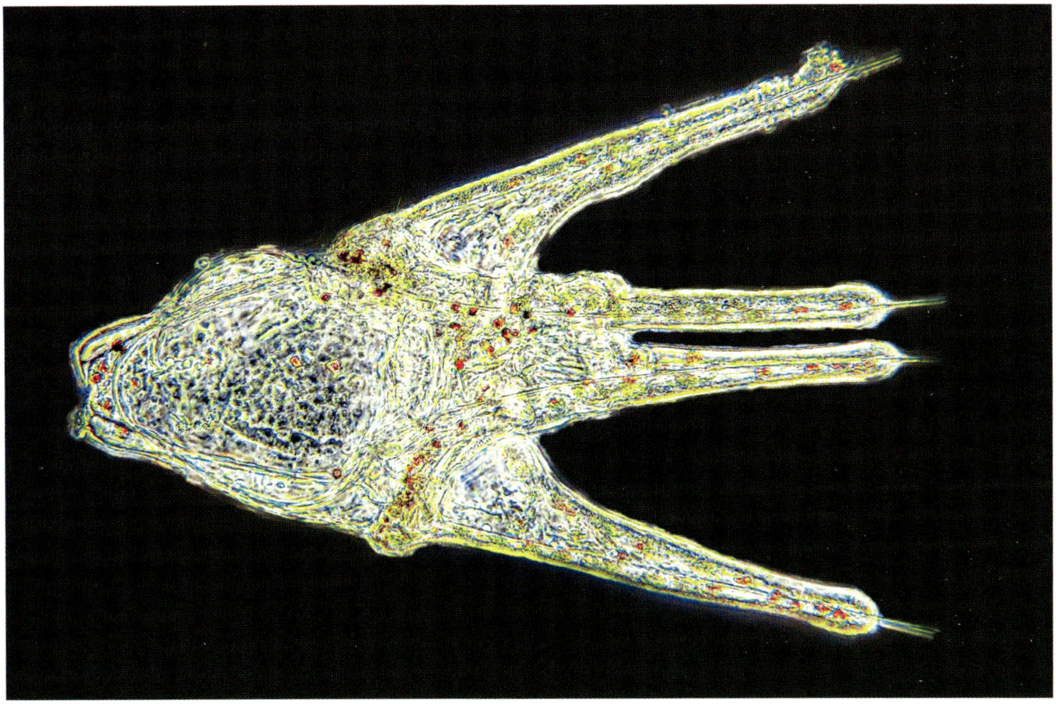

FIGURE 17.11 The sea urchin echinopluteus, phase contrast. 600 μm.

FIGURE 17.12 Late-stage echinopluteus. 3.5 mm.

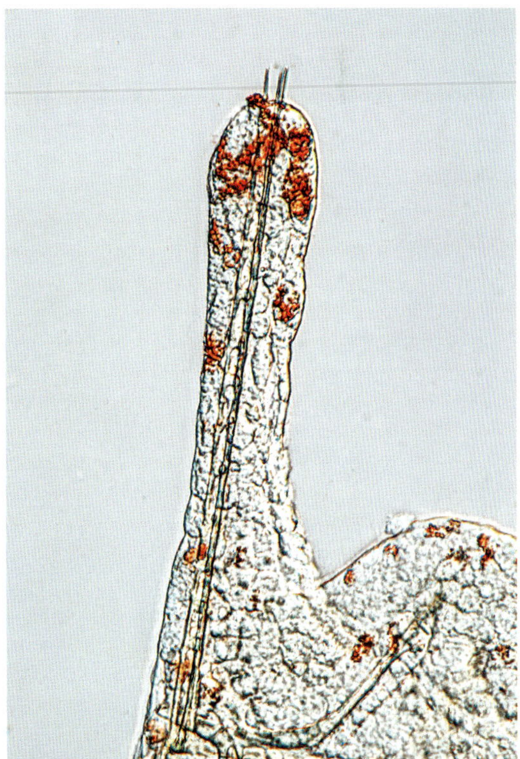

FIGURE 17.13 Arm of an echinopluteus, 350 μm.

Holothuroidea: the sea cucumbers

Several species of sea cucumber may be found at extreme low water on rocky shores, the larger Cotton Spinner *Holothuria forskali* and smaller, paler Sea-gherkin *Pawsonia saxicola* among them. The former is more likely below the shore and the latter has a more westerly distribution and is best seen around *Laminaria* holdfasts. *P. saxicola*'s symmetry looks atypical, as it is tubular, with a mouth at one end surrounded by tentacles and an anus at the other end. It is radially structured

FIGURE 17.14 The holothurian sea gherkin *Pawsonia saxicola*. 5 cm.

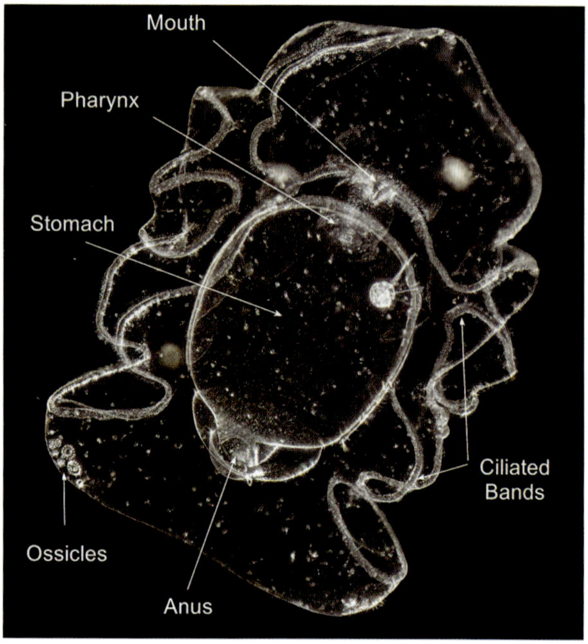

FIGURE 17.15 Sea cucumber: auricularia larva with a trochophore in the stomach. 1 mm height.

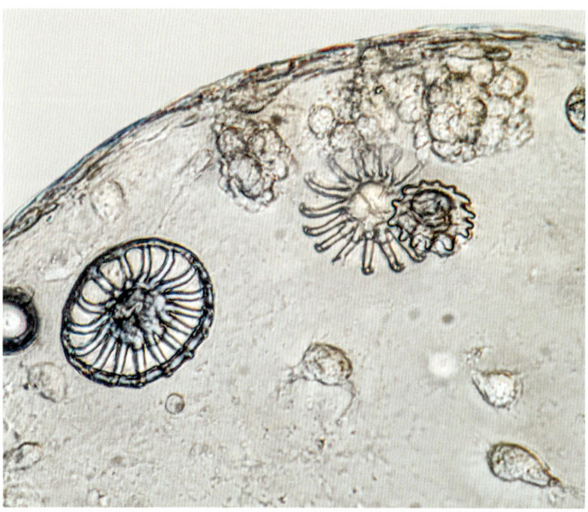

FIGURE 17.16 Sea cucumber: ossicles in the auricularia. 60 µm.

FIGURE 17.17 Sea cucumber: a developing pentactula larva; note the ossicles developing at the base of the tentacles. 4 mm.

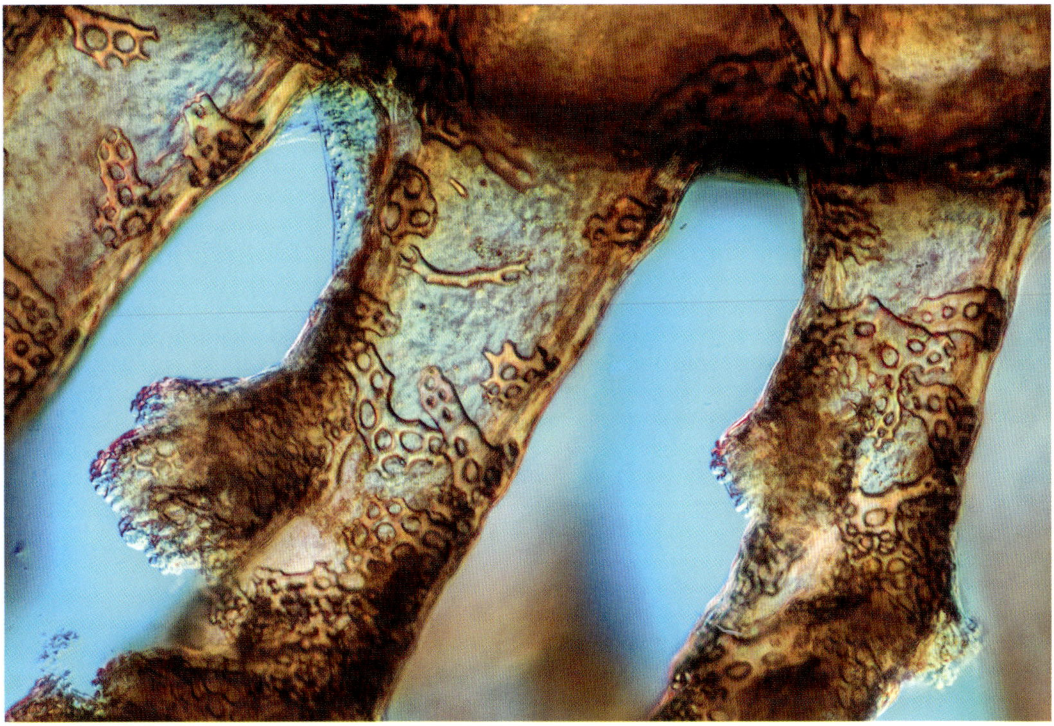

FIGURE 17.18 Sea cucumber: the ossicles on the tentacles of the pentactula larva.

around the mouth and divided lengthways into five sections marked by columns of tube feet along the length of the body. Food is trapped on the tentacles and transferred to the mouth.

While there are some hermaphrodite and brooding species, the common European species have separate sexes with external fertilisation after the release of the gametes. Zygote development takes several days to reach the larval stage known as the auricularia. This is a more specialised version of the starfish bipinnaria, with ciliated, multiple lobes. However, the ciliary band is continuous, folding in and around the entire body. The auricularia is also significantly larger than the bipinnaria, at around a millimetre long. The gut is visibly dominated by the stomach in the middle of the body and the calcite skeleton is reduced to a few ossicles at the bottom of the body. These are species specific, especially later in development.

For some weeks, growth produces a series of somatocoels at the side of the gut that will go on to create the water system, that enables locomotion, feeding, and respiration. Once this has occurred the auricularia metamorphoses into a barrel-shaped form, the doliolaria larva. This looks more like a juvenile holothurian. The original ciliary band now divides into several that go around the new larva. Ossicles develop around the upper gut and there are five tentacles, one for each of the five radial sections. The change continues with the loss of the ciliary rings as the larva, now known as a pentactula, settles on the shore. Calcite plates now begin to develop from the ossicles around the gut and these can be seen clearly under the microscope before the epidermis grows over.

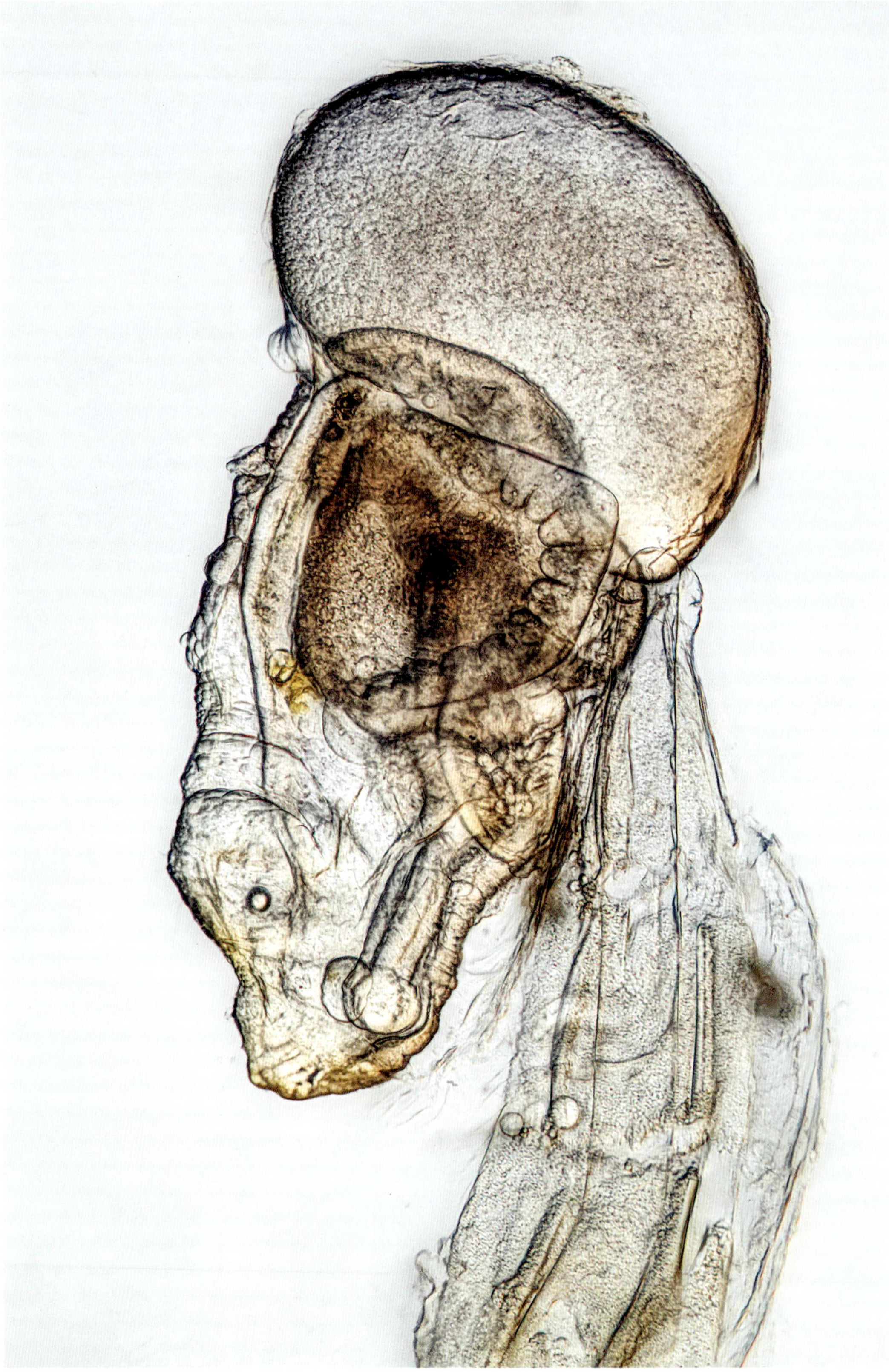

FIGURE 18.1 The main trunk of *Oikopleura*, a tunicate and chordate.

CHAPTER 18

Chordate and hemichordate relatives

Chordates are animals that possess, among other characteristics, a rod called a notochord and a spinal nerve cord that run together along the length of the body; fish and mammals, for example. Mature hemichordates do not have these, but their larval stages do, and thus it is thought that chordates and hemichordates have a common ancestor. While it might be thought that chordates would not be prominent in the plankton, some chordate groups make appearances at a variety of life stages. Fish eggs and larvae develop and feed within the plankton, becoming nektonic only later. Another group is the subphylum Tunicata: a strange bunch that are very much involved with the plankton and the main topic of this chapter.

Acorn worms

First, however, we will look at one of the planktonic hemichordates. As we have seen, there are plenty of so-called worms that are not, and in this case it is the acorn worm, Enteropneusta. These animals have a proboscis that, with some imagination, is slightly acorn-shaped and is narrowly attached to a collar. Between these is the mouth and behind the collar a gill region, followed by a tapering abdominal trunk. They live burrowed into sediment, especially sand with finer mud and detritus particles, where they feed on organic matter. In Europe there are only a few species and of those likely to be seen in the extreme lower shore *Saccoglossus* is the main genus. *Glossobalanus* (*Balanoglossus*) *sarniensis* is found in sublittoral sediment, although its distribution is unclear. The latter produces the most exquisite larvae called tornariae, which are completely mesmerising to watch swimming in a dish. As it matures it takes on a lantern shape, with an apical tuft and ciliary bands around the body. The mouth is in the middle and the gut develops into the lower portion, with an anus at the very bottom. Towards the final stage of development the top half extends to look like an acorn, which, with metamorphosis to the adult, will extend to become the proboscis. Behind the central mouth the gut portion begins to truncate, pushing the ciliated band to the very

FIGURE 18.2 The acorn worm *Saccoglossus*. 6 cm.

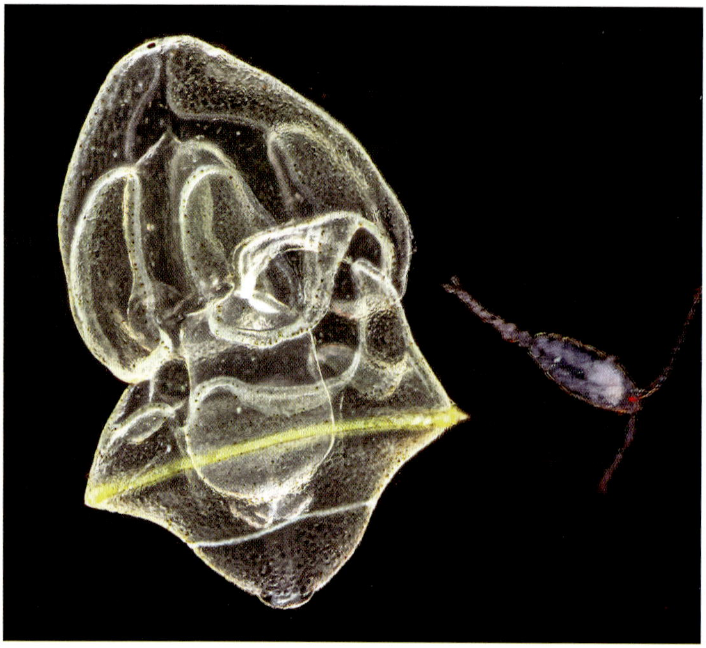

FIGURE 18.3 The beautiful tornaria larva of *Glossobalanus sarniensis*. 3 mm. NB. A copepod is also present in the photograph.

end, and a collar forms around the mouth. Tornariae can live in the plankton for several months, feeding on particulate matter brought by the ciliated bands to the mouth. During this time, if you are fortunate, dispersal may just bring one to your sampling point.

Tunicates

Tunicata, or Urochordata, is a very exclusive group. To be a member, uniquely to animals, organisms must produce cellulose. The particular form of cellulose is called tunicin and is used in the manufacture of the tunic, or protective coat, that covers the body. Adults typically have an inlet and outlet siphon to exchange water for filtration, as most species are suspension feeders. Several classes exist, including Thaliacea, which are pelagic tunicates, or salps. The most common representative is *Doliolum* and in southern Europe it can be prolific, growing at extraordinary rates;

FIGURE 18.4 A small colony of the compound ascidian *Botrylus* attached to a fucoid seaweed. 2 cm width.

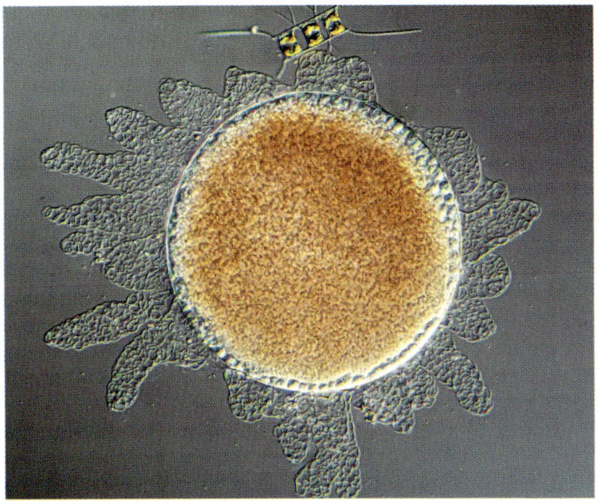

FIGURE 18.5 A tunicate egg. Diameter approximately 300 μm.

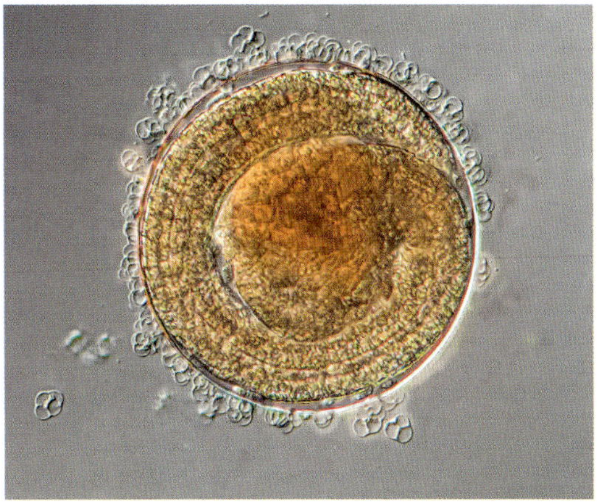

FIGURE 18.6 A tunicate embryo in the egg. The notochord is just visible. Diameter approximately 400 μm.

FIGURE 18.7 An ascidian tadpole larva. The prominent adhesive papillae are at the front of the head, which has a pigmented ocellus. From here the notochord runs down the trunk to the tail. On each side of the notochord are muscles and the entire animal is covered with a transparent tunic coat. 2.5 mm.

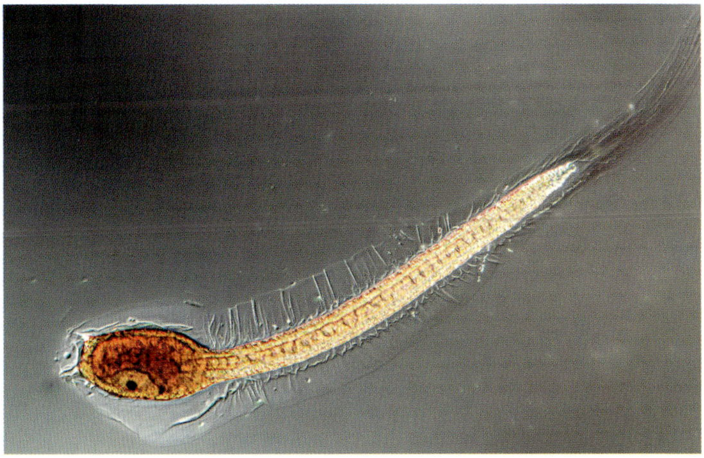

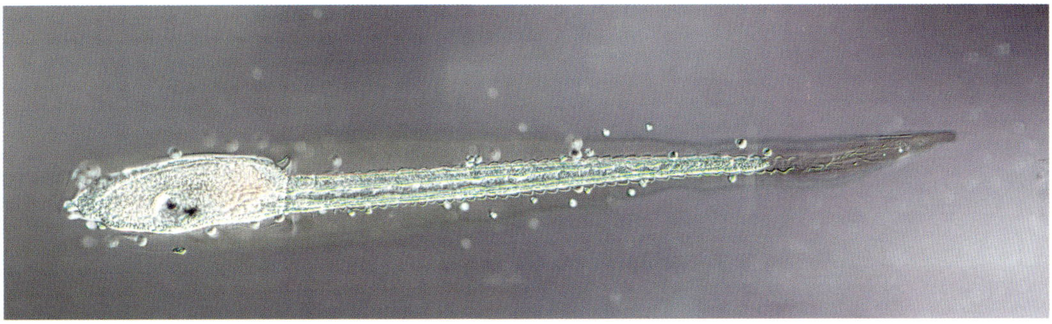

FIGURE 18.8 An ascidian tadpole larva, possibly of *Ciona*; note the yolk cells. 3 mm.

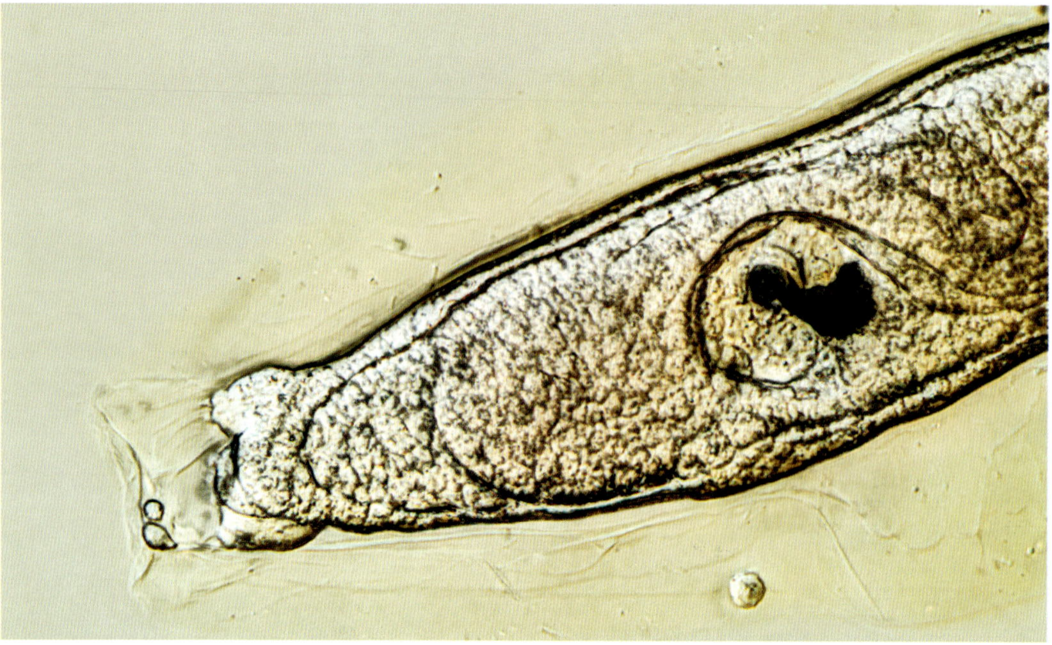

FIGURE 18.9 Close-up of the head of a tadpole larva. The central black dot is a statocyst, behind which is the ocellus. The adhesive papillae are at the end of the head, on the left.

an increase in body length of 10% an hour has been recorded. The tunic is encircled by a series of muscle bands that, on contracting, pump water through the body. Filtration removes food material and can be used to propel the body forward. On occasions they produce huge, dense concentrations of animals, but they are not neritic and unlikely to occur in shore samples.

The sea squirts

Class Ascidiacea contains what are commonly known as sea squirts, sessile creatures that are found on rocky shores, piers, and fouling the underside of boats and ships (see Figures 20.2–20.4). Some of these, such as *Ciona*, are solitary individuals with a bag-like, bilaterally symmetrical body. Others, the social ascidians such as the Lantern Sea Squirt *Clavelina lepadiformis*, form clumps and groups. Compound ascidians are small zooids that live together within a gelatinous matrix, often attached to seaweeds and holdfasts.

Sea squirts produce a non-feeding lecithotrophic larva called a tadpole. These have a similar body plan, with a long trunk and tail. At the core is the notochord, surrounded by muscle blocks for locomotion. The head has both an ocellus for light detection and a statocyst. At the very front of the head are several adhesive papillae. When the larva settles on a substrate to metamorphose into the adult these glue it to the surface so that it is attached by the head, with the tail uppermost.

Large, solitary sea squirts fertilise their eggs externally and the tadpole may be in the plankton for a few days. Compound ascidians may brood the eggs and release the larvae when they are almost ready to metamorphose, which can result in the larva spending only minutes in the plankton. The sessile, seashore adults can be common, but the larvae are not abundant in shore plankton samples due to the short period of free swimming. Yolk cells remain attached to the larva while in the plankton and can be observed around the tadpole, especially on the transparent tunic, which is difficult to see. There is no obvious gut, as this does not develop and function until after metamorphosis.

The Appendicularia

Class Appendicularia (also known as Larvacea) are free-swimming holoplanktonic tunicates. They are dominated by the genus *Oikopleura* and are a very important component of the zooplankton. In fact, their overall abundance is second only to that of copepods. They are between a few millimetres to a centimetre long and can reach very high densities – up to 50,000 per cubic metre of water. While the total number of sessile, shore-living tunicate species is around 3,000 there are fewer than 70 appendicularians known worldwide, and identification is difficult. *Oikopleura dioica* is the most common and widely distributed species. The body consists of a trunk on top of a long tail with a distinct notochord and limited musculature. The body plan is that of a modified sea squirt tadpole larva retained into the adult phase, which gives rise to the name Larvacea. They never become sessile or benthic, but keep their pelagic way of life. The most remarkable feature of *Oikopleura* is their exceptionally abbreviated lifespan: they grow, develop, reproduce and die within an average of five to six days.

Figure 18.11 shows the principal structures within a mature adult, including a well-defined gut and brain. Near the mouth there are two important components: the endostyle and buccal gland. These can produce copious amounts of mucus that bring food particles into the mouth. At the apex a large gonad forms. The epithelium of the trunk is a single layer of cells that secretes a rudimentary 'house' which expands to produce an enclosure around three times the length of the organism, a transparent and fragile tunic that both provides protection and enables the

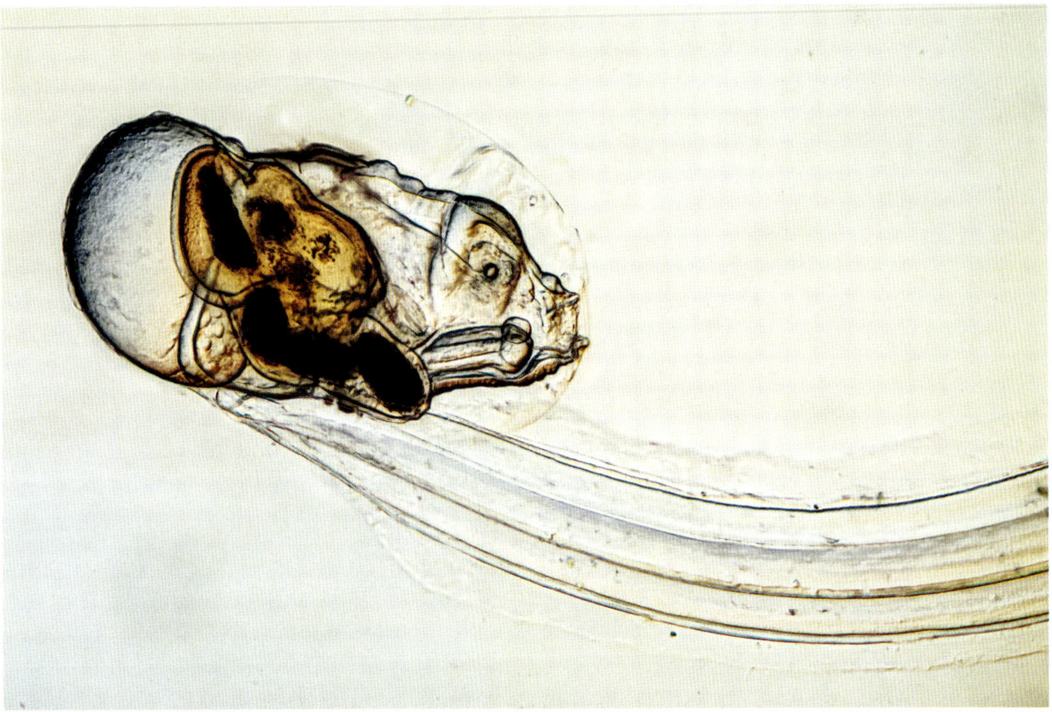

FIGURE 18.10 Mature male *Oikopleura* with the main trunk and tail, 2 mm.

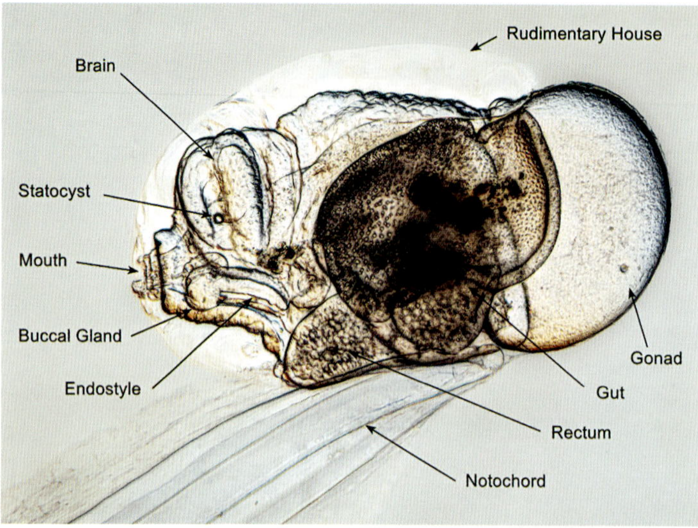

FIGURE 18.11 The structure and organs of a mature *Oikopleura*.

collection of food. It is made from the cellulose-like substance tunicin, mentioned above, and a unique protein called oikosin. The proteins are secreted by specialised cells called oikoplasts. It is thought that some kind of gene transfer occurred during the evolution of Appendicularia, allowing *Oikopleura* to manufacture a slightly more complex form of cellulose than plants. These houses are the most complex extracellular structures of any organism and are completely non-cellular. There are several outlet siphons and a large inlet with an ingenious mucus filter to remove particles (mainly nanoplankton) from the water. This filter often comes in two parts, the first to keep out particles that are too large and the second to remove the desired size of material from the water. The mouth is closely attached to the second filter, with mucus streaming the food into the gut.

To feed, the organism gently moves to bring water, and the food it contains, into the 'house'. While there is plenty of food arriving the animal moves slowly, but when the particulate matter in the water reduces the tail action speeds up. This quickly draws in more water that can be expelled to provide jet propulsion, moving it to a new location and, possibly, more food. As it feeds so it will be secreting a new, rudimentary house around the trunk. After a few hours, the filters become clogged with debris and *Oikopleura* responds by swimming out to make a new house. Having lost the original house it now expands the rudimentary one, which takes around ten minutes to reach its full size. Up to 16 of these houses can be created in a 24-hour period. The house may also be abandoned, probably as an escape response, if the animal is threatened by a predator.

In plankton samples it is rare for the house to survive intact with the animal inside. Apart from the innate fragility of the structure, the animal probably considers the plankton net a predator, and jettisons the house. When rejected, the empty housings, full of detritus, sink to the benthos, where they are an important source of food for bottom-dwelling animals, especially in the deep sea. *Oikopleura* therefore plays an important role in the removal of carbon from the upper layers of the ocean, as its used accommodation and its contents sink to the bottom of the sea. This process not only sequesters carbon but also acts to regulate virus populations. Viruses of the family Phycodnauridae, 0.2 microns in size, are filtered and pass through the gut of *Oikopleura*, ultimately being removed to the ocean depths as the houses are discarded. Between spring and autumn, plankton samples often have membranous debris full of detritus and nanoplankton, some of which is likely to be the shed houses of *Oikopleura*.

Oikopleura is quite seasonal and is found around the coast from May to October. The lifecycle, though short, is complex and completely tied in with temperature and food availability. The female produces between 40 and 400 eggs, depending on the amount of food present. The maturation time from egg to adult depends on the temperature, taking approximately two days at 29°C, four days at 20°C, six days at 15°C and ten days at 13°C. There are four main phases. At 13°C the developing embryo hatches from the egg into a larva after six hours; this larva becomes a juvenile at 18 hours, which grows in the plankton for 10 days before maturing into a sexual adult.

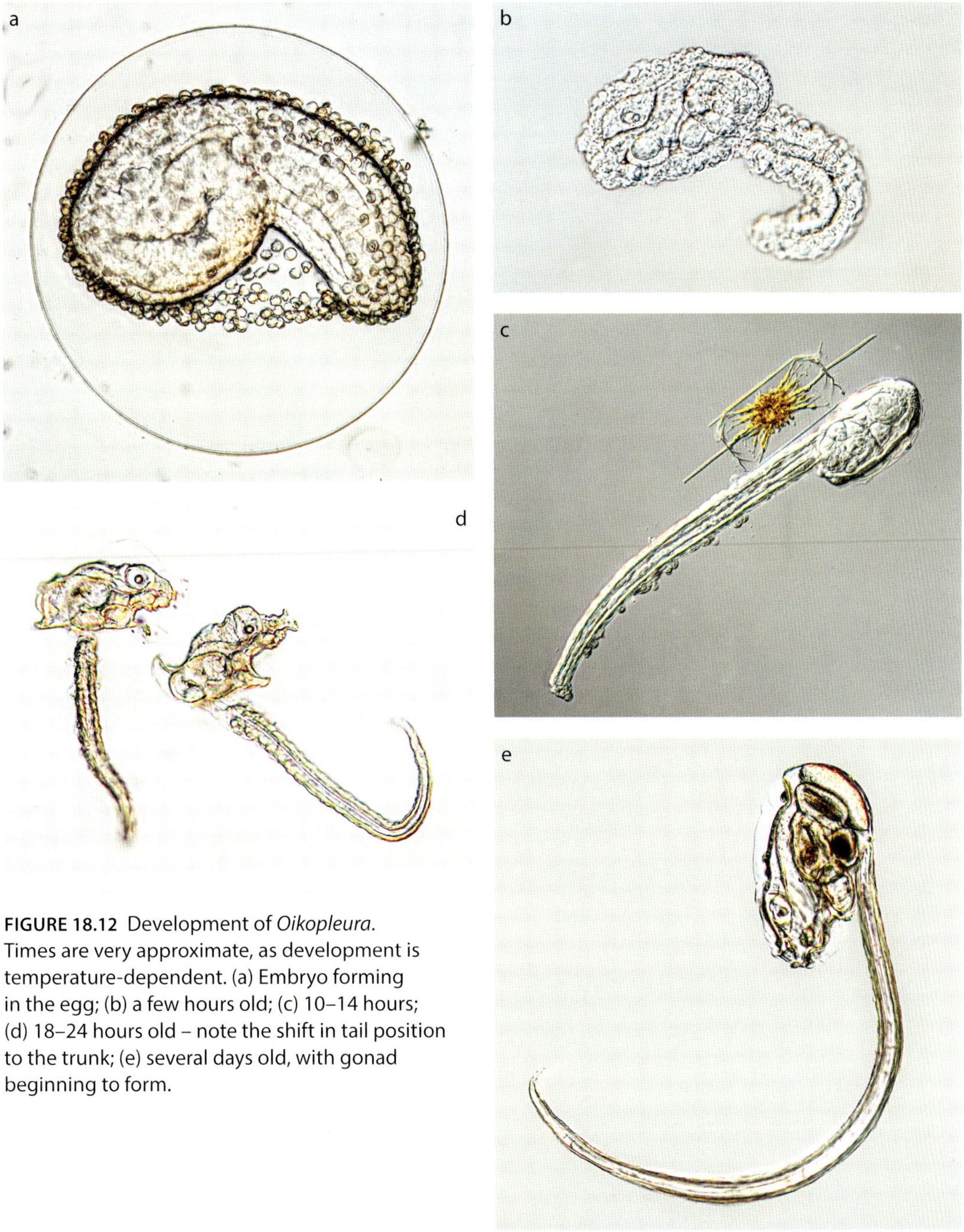

FIGURE 18.12 Development of *Oikopleura*. Times are very approximate, as development is temperature-dependent. (a) Embryo forming in the egg; (b) a few hours old; (c) 10–14 hours; (d) 18–24 hours old – note the shift in tail position to the trunk; (e) several days old, with gonad beginning to form.

Fish eggs and larvae

The Blenny or Shanny *Lipophrys pholis* is probably the most ubiquitous seashore fish, adapted to survive on the shore even when the tide recedes by living in rock pools or damp crevices and under stones. The female lays yolky eggs that adhere to weed in a nest on the shore, where the male fertilises and protects them. One female may produce as many as 8,000 eggs. Various pipefish species are also associated with shores; in these species the females lay a batch of 150 eggs that she attaches to a groove on the underside of the male for him to protect. Once the eggs of shore fish hatch, the larvae enter the plankton for dispersal. Determining larval duration during this time, as well as their size at settlement, is achieved by analysis of the microstructure of the otolith, a tiny calcium carbonate structure in the inner ear that, when sectioned, can provide information about the organism's entire life up to that point. Pelagic larval duration estimates up to the moment of settlement for the Blenny range from 30 to 75 days, with rates being lower in the south of Europe and higher

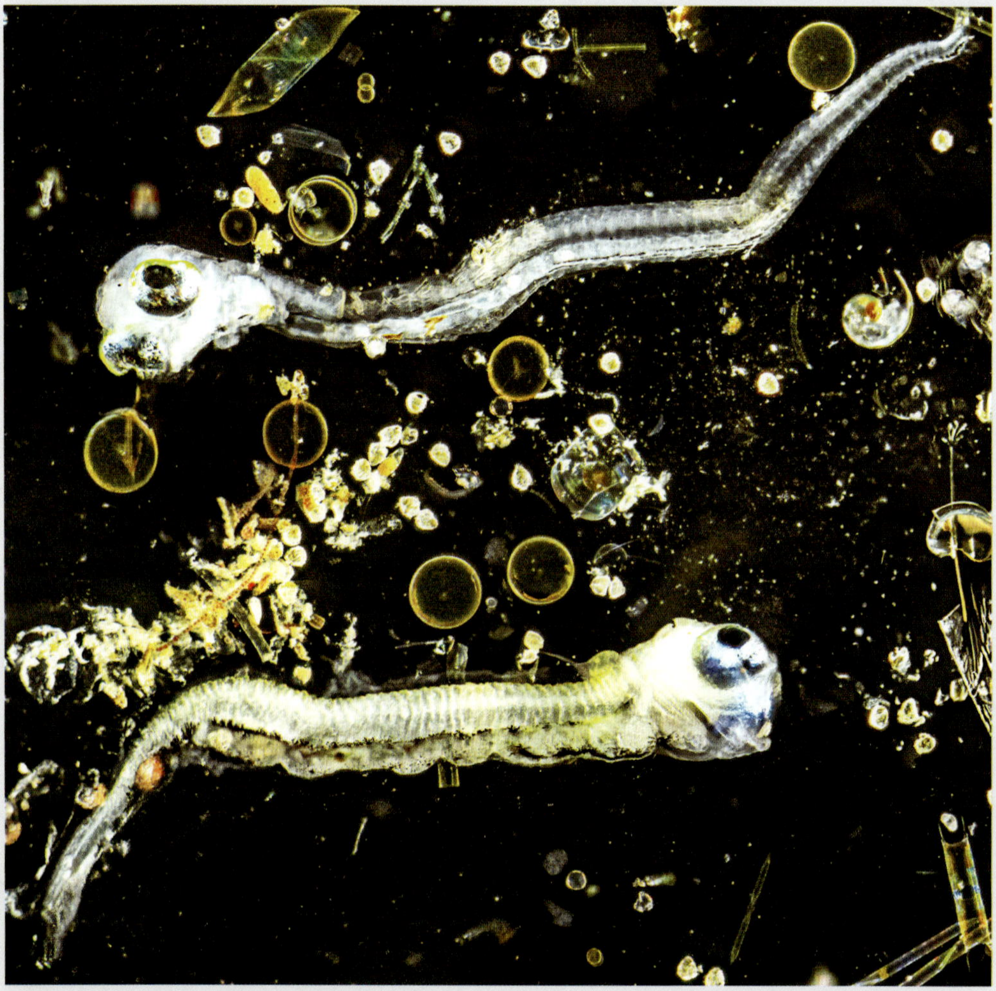

FIGURE 18.13 Fish larvae collected in a plankton net and slightly damaged by the process. 5–7 mm.

FIGURE 18.14 A cluster of developing fish eggs. The outer membrane is the chorion, within which is the perivitelline space. Attached to the yolk sac is a single drop of oil. Egg diameter 1 mm.

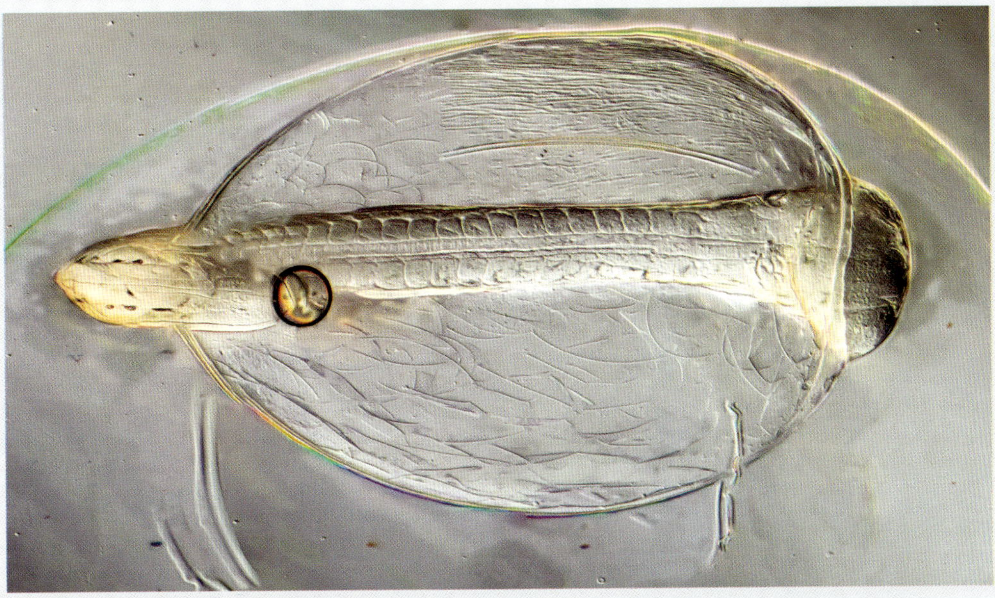

FIGURE 18.15 A single developing egg, at this stage just a germ band lying curved around the yolk.

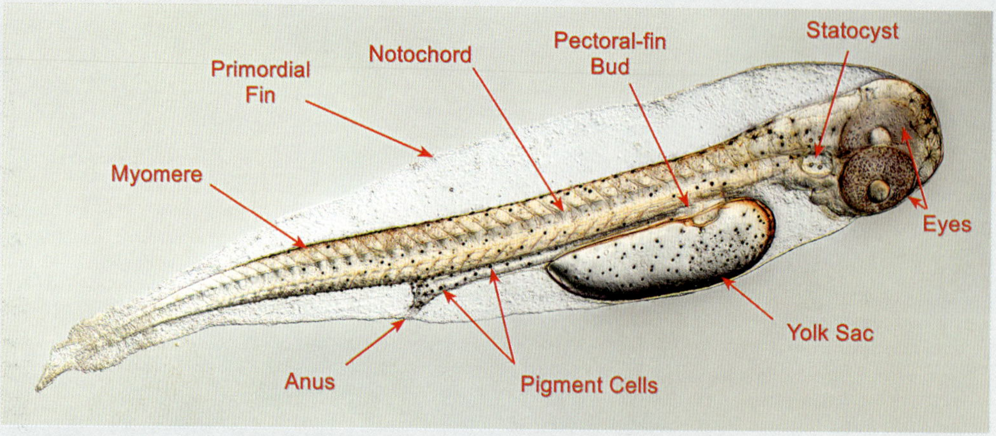

FIGURE 18.16 A yolk-sac fish larva, undamaged by the plankton net, showing detailed structure. Pigment cells are amoeboid and capable of moving to different locations creating species-specific larval patterns at the end of the yolk-sac stage. With a lack of these and other changes it is difficult to determine species but is likely a Sand Eel *Ammodytes* sp. 5 mm.

in the north. They carry little yolk and begin eating planktonic matter within 24 hours. This dispersal period will sweep them along the coast and offshore.

The reverse also occurs, where the progeny of deep-water bony fish living offshore will be brought in on tidal currents. Offshore species, unlike those living on the shore, are more likely to be broadcast spawners, releasing huge amounts of sperm and eggs. Of the latter, a hundred million from a single female is not unknown. Eggs are spherical and contain at least one drop of oil inside, which provides buoyancy. After hatching, the yolk and oil will remain for a short while and once it has been absorbed the young will then feed on planktonic organisms. It is not unusual, however, for both eggs and larvae to be absent from shore samples, and those that do occur will be found in late summer in small numbers. Eggs invariably float to the surface in the sample dish, while larvae, having usually been damaged by the net during collection, will invariably sink to the bottom.

During the juvenile expansion stage it does not increase the number of cells, but grows in volume. Once the trunk is fully developed the larva undergoes 'tail shift', in which the tail moves from a rear position to a ventral orientation and twists 90 degrees relative to the trunk (Figure 18.12d). Following tail shift, the larvacean begins secretion of its first house.

The population dynamics of this genus follow a boom-and-bust pattern, with high grazing rates followed by fast reproduction with short generation times producing huge blooms. Reproduction can be very fast when conditions are favourable and decline can happen just as rapidly. For example, the biomass of *Oikopleura* can increase tenfold in just 24 hours. By contrast, in the same amount of time the copepod *Paracalanus* has a maximum increase of 0.67%.

This extraordinary growth rate has led to an increasing recognition of their importance as a food source, where they outstrip calanoid copepods by a factor of ×2.4. Early in the life of the Flounder *Pleuronectes flesus*, for example, the larvae feed on copepods, but later they switch to a new favourite: members of the Appendicularia. Plaice *P. platessa* larvae are initially pelagic and in the southern part of the North Sea feed exclusively on *Oikopleura*, feasting on them during the summer population peak. Different-sized fish larvae will feed on the different sizes or ages of *Oikopleura*. Smaller organisms also make use of them; with such high population densities it is inevitable that they become parasitised by dinoflagellates and *Oodinium* is the most likely ectoparasite.

Oikopleura population density begins to rise in early summer, usually just after there is a phytoplankton bloom in April and May. When diatoms are at their peak this coincides with a high concentration of pico and nanoplankton biomass and it is this that stimulates the *Oikopleura* to rise in numbers. Their growth rate and recruitment, however, will decrease as the general zooplankton increases: competition with other members of the plankton, rather than the availability of food, controls their population. For growth and reproduction *Oikopleura* need a lifetime of steady, consistent access to food, rather than high food concentrations at any one time, as their food requirement threshold is low. That is good for their food security and regulation of the population. A bloom of diatoms, equally, may have an adverse effect on *Oikopleura* density, possibly because of biotoxins produced by the diatoms, which are known to inhibit embryo development. It is thought that ocean warming will only intensify this effect.

Physonect siphonophore juvenile colony, 1 cm in length. The pneumatophore (float) at the top has a red apex and below is the gas gland. The yellow arrows indicate the gastrozooids for feeding, with sting batteries at the base. Blue arrows are the reproductive gonozooids. Gelatinous plankters like this and other cnidarians are increasing as sea temperature rise.

PART III

The marine plankton environment

FIGURE 19.1 The Atlantic Puffin *Fratercula arctica* with sand eels *Ammodytes* sp.

CHAPTER 19

Cycles and recycles

Storm Ciarán struck north-western Europe at the end of October 2023 with reported gusts of 290 km/hour. In the calm a few days prior to this, plankton samples around Milford Haven were rich in diatoms and dinoflagellate species. Zooplankton samples contained arrow worms, hydromedusae and polychaete larvae, but few copepods. The content in November samples was very low. Species richness was a quarter of the previous month, with debris, faecal pellets and particulate matter dominant. Any horned dinoflagellates, *Tripos*, were badly damaged, and large numbers of resting stages of these and diatoms were present. Bacteria levels were very high, as the wealth of dead and decaying material provided them with huge amounts of food.

Six weeks after the storm the quantity and diversity had returned to levels found beforehand, although the composition was very different. The exception was the tintinnids, which were the first species to recover. Dinoflagellates would not recover until the spring, when their resting spores developed, but winter diatoms were back, along with a high density of mollusc veligers, particularly periwinkles. By March 2024 diatoms that tend to be common in winter, such as *Bacillaria* and *Odontella mobiliensis*, started to decline, disappearing in late May and reappearing again only in the autumn. Instead, diatoms such as *Chaetoceros*, *Ditylum* and *Rhizosolenia* began to bloom, continuing well into the summer.

Rainfall from late winter 2023 through the first few months of 2024 was unusually high in much of western Europe, causing severe floods and late sowing of crops. Rivers in spate brought substantial material down from the hills and the spring diatoms were joined by freshwater desmids such as *Staurastrum*, *Closterium* and *Micrasterias*, which became common constituents of the phytoplankton for some weeks. I normally see a June plankton dip, but in 2024 all trends seemed delayed by about a month, so it was not until late August that the late summer rise in gelatinous plankton began. Autumn 2024 saw an explosion in hydromedusae, ctenophores, arrow worms and the siphonophore *Muggiaea*. Phytoplankton bloomed, with a wealth of dinoflagellate diversity as well as polychaetes and bryozoan larvae. That was until Storm Darragh.

In December 2024 Storm Darragh devastated the plankton in the Irish Sea and west Wales. For weeks plankton levels were unusually low, even compared to the year earlier. A slow recovery during January 2025 was followed by a spring that was the reverse of 2024's: probably the warmest and driest on record. In March the first diatom bloom occurred: *Lauderia*, *Skeletonema* and *Chaetoceros*. Large diatoms disappeared and dinoflagellates were few. April saw these populations crash, to be replaced by the tiny *Cerataulina*, which along with a rise in *Phaeocystis*, turned the water smelly and green. Water temperatures were several degrees higher than normal. Development of the zooplankton stalled and only in mid-May did the bloom subside and the dinoflagellates and zooplankton proliferate. Swarms of comb jellies arrived. Unlike the previous year, the June dip for zooplankton did not occur and some summer populations bloomed early like copepods and appendicularians.

This is a brief, anecdotal summary of what was happening in Milford Haven, Pembrokeshire, over the course of a couple of years. It illustrates that events, while cyclical to a degree, change and will keep changing. Some are predictable, but many are not. Ten years ago the amount of

Adaptations to planktonic life

Plankters need to maintain their position in the water as they drift and without some form of adaptation, unless they are naturally neutrally buoyant, they will sink. The extent of sinking depends on various factors, such as the viscosity of the water. A higher temperature makes it less viscous, and organisms more prone to sinking. Furthermore, the size of the organism matters. Smaller organisms have a greater relative surface area to body mass, so the more microscopic forms are less likely to sink. Femtoplankton and picoplankton, therefore, have few adaptations to prevent sinking. For bigger organisms creating drag helps: to increase surface area, the most widespread method is to produce hairs, chaetae or extensions. The diatom *Chaetoceros* is very hairy, as are the larvae of polychaetes. Crab zoea have long, extended spines. Mollusc larvae can produce threads of mucus and arrow worms have flattened bodies with extended fins. The wide arms of echinoderm pluteus larvae become more numerous and extended the older and larger they become.

Frictional resistance alone will not stop organisms sinking, but it slows the process before other techniques can increase buoyancy. Flotation chambers are one, like the siphonophores, with a gas-filled float. Oil droplets further increase buoyancy, as seen in fish eggs, copepods and many polychaete larvae. Or a combination of methods may be employed. The beautiful radiolarian type *Acanthometra* employs an extensive frothy layer between its 20 spines, as well as oil droplets around the cell. *Muggiaea*, with its gelatinous bell float, also has oil droplets present. Carbon dioxide is typically the gas used to make froth and air pockets. By changing the metabolism around the float the concentration of gas can be adjusted to increase or decrease buoyancy, allowing vertical movement. Oils are useful energy storage compounds, but most plankters use these lipids for both food and buoyancy.

Perhaps one of the most amazing adaptations is a lighter body. To achieve this, heavy components are replaced by lighter ones. Shells of mollusc larvae and pelagic adults may have calcium carbonate replaced by protein. Gelatinous bodies can be diluted to be more watery.

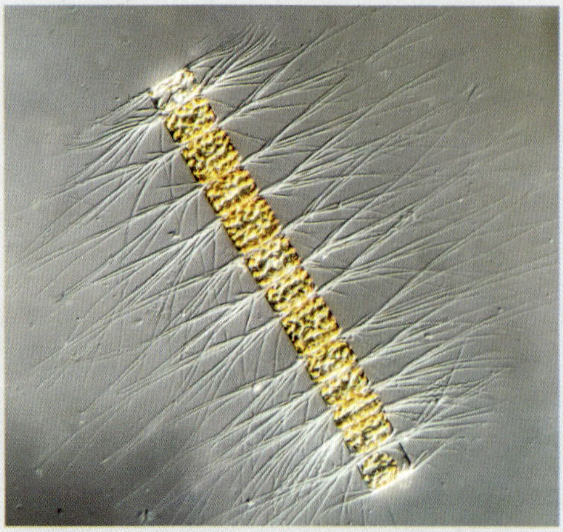

FIGURE 19.2 A *Chaetoceros* diatom chain lit to emphasise the huge density of setae that reduce sinking. Cell width 50 µm.

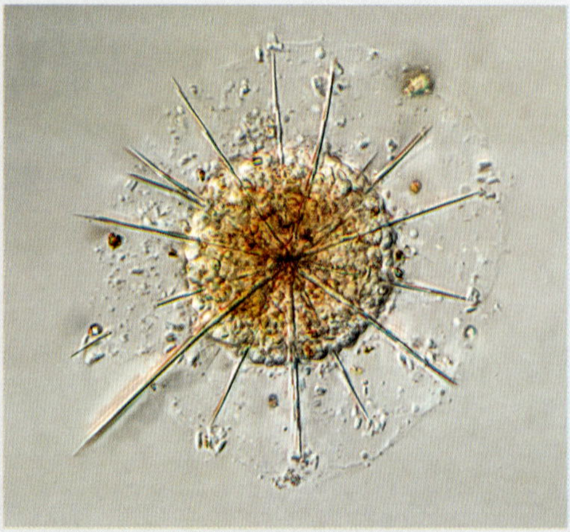

FIGURE 19.3 The large radiolarian *Acanthometra pellicida*, which has a frothy cell and a large surface area to reduce sinking. Diameter 500 µm.

FIGURE 19.4 Zoea of the Broad-clawed Porcelain Crab *Porcellana platycheles*. 2 mm.

Heavy, dense ions in the cytoplasm, such as magnesium and sulphates, can be replaced by lighter ones. This will reduce the efficiency of the metabolism, but if less energy is expended to stop sinking then the overall effect will be positive. The benefit of buoyancy is that passive transport within the plankton is possible, using the natural movement of the water.

Living in an open, three-dimensional medium with nowhere to hide makes it difficult to avoid predators, but, handily, adaptations to reduce sinking might also help organisms to avoid being eaten. A covering of hair, like that of the *Chaetoceros* diatoms, will make organisms less palatable, for example. Cnidarians might use sting cells as protection as well as for catching prey. This consideration is just as important for a carnivore stalking prey. In many cases the problem is not being seen, as light levels are often low, but prey can detect movement, which creates vibrations in the water; copepods, for instance, have an array of sensors across the entire width of the antennules, held out in front of the body, with which to sense potential predators. Most of the zooplankton have some sense receptors such as hairs, which pick up the slightest water disturbance. Therefore, ambush predators such as arrow worms remain static in the water, waiting for prey to pass, accurately detecting disturbances caused by their prey using their own sensory tufts along the body.

Planktotrophic larvae spend more time in the plankton than their lecithotrophic counterparts. As already described, a longer pelagic duration results in increased dispersal and gene flow, but also a high mortality. By producing huge numbers of progeny, counted in millions of offspring, the chances of some surviving are good. But the efficiency of this must be balanced with supplying the eggs with an absolute minimum of stored food. Lecithotrophic larva, on the other hand, can spend as little as an hour or two in the plankton before settling. Mortality will be lower and far fewer eggs need to be released, but a high-energy yolky food store will be needed to promote rapid development; lecithotrophic larvae do not even need to waste resources on growing a gut. However, they are more likely to settle within the existing population, thus reducing genetic variation.

gelatinous plankton was much less in late summer, but in 2023 the highest sea temperature ever recorded was seen in Pembrokeshire, and we know that warming seas favour the jellies.

Finding a mate in the vast expanse of water is another problem. Some unicellular organisms solve this problem by using asexual reproduction to produce dense populations followed by conjugation with another individual at a time when they are more likely to find one. Conjugation will shuffle the genes, leading to variation. Hermaphrodite adults are a common theme, especially when both gonads mature at the same time to give a chance of self-fertilisation. If not, then the production of a large number of gametes for external fertilisation increases survival. This may require some form of synchronisation, so that the two sexes release gametes at the same time. Although there are cases of chemical (pheromone) triggers enabling sexes to find each other, synchronisation by the phases of the moon is a method found in polychaetes and some molluscs.

Food webs

Estuarine and coastal marine food webs are dominated by plankton. Diatoms and dinoflagellates fix the sun's energy and are, in turn, consumed by the zooplankton, such as copepods and planktotrophic larval forms from the shore. These too will be preyed upon by cladocerans, hydromedusae and polychaetes such as *Tomopteris*. Fish such as the Atlantic Herring *Clupea harengus* will feed on phytoplankton, initially when young but later shift to consuming copepods, which are an essential food source for most fish. This includes a number of widespread sand eel or sand lance species belonging to the family Ammodytidae, such as *Ammodytes tobianus*, which are a major part of the diet of sea birds such as the Atlantic Puffin and Razorbill *Alca torda*. There is considerable concern that fishing for sand eels will impact seabird populations. These fish are collected not for human consumption, but to produce fertiliser and animal feed, and over the last few decades their populations in the North Sea have declined as a result of fishing. Birds such

FIGURE 19.5 The English Channel and North Sea at the height of a phytoplankton bloom near the surface of shallow coastal water. As well as these microorganisms, there is great deal of nutrient and sediment visible in areas such as the German coast, Bristol Channel, the Wash and the Humber Estuary. (NASA)

FIGURE 19.6 The world's largest fish species, Whale Shark *Rhincodon typus*, in plankton-rich water. (Cameron Witney)

as the Puffin begin their breeding season well, with developing young, and then a sudden crash occurs as insufficient food exists to complete chick fledging. Some countries are beginning to act to reduce sand eel fishing and currently populations of these birds on the west coasts of Europe are faring well, with relatively stable numbers, although what effect bird flu might ultimately have is another story.

Every plankton sample will have a high percentage of organic matter that feeds the smallest inhabitants, such as bacteria, the microbes that degrade organic matter. This picoplankton will be the food source for many protists and invertebrate larvae. *Oikopleura*, a large holoplanktonic tunicate, also feeds on this minute material and forms an important trophic shortcut to the top of the food web, where there are fish and their larvae. *Oikopleura* is a keystone species with major roles in both the coast and ocean systems.

Whether it is copepods feeding on diatoms or *Oikopleura* consuming smaller plankton, between the two groups they are the main supporters of fish populations. Incredibly, these short food chains support some of the largest organisms on the planet. The baleen whales filter plankton, mainly larger crustacean species like krill, although they do not commonly feed close to the shore. However, the Basking Shark *Cetorhinus maximus*, the second-largest living fish, often arrives in Europe to coincide with the spring blooms, where it feeds by filtering the plankton in coastal waters. They can be a common sight on the south-west coast of the UK. The largest fish species, at 12–18 m in length, is the Whale Shark *Rhincodon typus*, which is entirely sustained by filtering plankton and small fish. Although not common in Europe it does have a large Atlantic population, which mainly keeps to warmer and tropical waters.

Vertical migration

Diel (diurnal) vertical migration, briefly explored in Chapter 13, is thought in zooplankton to be a predator avoidance strategy, utilising the hours of darkness when fish such as herring cannot use their sight to locate feeding crustaceans near the surface. In the early 1940s, during the development of sonar, the Royal Navy believed that they had discovered a large submarine that rose to the surface at night. What they had actually found was the migration of copepods and other zooplankton rising to feed at night and then descending to the depths by day.

In fact, most plankters undergo some form of daily vertical migration and this strategy is found in all aquatic systems, including fresh-water ones. Phytoplankton has a passive tendency towards vertical movement in search of light; during the day photosynthesis produces oxygen, some of which is retained in the frustule and provides a degree of buoyancy. The more the organism photosynthesises, the closer it can get to the surface. As darkness occurs, so does the beginning of a descent through the water column. If the plankter descends too rapidly, especially in winter when the photoperiod is short, it may drop too low and be out of the photic zone, resulting in no photosynthesis. It will then be completely lost and descend to the benthos.

Daily cycles of vertical plankton movement occur less frequently near the seashore, as movement here will be governed more by the tides. However, the composition of samples taken on a high spring tide can vary depending on the time of day, so that, for example, collections at dawn will contain more copepods than a collection taken in the afternoon.

Coastal nutrients, the carbon cycle and the biological pump

Coastal regions are complex, highly dynamic and productive systems, to a degree greater even than oceanic systems. The coast is where 40% of the world's human population lives. It is also a marine ecosystem that has varying inputs of terrestrial organic matter, nutrients and sediment. Its salinity can undergo considerable variation as a result both of fresh water flowing from the land and, in summer, the warming of very shallow areas, increasing evaporation. Changes in light, temperature, salinity and nutrients will be substantial over a year and from year to year. In the deep oceans, by contrast, there is a chemical and physical stability that is missing on the coast, where the environment is much less predictable, and there planktonic system dynamics can be relatively easily predicted, with blooms occurring at known intervals along with the steady degradation of organic matter.

Sheltered waterbodies such as Poole Harbour and Milford Haven have sea temperatures that are higher than on the open coast. In addition, nutrients, especially human effluent and farm runoff arriving via streams and rivers, are more likely to be contained for longer periods in one area than in open areas swept by the tide. In these shallow waterways light may reach the benthos, although the water can be more turbid with silt, especially when wind disturbance occurs. The predictability of seasonal gales and storms has changed with a warming planet, such that strong wind or calm periods can happen at any time. In the last few years at least one or two gale-force winds have hit the west coast of northern Europe in mid-summer. Consequently, some shore plankton populations may be triggered into growth at any time of the year if, for example, nutrients such as nitrogen are flushed from farmland at the same time that benthos disturbance occurs, bringing the essential silica and resting stages up from the bottom. This occurred multiple times during 2023 in Milford Haven, giving rise to a pronounced bloom in December, as discussed above in this chapter. In many ways, the unpredictability of what may suddenly turn up is part of the joy of looking at shore plankton.

Turbidity is not just inorganic silt. Shore plankton samples contain a tremendous amount of detritus and organic material, much of which is coalesced particulate matter and faeces. Copepod faeces are abundant and, when excreted, can take many hours to slowly descend through the water, while wind action quickly brings them back into the plankton. Unlike in the open sea, a considerable amount of organic matter is derived from the seashore. Each tide will not only deposit seaweed on the strandline but also collect decayed material and take it back into the water column. As well as the visible particulate organic matter (POM), dissolved organic matter (DOM), a cocktail of water-soluble compounds containing organic carbon from a variety of biological and geological sources, makes up a very important component in sea water. For many of the smaller zooplankton, such as ciliates, as well as some larval forms, life in the plankton is akin to living in a nutritious soup, as DOM passes directly into the body as food. However, no organisms utilise DOM as fully as do bacteria, which, being so small, have a higher surface area to volume ratio than zooplankton. This is the base layer of the microbial loop, described in Chapter 6. DOM is also an essential part of the biogeochemical cycling of carbon, the energy currency of the sea.

The marine carbon cycle is an essential earth system, even within the coastal zone. In marine environments carbon originates both from organic sources via microbial recycling and from the atmosphere, as carbon dioxide dissolves in the water surface to produce bicarbonate ions (HCO_3) that are used by photosynthetic phytoplankton in the illuminated shallow water. Here the bicarbonate is metabolised into organic materials for the growth and development of these producers, which release oxygen in the process, most of which is retained within the aquatic system and used by the organisms there. Phytoplankton are then consumed by zooplankton, and much of the carbon ingested by them will be utilised within cell walls, as calcium carbonate in shells and hair, setae and chaetae. Mostly these consumers are very messy eaters that spill a tremendous amount of POM into the water, as well as releasing faecal material. The phytoplankton fix the carbon into molecules that now, in particulate form, end up sinking to the bottom. Some of this aggregates together on the way to the benthos. For instance, the 'house' of the tunicate *Oikopleura*, which is produced and released many times a day, takes huge amounts of POM and microorganisms to the seabed, making it a vital sequester of carbon. Although it is not so significant along the coast, in ocean systems this blizzard of descending POM is referred to as 'marine snow' and is part of what is known as the biological pump mechanism: the sequestering or removal of carbon from the atmosphere to lock it in the seabed. A lack of oxygen in the depths can leave this material undecomposed for thousands of years, although disturbance is more likely in the very shallow water of the neritic zone. Diel vertical migration contributes to this process, as many of the organisms that descend during the day never recover and continue to sink to the bottom.

Annual cycles

Environmental stresses to planktonic organisms include temperature and light, along with insufficient nutrients. When conditions are ideal they bloom, but during stressful periods the population may survive through the formation of cysts or resting stages, as we have seen in tintinnids, protists and copepods.

Although there will be variation between locations, organisms show distinct cycles in response to these factors. There are patterns across the year in diatom blooms, for example, each of which have their own specific needs and tolerances. The large species, such as *Bacillaria paxillifer* and *Odontella sinensis*, usually disappear in the summer and dominate in the winter. *Thalassiosira gravida* can be abundant in spring and into the summer, while *Rhizosolenia* (*Guinardia*) *flaccida* is common from early summer into the autumn. In some years the long, needle-like diatom *Rhizosolenia hebetata* can completely dominate for months through the summer.

Patterns can be seen in zooplankton, as well. Chaetognaths, cladocerans, appendicularians and hydromedusae have their highest abundance in late summer and autumn, when the temperature is highest (16–18°C). This is reflected in an intense reproduction period producing a large number of juveniles. A stable weather period during winter can also promote the increase of zooplankton, with the release of mollusc veligers and barnacle nauplii.

FIGURE 20.1 A swarm of Moon Jellyfish *Aurelia aurita*. In large numbers these can become a pest, affecting beach swimmers and blocking machinery inlets. This is an increasing problem with warming seas. (Michal Mikulec)

CHAPTER 20

Plankton, humans and climate

My favourite beachcombing shore is a rocky, wave-cut platform in Dorset that projects from the coast, catching all manner of strange objects. Sifting through both natural and human detritus and figuring out where it might have come from is part of the fun. Sometimes the source is obvious, like the four-pack of unopened American beer (only three cans left) caught in a rock pool. A perfect discovery on a warm day, which might have arrived after a journey across the Atlantic. On the same shore a fully intact coconut with its fibrous outer coat intact was not an unusual find, as it has adapted to drift in ocean currents. There are many different tropical beans that can appear on western European shores, although the only one I have found was a horse-eye bean, *Mucuna* sp., which washed up on the Dorset shore. Unusual items like these are moved on the world's extensive rotating currents, forming what are called gyres, on which our survival depends (Czerski 2023). The North Atlantic Gyre includes the Gulf Stream, which travels up the coast of USA from the Caribbean and then across to Europe, and the slightly weaker Canary Current, which goes south, past Portugal and North Africa, before reaching the Equatorial Counter-current, the upper part of which crosses west to Central America. Plankton on your local seashore is part of an amazing interconnection with other marine environments across the globe. You could be observing on your patch the progeny of organisms with their origin many thousands of miles away.

The globally interconnected nature of the currents affecting plankton is the main topic of this chapter, from localised eutrophication to global issues of climate change, invasive species, microplastic pollution, chemical and oil pollution and overfishing. All are stressors on marine plankton and the health and functionality of marine ecosystems, which then has major implications for human well-being, not least in terms of food and carbon storage.

Shipping and invasive species

The large diatom *Odontella sinensis* (Figure 7.5) is widely distributed in European waters and is often the most abundant species in neritic phytoplankton. Originating in the Indo-Pacific, it has a wide tolerance of temperature and salinity variation, enabling it to survive and spread in a variety of situations. Prior to 1903 it was unknown in the Atlantic; it arrived suddenly off the Danish coast and by 1908 had spread through the North Sea to the Irish Sea. It reached the Isle of Man in 1909 and at about this time biologists started using it as an indicator of current direction and speed. How it made the leap from one side of the world to the other is perhaps unsurprising: it came via shipping to the Elbe in 1903.

Shipping carries marine organisms in two main ways: in ballast water and as fouling organisms attached to the underside of the hull. Water taken on board to fill ballast tanks at the point of origin is eventually discharged at the destination port. Many adults and larvae would not survive the journey, but some adults do and then produce larvae, which can disperse in the new locality if the habitat is suitable. A wide range of tintinnid ciliate species have been moved around the world in ballast-tank water and for so long that keeping track of the changes has been difficult. As a unicellular species with no known pathogens associated with them these ciliates are thought to have only minimal effects on local communities. However, this is not true of all species, and many non-native marine organisms can live within the algae and sessile creatures that cause

FIGURE 20.2 Fouling on the hull of a boat. Kelp and red algae as well as encrusting animals. White arrows indicate sea squirts and red arrows bryozoans.

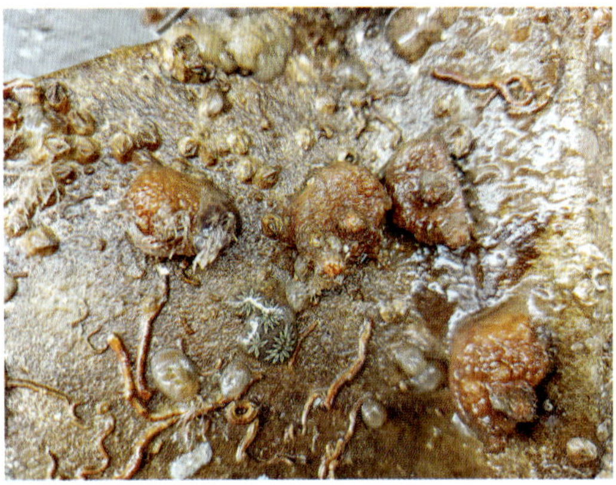

FIGURE 20.3 A close-up of boat fouling, showing sea squirts and a star ascidian colony.

FIGURE 20.4 The invasive Leathery Sea Squirt *Styela clava* from Asia, now a common member of the marine fauna in many parts of the world. Note the inlet and outlet siphons.

fouling under ships, such that entire communities can develop if ship hulls are not kept clean. In some places, shipping may even deliver new organisms to locations made more suitable for them by human activity. For example, ports can be located near power generation plants which need water to cool the generators. The warm outflow water then raises the temperature of the coastal water around the plant, perhaps just a degree or two, but sufficient to provide a new niche.

Odontella sinensis is an early example of the known movement of exotic species, although shipping in centuries past may well have had similar consequences. There are certainly many documented examples since, and the number of invasive species has risen significantly as shipping increases. For instance, in 1953 the Leathery Sea Squirt *Styela clava*, a sea squirt or tunicate with a rough coat that is native to the north-west Pacific, was found in Plymouth, on the south coast of England. At the end of the Korean War, in 1951, the military ships used in the conflict returned to their home port, Plymouth, which turned out to be the perfect environment for the larvae to develop and spread. The spread was helped by harbours warmed by outflows from power generators, like the Solent.

In the 1970s the Red Ripple Bryozoan *Watersipora subatra* arrived in Europe, also from the Pacific. Subsequently, the Oriental Shrimp *Palaemon macrodactylus*, from the north-east Pacific, where it is an important human food, arrived in the Thames Estuary plankton in 1992, probably via ballast water. My own find of juveniles within the Cleddau Estuary in summer 2024 is thought to be the first record for Wales. As with so many invasives, this species is highly tolerant of variations in temperature, salinity and oxygen levels and could thus spread anywhere within Milford Haven. There is concern that it may impact local shrimp species, although at this stage very little is known.

In the 2010s the Pacific nemertean *Cephalothrix simula* arrived in Europe and continues to spread via its planktonic larva. While most nemerteans are predatory, consuming prey with their proboscis, this particular species stuns its prey with the neurotoxin tetrodotoxin, also associated with pufferfish, which is produced by the large biome of bacteria living inside it, including *Alteromonas* spp., *Vibrio alginolyticus* and *Pseudomonas luteola*. This nemertean is now common on the south coast of the UK and, given the currents in the Bay of Biscay and the Channel, it is likely that its pilidium larvae will disperse it further. Tests conducted on UK specimens have shown high levels of the toxin present. Only 5% of the world's nemertea species have had their toxins studied. Some are now being looked at for medical use as well as their persistence through food webs. For the most complete analysis of toxins, download the comprehensive paper 'Toxins of Nertinea Worms' (Göransson U., et al. 2019).

Invasive species may also expand their range via aquaculture. Pacific Oysters *Magallana gigas* have been transported worldwide, as have many other examples of shellfish for human consumption. If care is not taken other species may be transported with them and, although

FIGURE 20.5 A juvenile Oriental Shrimp *Palaemon macrodactylus* showing a characteristic spine on the back.

FIGURE 20.6 A dead Pacific Oyster *Magallana gigas* on a Devon shore. This shows the lower of the two valves that make up the shell attached to the rock surface. Despite the animal being dead, it has crowded out other local species with this valve preventing the normal community from being able to develop.

some countries have rigid biosecurity, some are less conscientious. Now *M. gigas* is found on UK seashores, where their range is expanding, as they create huge numbers of planktonic larvae. In England and Wales it is regarded as an invasive species. They grow attached to rock, out-compete local species for space and are having an impact on seashore communities.

Microplastics

Plastic pollution has been headline news for a number of years now. In 2020 the World Health Organization concluded that too little was understood about the health impact of passive microplastic consumption by humans and that efforts to reduce plastic pollution were paramount. Microplastics are derived from the physical break-up of plastic waste, as well as microscopic beads used in cosmetics and other products. Much can be seen on the strandline as colourful arrays of tiny plastic pieces. Some of these will be regular spheres called nurdles that are used to make plastic products, but are lost during transport and handling. These are quite large as microplastics go, at around 4–5 mm. Most of the concern around plankton is with the much smaller particles, measuring less than 10 microns in size, which can be consumed by plankters and observed under the microscope by using epifluorescence techniques, which make the microplastics show up as distinct patchworks around planktonic organisms.

Zooplankton have their highest productivity on the coast, which also has some of the highest microplastic concentrations, coming both from the sea and the land. Cardiff University research discovered that 86 trillion to 710 trillion microplastic particles annually contaminate farmland in the UK (Cardiff.ac.uk 2023). As well as entering the human food chain, many of them enter rivers and ultimately end up at the coast. Microplastics have been found in the bodies of most examples of zooplankton studied, and are thought to have the greatest impact on copepod and krill populations. They negatively affect feeding behaviour, growth, development and reproduction. Larger crustaceans in the plankton, such as krill, will readily ingest microplastics, fragmenting them into smaller particles. Crustaceans that eat nanoplastics may not be killed directly, but their swimming

FIGURE 20.7 Nurdles and microplastics on a sheltered shore strandline.

ability is impaired. Juveniles may fail to shed the exoskeleton. Medusae and hydromedusae are known to feed on microplastics and have them caught on the tentacle surfaces.

Like other toxins, microplastics are concentrated up the food chain, so that fish with significant microplastic loads are being consumed by humans. The spread of toxin-producing bacteria – mentioned above – is exacerbated by their adherence to microplastics that then collect in the strandline. All the while, more microplastics are entering the ecosystem from an already huge amount of macroplastic in the environment.

Acidification and eutrophication

An increase in atmospheric carbon dioxide levels has led not just to an increase in global temperatures but to a slow acidification of the oceans as it dissolves into seawater. The current pH of the oceans averages 8.1, a drop of 0.1 since the Industrial Revolution. This does not sound like a big shift, but it is already having an effect. In copepods acidification has shown to reduce egg viability, and nauplii may fail to develop. Surface ocean acidity is projected to increase by about 38–41% by the end of the century. That is a substantial change.

For molluscs in the plankton, attempting to produce a shell of calcium carbonate will become an impossibility if that degree of acidification occurs. The same problem will occur in echinoderms, particularly the sea urchin manufacturing a test or the ophiopluteus of brittlestars creating calcite. In fact, any plankter needing to use calcium carbonate will meet new problems.

Climate change

The impact of climate change on ecosystems is complex and often no direct relationship can be seen. However, the sea is clearly warming, as summer temperatures at coastal sites around Europe regularly break records. For example, the sea in west Wales in August 2023 exceeded 18°C. Along with increases in nutrient levels this can give rise to harmful algal blooms (HABs). Such blooms can clog the gills of fish and release dangerous toxins. To reduce grazing by copepods, diatoms can produce toxins such as oxylipin, which reduces copepod reproduction. It can also affect sea urchins, ascidians, molluscs and annelids, where it can also reduce egg and sperm production. Some dinoflagellate blooms release toxins that directly affect humans and domestic animals. Often, we hear warnings to keep dogs away from certain waterbodies because of these toxins. Bacteria are likely to increase faster than diatom populations with warming seas, and the types of bacteria present may also change. Certain human pathogens might flourish relative to other bacterioplankton, impacting humans. As zooplankton is affected slightly differently by increased water temperature than the phytoplankton there is a chance that these two components of the plankton will move out of synchronisation with each other.

The changing species balance caused by warming seas will affect prey–predator relationships, so that only those able to adapt to new food sources and environmental regimes will survive. For example, with warming seawater calanoid copepods that diapause overwinter may not need so much time to suspend development, becoming active at a different time. This may have serious consequences if their emergence is at a different time to that of the food source with which these species are currently synchronised. Ctenophores are developing a month earlier than they did 40 years ago. If they are now feeding on copepods a month earlier, what effect will this have long term? Currently, we can only speculate. While warming seas reduce crustacean populations such as copepods we are seeing an increase in jellyfish and siphonophores, the gelatinous plankton. Gigantic swarms of jellyfish have been detected in the Mediterranean in summer, while siphonophores such as *Muggiaea* are appearing in high densities on the seashore in late summer. Both have negative impacts on tourism and fisheries.

If global temperatures continue at their current rate of increase there may be an overall rise in plankton species richness caused by the warming sea. There has been a noticeable appearance of certain plankton species further and further north. *Rhizosolenia* (*Neocalyptrella*) *robusta* is a large diatom (Figure 7.6) that was originally considered subtropical, with a northerly limit near the English Channel. In the last few decades it has become more common up the Irish Sea to Scotland when the temperature is at its maximum. The shift in plankter species' distributions is moving towards the North Pole at a median speed of 35 km per decade. The richness of phytoplankton is projected to increase by more than 16% over most regions except the Arctic Ocean. The biodiversity and abundance of zooplankton is likely to change in the future with a decline in the tropics but a substantial increase in temperate regions. It is thought that nearly 40% of the plankton community structure in temperate latitudes will be replaced by populations that are shifting north.

FIGURE 20.8 Tourism. In large numbers, jellyfish – here Compass Jellyfish *Chrysaora hysoscella* – can have a strongly negative effect on tourism. (Kenny van Roosbroek)

With the increase in mild winters, rainfall patterns have changed. Long periods of precipitation increase flooding and outflow from rivers, taking with it fresh-water algae and nutrients into estuaries. Agricultural slurry and human effluent from overflowing sewage systems contribute large quantities of nutrients that trigger algal blooms. If the estuary flows directly into the open sea the effect can be minimal, but discharge into contained bays and harbours will increase the likelihood of eutrophication and algal blooms. Nitrogen and phosphorous from the organic matter will be an important trigger and may cause HABs, as we have already discussed. But the problem is exacerbated when these enormous numbers of algae die because the bacteria that decompose them severely deplete the ecosystem of dissolved oxygen in the process. This hypoxic state, or anoxia, then suffocates other organisms, creating dead zones along the coast.

Plankton and humans

In Norway a new industry has begun to manufacture omega-3 oil from zooplankton such as calanoid copepods, rather than from fish with the desire to make production more sustainable. Zooplankton is also widely collected and used as fish food. In Asia and Central America combined, more than 900,000 tonnes of jellyfish are landed for human consumption every year (Behera et al. 2020) and, because of the high collagen levels in the bell, for use in the medical industry. In a few cases, when large swarms of jellyfish occur they have been gathered and used for fertiliser. Jellyfish are also employed in medical research, particularly in genetic studies; Green Fluorescent Protein, used as a gene marker, was first extracted from the jellyfish *Aequorea victoria*. It has now been isolated from sea anemones and copepods, too.

The exploitation described above has immediately obvious applications and benefits for humans, but perhaps the most significant benefit for people in the long run is plankton's ability to sequester carbon through its role in gathering carbon dioxide from the air and, ultimately, removing it to storage at the bottom of the sea as discussed in Chapter 19. The chances are that exploitation of components like *Calanus* would have strong implications. The most significant human influences on plankton will be the indirect factors of climate and carbon dioxide.

Glossary

Amictic incapable of being fertilised; in amictic eggs, the embryo develops without fertilisation (parthenogenesis). See Mictic.

Amphiesma multiple-layer cell wall covering found in dinoflagellates.

Auricularia early larva of holothurians or sea cucumbers.

Benthos/benthic the bottom of the sea, below the water column and hence the opposite of pelagic.

Binary fission simple cell division into two daughter cells as found in prokaryotic organisms like bacteria, enabling rapid growth.

Bipinnaria is the first distinct larval stage of starfish. See Brachiolaria.

Biramous an invertebrate appendage that forks into two.

Blastostyle a zooid in a hydroid that produces medusae.

Brachiolaria is the last larval stage in starfish development. See Bipinnaria.

Bright field illumination is the simplest form of microscope lighting with a light source coming from the base.

Cephalisation development of a distinct head region.

Cephalosome head of an arthropod, with particular relevance to copepoda.

Cercaria second stage larva of parasitic, trematode flatworms. See Miracidium.

Chelae the pincers on the front appendages of crustaceans such as crabs.

Chromatophores specialised cells containing pigment that with contraction and expansion can produce colour change in animals.

Ciliate a single-celled protist with cilia on the surface for locomotion and/or feeding.

Coelom the main body cavity found in animals surrounding the gut. Visible and distinct in polychaetes, it becomes less obvious in higher animal forms like chordates.

Coenobial a colony that has a set number of cells arranged in a specific manner, and which may be surrounded by a membrane.

Coenosarc a series of thin tubules that distribute material, i.e. food, in hydroid colonies.

Commensalism a relationship between two organisms where one benefits but the other neither gains or loses.

Compensation point where the amount of light available means that the rate of photosynthesis is equal to the rate of respiration, i.e. the amount of energy and oxygen being produced or used is equal and so there is no net gain. This also means that the amount of CO_2 being used or produced is the same.

Consumer in this context refers to an organism deriving its nutrition by eating another organism.

Copepod small aquatic crustacean, commonly considered 'insects of the sea', forming an essential link between producers and fish.

Cormidium clusters of three different types of zooid forming a colony of polyps found in siphonophores (hydroids) that chain together with other cormidia to create supercolonies. See Eudoxid.

Coronate a larval form of some bryozoa but which dwells in the plankton for a short time. See Cyphonautes.

Cyphonautes a common larval form of bryozoa living in the plankton for weeks. See Coronate.

Cytochrome an iron-containing catalyst, important for electron transfer.

Cytopharynx found in some protists, it is a passageway into the cell to carry food from the mouth or cytostome. See Cytostome.

Cytoplasm is the semi-liquid material in a cell that is within the cell membrane and outside the nucleus.

Cytostome in some protists like ciliates it is equivalent to the mouth. See Cytopharynx.

Dactylozooid specialised polyp in hydroids, in this case for defence.

Dark field illumination a technique of angling the light passing into a microscope so translucent specimens have better contrast.

Diapause a non-feeding period, such as overwintering, found in some arthropods, where the organism remains active but does not develop.

Diatom found as single cells or in chains, the cell wall is made of silica; these are the main component of phytoplankton, fixing carbon and forming the basis of food webs in the sea.

Diel vertical migration diel is used in biology to suggest a cycle of something occurring over a 24-hour period. During this time the organisms, like copepods, move vertically through the waterbody.

Differential interference contrast (DIC) a complex lighting technique used to create high contrast in transparent specimens under the microscope.

Dinoflagellate a protist that has two dissimilar flagellae.

Doliolaria a barrel-shaped larval stage in the development of certain echinoderms.

Ecdysis the process of arthropods shedding their exoskeleton to grow.

Echinopluteus the specific form of pluteus larval stage found in sea urchins.

Endostyle – referring to appendicularians – a ciliated organ on the wall of the foregut used for feeding to produce mucus collecting fine food material.

Entolecithal the ability to have sufficient yolk material in an egg for development of the embryo.

Epifluorescence a microscopy technique that can be used to observe microplastic pollution in samples. A specific wavelength of light is directed into the microscope to stimulate electrons in the specimen. This causes certain parts of the specimen to fluoresce, and brilliant streaks of light appear where tiny fragments of plastic are present.

Epineustonic organism living above the surface of the water.

Epitoky found in some polychaetes, e.g. syllids, this is a way of finding members of the same species for reproduction by producing a sexual phase that enters the plankton.

Epizoite an organism that lives on the outside of an animal, using it just for attachment and not for nutrition.

Eudoxid a sexual stage in the development of siphonophores and released from a cormidium.

Eutely where an organism, e.g. rotifer, has a fixed number of cells in the body that once mature never develop any more – hence mitosis ceases.

Exuvia the shed exoskeleton of a developing arthropod, e.g. barnacle.

Femtoplankton plankters with a size between 0.02 and 0.2 μm, e.g. viruses.

Fultoportula a tubular process found in centric diatoms that creates a pore on the valve surface.

Furca a structure that is forked. Can be found at the end of the posterior of some animals.

Gastrozooids specialised polyps in hydroids, in this case for feeding.

Gene flow where genetic material is transferred from one population to another, possibly through dispersal of larvae.

Gonophores the reproductive structures on a gonozooid for making gametes. See Gonozooid.

Gonozooids specialised polyps in hydroids, used for reproduction. They contain structures called gonophores that create the gametes.

Haemocoel a blood-filled space in arthropods and molluscs that bathes the organs. The blood is called haemolymph.
Haptocyst a toxic tip to a tentacle found in suctorian protists and deployed in catching prey.
Harpacticoid see copepods.
Heterotroph an organism that derives its nutrition by ingesting another organism.
Holoplankton permanent members of the plankton, living their entire life there.
Hyaline a layer of translucent material made from protein.
Hydrocaulus the 'stems' of hydroids, holding the polyps growing on the shore.
Hydrorhiza the root-like structures of hydroids growing on the shore.
Hyponeustonic of an organism living just beneath the water surface.
Infraciliature collection of complex structures that lie beneath the cilia of ciliates.
Instar a stage in the larval development of arthropods between shedding the exoskeleton. See Ecdysis and Exuvia.
Kentrogon specialist female larval stage of the parasitic barnacle, *Sacculina*. See Vermigon.
Kinetosome a microtubular structure found at the base of protist cilia and flagella.
Lecithotrophic refers to eggs which contain large quantities of stored food, allowing the larva to develop independently of the rest of the plankton. Growth is quick and the larva drifts in the sea for a short time, perhaps just a day or two.
Lorica the shell, or test, secreted by tintinnids and some other ciliates as protection – may be clear or studded with minerals.
Macroplankton plankters with a size between 20 and 200 mm, e.g. arrow worms.
Manubrium a tubular extension of the mouth in hydromedusae, for collecting food.
Megalopa the final-stage larva found in crabs when they settle out of the plankton to metamorphose into an adult.
Megaplankton plankters with a size between 20 and 200 cm, e.g. large jellyfish.
Meroplankton organisms that live only part of their lifecycle, typically as a larva, in the plankton
Mesenchyme is simple undifferentiated tissue, often considered as a connective tissue found between the outer (ectoderm) and inner (endoderm) tissue and which in cnidarians may produce mesoglea.
Mesoglea a gelatinous, extracellular layer found in cnidarians, particularly distinct in jellyfish.
Mesoplankton plankters with a size between 0.2 and 20 mm, e.g. copepods.
Metanephridia excretory tubules found in certain invertebrates such as arthropods and annelids.
Metasome the second-foremost body part of copepod.
Metazoa the name given to multicellular animals. Cf. Protist
Microbial loop the recycling of dissolved organic matter by microbes like bacteria and fungi.
Microplankton plankters with a size between 20 and 200 μm, e.g. tintinnids.
Mictic sexual reproduction in the context of rotifers where females produce haploid eggs which, if they are fertilised by a male, will produce a resting egg. Unfertilised, they can produce males. Amictic females create diploid eggs that develop into more females without fertilisation.
Miracidium the first stage larva of parasitic, trematode flatworms. See Cercaria.
Mitosis cell or growth division where the number of chromosomes in all cases remain the same.
Myoneme a contractile filament found in the stalk of certain protists, e.g. *Vorticella*.
Nanoplankton plankters with a size between 2.0 and 20 μm, e.g. flagellates.
Nauplius (pl. nauplii) the first larval stage of some crustaceans, e.g. barnacles.
Nectophores swimming bells found in some siphonophores.

Nekton pelagic organisms that can swim independent of currents.

Neritic the shallow water around the coast.

Neuston organisms living in, on or just under the surface of the seawater.

Ocellus/ocelli simple light receptors found in many invertebrates.

Operculum the hard 'trapdoor' that closes over the aperture in the shell of a gastropod mollusc when it retreats back inside.

Ophiopluteus the specific pluteus larva found in brittle star echinoderms.

Ossicles calcareous parts of the echinoderm endoskeleton, providing rigidity.

Parapodia extensions on either side of each body segment of polychaetes.

Pelagic living in open water or the sea.

Pellicle a thin layer or film covering that may be composed of tissue, protein or other organic materials.

Pentacrinoid larval stage found in crinoids or feather stars. See Vitellaria.

Pentactula final larval stage before an adult holothurian or sea cucumber forms.

Perisarc made from protein and chitin, forms a sheath or coating over the stalk of hydroids.

Phase contrast a form of microscope illumination that increases contrast and has been known as 'optical staining'.

Phagotrophism ingestion of organic matter and small organisms by engulfing them rather than through an oral cavity.

Phototaxis an organism's behavioural response when attracted towards light or being stimulated to move away.

Phytoplankton primarily diatoms and most dinoflagellate species; both carry out photosynthesis. See Producer and Primary production.

Picoplankton plankters with a size between 0.2 and 2 µm, e.g. bacteria.

Planktotrophic eggs contain little stored food and on hatching larvae will depend on finding nutrients in the plankton. This reproductive strategy is more uncertain than lecithotrophy, and growth can take weeks.

Planula/planulae the larval phase found in cnidarians. The most commonly found are those of sea anemones.

Plastid is a membrane-bounded organelle found in algae and plants and the site of photosynthesis. There are different types, of which chloroplasts are one.

Pleopod usually five pairs of these forked appendages, also known as swimmerets, are attached to a crustacean abdomen.

Pleuston organisms floating at the surface of water and affected by wind. See Neuston.

Pneumatophore flotation devices found in some siphonophores.

Praniza a larval stage unique to *Gnathia maxillaris*, an isopod crustacean.

Primary production carried out by the phytoplankton using photosynthesis to fix carbon and energy.

Producers phytoplankton that produce energy via photosynthesis

Prosome term used in crustacea for the anterior of the body and in copepods specifically to refer to the combined cephalasome and metasome.

Prostomium the first segment of an annelid body; in polychaetes it can be well developed/cephalised, with mouth, palps and other structures.

Protist unicellular, primitive animals. Cf. Metazoan.

Protonymphon larval stage of sea spiders.

Prototroch is the main ciliary band that goes around the equator of a trochophore.

Pseudopodia extensions of cytoplasm that a unicellular organism uses for feeding and/or locomotion.

Pyrenoid structures within algal plastids that fix and concentrate carbon dioxide to ensure maximum photosynthesis is maintained, often difficult in water-based environments.

Reticulopodia found in some protists, particularly foraminifera, these are fine extensions of the cytoplasm for searching out food.

Rhabdome a rod-like structure in the complex eyes of crustacea and insects that has photoreceptors to trigger nerve impulses.

Rhopalium sense organs in medusae and ephyra of cnidarians that can detect light and gravity.

Rosette glands yellow spots, which produce bioluminescence, found on the parapods of the polychaete genus *Tomopteris*.

Scopula a 'glandular' region on some suctorian protists that create the stalk.

Scyphistomae a polyp stage in the lifecycle of jellyfish, found on rocky shores.

Sensilli hairs on the surface of crustaceans and other arthropods that form part of the sensory system, picking up information such as temperature and taste.

Setae stiff hairs used to assist locomotion and/or feeding.

Soft microplankton the unicellular organisms in microplankton lacking any hard protective coat, e.g. lorica in tintinnids.

Somatic ciliature collective term referring to cilia arranged in rows.

Somatocoel part of the coelom, relates particularly to echinoderm development.

Somite the term varies in its use depending on the organism, but typically refers to a segment.

Statocyst a sense organ for determining gravity and orientation in some aquatic invertebrates. These are tiny bulb-like chambers with bristles. In a central space are statoliths, calcareous particles that when moved by gravity during locomotion touch the bristles and trigger the sensory cells.

Stomodeum a space in an embryo that will develop into a mouth.

Strobilation as jellyfish polyps grow on the rocky shore they split, the process of strobilation, to produce a stack of strobila like a stack of plates. The ones at the top drift off into the plankton, called ephyra, to form new jellyfish. See Scyphistomae.

Suspension feeders animals, particularly sedentary polychaetes, that feed on organic matter suspended in the water.

Symbiont an organism involved in a symbiotic relationship.

Telotroch a group of cilia found on certain larval forms, e.g. trochophores; also the name given to a swimming stage of some ciliate protists.

Tentillum used in relation to cnidarians it is a small tentacle branch off a larger one.

Tornaria the larva of *Glossobalanus sarniensis* (acorn worm).

Trichite membrane-covered rods used in some protists as a method of collecting food and/or defence, as they can be extruded from the cell.

Trichocyst found on the ciliate or flagella surface, these are organelles used for defence or collecting food. Several different types exist.

Trochophore first-stage larva of several phyla but primarily polychaetes. Others include molluscs.

Tychoplankton temporary visitors to the plankton, maybe just for a day or two.

Uropod collectively these are like a tail fan on some crustaceans, e.g. shrimp; the appendages found on the last segment of the abdomen.

Urosome the tail segments of copepods.

Vacuole cell organelle that is a fluid-filled pouch or sac. In many cases produced for the digestion of food that has been ingested, but are also used for storage. Particularly relevant for protists.

Velum a two-lobed ciliated organ in gastropod larvae (veliger) used for locomotion; in bivalves it is usually a single lobe.

Vermigon a specialist secondary female larval stage of the parasitic barnacle, *Sacculina*.

Vertical migration of plankton in the water can occur during a 24-hour cycle, typically for phytoplankton moving up towards the light (sinking at night) and zooplankton, such as copepods, moving down by day to avoid predation and then moving up at night.

Vitellaria primitive larval stage found in crinoids or feather stars, later developing into a pentacrinoid larva.

Zoea larval stage of crustaceans, especially decapods.

Zooid small individuals that collectively form a colony. In some cases the zooid develops specialised functions such as feeding or reproduction.

Zooplankton collective term for all the individuals in the plankton that are heterotrophic, typically animals.

Resources and bibliography

Resources

Microscopy

The Royal Microscopy Society is based in Oxford: www.rms.org.uk
The Quekett Microscopical Club: www.quekett.org

Equipment supply

NHBS Ltd: www.nhbs.com

Other useful websites

www.algaebase.org
http://diatoms.org
www.keweenawalgae.mtu.edu
https://phytoplankton.eoas.ubc.ca
www.planktonmilfordhaven.org
https://groups.io/g/planktonnet
Helgoland Marine plankton: https://av.tib.eu/series/1151

Bibliography

The following is a limited list of books, papers and articles. Books, particularly those older than 75 years, are often available free of charge as digital versions on sites such as the Internet Archive (archive.org). Online papers may be available from sources such as Researchgate (https://www.researchgate.net/search) and ScienceDirect (https://www.sciencedirect.com/).

Identification guides

Conway, D.V.P. (2012a) *Marine zooplankton of southern Britain. Part 1: Radiolaria, Heliozoa, Foraminifera, Ciliophora, Cnidaria, Ctenophora, Platyhelminthes, Nemertea, Rotifera and Mollusca* (ed. A.W.G. John). Marine Biological Association of the United Kingdom 25. Plymouth.

Conway, D.V.P. (2012b) *Marine zooplankton of southern Britain. Part 2: Arachnida, Pycnogonida, Cladocera, Facetotecta, Cirripedia and Copepoda* (ed. A.W.G. John). Marine Biological Association of the United Kingdom Occasional Publication 26. Plymouth.

Conway, D.V.P. (2015) *Marine zooplankton of southern Britain. Part 3: Ostracoda, Stomatopoda, Nebaliacea, Mysida, Amphipoda, Isopoda, Cumacea, Euphausiacea, Decapoda, Annelida, Tardigrada, Nematoda, Phoronida, Bryozoa, Entoprocta, Brachiopoda, Echinodermata, Chaetognatha, Hemichordata and Chordata* (ed. A.W.G. John). Marine Biological Association of the United Kingdom Occasional Publications 27. Plymouth.

Hoppenrath, M., Elbrachter, M. and Drebes, G. (2009) *Marine Phytoplankton: selected microphytoplankton species from the North Sea around Helgoland and Sylt*, Kleine Senckenberg-Reihe 49. Stuttgart: E. Schweizerbart'sche Verlagsbuchh.

Kraberg, A., Baumann, M. and Durselen, C. (2010) *Coastal Phytoplankton: Photo guide for northern European seas*. Munich: Verlag Dr. F. Pfeil.

Larink, O. and Westheide, W. (2011) *Coastal Plankton: Photo guide for European seas*, 2nd edn. Munich: Verlag Dr. F. Pfeil.

Rodríguez, J., Alemany, F. and García, A. (2017) *A guide to the eggs and larvae of 100 common Western Mediterranean Sea bony fish species*. Food and agriculture organization of the United Nations

Tomas, C.R. (ed.) (1997) *Identifying Marine Phytoplankton*. Academic Press. NB. A comprehensive 860 page book available as a free download.

Other useful identification material

Bishop, J.D.D., Yunnie, A.L.E., Baxter, E.J. and Wood, C.A. (2017) *Guide to the Early Post-settlement Stages of Fouling Marine Invertebrates in Britain (Version 2)*. Marine Biological Association of the United Kingdom Occasional Publication 29. Plymouth.

Conway, D. (2024) *Identification of Copepod Eggs, Nauplii and Copepodites, and Facetotectan Nauplii and Cyprids*. Marine Biological Association of the United Kingdom Occasional Publication. Plymouth.

Hayward, P.J. and Ryland, J.S. (2012) *Handbook of the Marine Fauna of North-West Europe*. Oxford.

Muxagata, E. and Williams, J.A. (2004) The Mesozooplankton of the Solent–Southampton Water system: A photographic guide. Southampton Oceanography Centre Internal document no. 97. Southampton.

Newell, G.E. and Newell, R.C. (1963) *Marine Plankton: A practical guide*. London: Hutchinson Educational.

Trewhella, S. and Hatcher, J. (2015) *The Essential Guide to Beachcombing and the Strandline*. Plymouth: Wild Nature Press. https://doi.org/10.1515/9780691232423

PHYTOPLANKTON

Bourrelly, P. (1970) *Les Algues d'eau Douce, Vols 1, 2 and 3*. Paris: N. Boubée & Cie. Volume 3 covers dinoflagellate identification.

Dring, M.J. (1982) *The Biology of Marine Plants*. London: Arnold.

Fritsch, F.E. (1965) *The Structure and Reproduction of The Algae*, 2 vols. Cambridge: Cambridge University Press.

Graham, L. and Wilcox, L. (2000) *Algae*. Upper Saddler River, NJ: Prentice Hall.

Henderson, Z. (2024) The dinoflagellates: a review. *The Quekett Journal of Microscopy* 44: 597–612.

Lebour, M.V. (1930) *The Planktonic Diatoms of Northern Seas*. London: Ray Society.

Round, F.S. (1975) *The Biology of the Algae*. London: Arnold.

PROTISTS/PROTOZOA

Archibald, J.M., Simpson, G.B. and Slamovits, C.H. (eds) (2017) *Handbook of the Protists*, Cham: Springer. https://doi.org/10.1007/978-3-319-28149-0

Hawes, R.S.J. and Mackinnon, D.L. (1961) *An Introduction to the Study of Protozoa*. Oxford.

Haynes, J.R. (1973) Cardigan Bay recent foraminifera (cruises of the R.V. Antur, 1962–1964. *Bulletin of the British Museum* (Natural History) zoology supplement 4. https://doi.org/10.5962/p.312693

Jones, A.R. (1974) *The Ciliates*. London: Hutchinson.

Murray, J.W. (1979) *British Near-shore Foraminiferids*, Synopsis of the British Fauna. London: Field Studies Council.

Sleigh, M. (1989) *Protozoa and other Protists*, Edward Arnold.

ROTIFERS

Pontin, R.M. (1978) *A Key to British Freshwater Planktonic Rotifera*. Ambleside: Freshwater Biological Association. Volume includes *Synchaeta* and brackish water species.

CNIDARIA AND CTENOPHORES

Cornelius, P.F.S. (1995) *Thecate Hydroids and their Medusae Parts 1 and 2*, Synopsis of the British Fauna. London: Field Studies Council.

Mayer, A.G. (1910) *Medusae of the World*, 3 vols. Washington DC: Carnegie Institution. https://doi.org/10.5962/bhl.title.159245

Schuchert, P. (2007) The European athecate hydroids and their medusae (Hydrozoa, Cnidaria): Filifera Part 2. *Revue Suisse de Zoologie* 114 (2): 195–396. https://doi.org/10.5962/bhl.part.80395

ARTHROPODS

Angel, M.V. (1993) *Marine Planktonic Ostracods*, Synopsis of the British Fauna. London: Field Studies Council.

Brady, G.S. (1880) *A Monograph of the Free and Semi-parasitic Copepoda of the British Isles*, 3 vols. London: Ray Society.

Green, J. and Macquitty, M. (1987) *Halacarid Mites*, Synopsis of the British Fauna. London: Field Studies Council.

Gurney, R. (1942) *Larvae of Decapod Crustacea*. London: Ray Society.

Hingley, M. (2021) *Microscopic life in Sphagnum*. Pelagic Publishing.

Huys, R., Gee, J.M., Moore, C.G. and Hamond, R. (1996) *Marine and Brackish Water Harpacticoid Copepods*, Synopsis of the British Fauna. London: Field Studies Council.

Jones, N.S. (1976) *British Cumaceans*, Synopsis of the British Fauna. London: Field Studies Council.

Marshall, S.M. and Orr, A.P. (1955) *The Biology of a Marine Copepod*. Edinburgh: Oliver & Boyd.

Miller, C. (2023) *Oar Feet and Opal Teeth*, Oxford: Oxford University Press. https://doi.org/10.1093/oso/9780197637326.001.0001

Stott, R. (2003) *Darwin and the Barnacle*. London: Faber and Faber.

POLYCHAETES

McIntosh, W.C. (1915) *The British Marine Annelids*, 9 vols with extra plates. London: Ray Society.

NEMATODES

Platt, H.M. and Warwick, R.M. (1983) *Free-living Marine Nematodes*, Synopsis of the British Fauna. London: Field Studies Council.

NEMERTINEA

Göransson, U., Jacobsson, E., Strand, M. and Andersson, H.S. (2019) *The Toxins of Nemertean Worms. Toxins (Basel)* 11(2): 120. https://doi.org/10.3390/toxins11020120

CHAETOGNATHA

Pierrot-Bults, A.C. and Chidgey, K.C. (1988) *Chaetognatha*, Synopsis of the British Fauna. London: Field Studies Council. https://doi.org/10.1163/9789004611412

BRYOZOA AND LOPHORATES

Hayward, P.J. and Ryland, J.S. (1998) *Cheilostomatous Bryozoa*, Synopsis of the British Fauna. London: Field Studies Council.

TUNICATES

Alder, J. and Hancock, A. (1912) *The British Tunicata*, 3 vols. London: Ray Society.

Fraser, J.H. (1981) *British Pelagic Tunicates*, Synopsis of the British Fauna. London: Field Studies Council.

GENERAL MARINE AND BIOLOGY

Archer-Thomson, J. and Cremona, J. (2019) *Rocky Shores*. London: Bloomsbury.

Barnes, R.S.K., Callow, P. and Olive, P.J.W. (2001) *The Invertebrates: A new synthesis*, Oxford: Blackwell Scientific.

Brusca, R.C., Giribet, G. and Moore, W. (2023), *Invertebrates*, 4th edn, Oxford University Press. https://doi.org/10.1093/hesc/9780197554418.001.0001

Cremona, J. (2014) *Seashores: An ecological guide*. Ramsbury: Crowood.

Czerski, H. (2023) *Blue Machine: How the ocean shapes our world*. London: Transworld.

Dipper, F. (2019) *The Marine World: A natural history of ocean life*, Plymouth: Wild Nature Press.

Fish, J.D. and Fish, S. (2011) *A Student's Guide to the Seashore*, 3rd edn. Cambridge: Cambridge University Press. https://doi.org/10.1017/CBO9781139035125

Fraser, J. (1962) *Nature Adrift*. London: G.T. Foulis & Co.

Grahame, J. (1987) *Plankton and Fisheries*. London: Edward Arnold.

Green, J. (1968) *The Biology of Estuarine Animals*. Sidgewick & Jackson.

Hardy, A. (1955) *The Open Sea: The World of Plankton*. Collins New Naturalist.

Jackson, T. and Parker, J. (2024) *Plankton: A worldwide guide*. Princeton, NJ: Princeton University Press. https://doi.org/10.1515/9780691256085

Little, C., Williams, G.A. and Trowbridge, C.D. (2009) *The Biology of Rocky Shores*, 2nd edn, Biology of Habitats. Oxford. https://doi.org/10.1093/oso/9780198564904.001.0001

Stal, L. and Cretoiu, M. (eds) (2022) *The Microbiomes of Humans, Animals, Plants, and the Environment Pt. 3 Marine Habitats*. Cham: Springer. https://doi.org/10.1007/978-3-030-90383-1

Suttle, C.A. (2007) Marine viruses: major players in the global ecosystem, *Nature* 5: 801–812. https://doi.org/10.1038/nrmicro1750

Waller, G. (ed.) (1996) *Sealife: a Complete Guide to the Marine Environment*. Mountfield: Pica Press.

MICROSCOPY AND PHOTOGRAPHY

Cremona, J. (2017) *Extreme Close-up Photography and Focus Stacking*. Marlborough: Crowood.

Cremona, J. (2018) *Beyond Extreme Close-up Photography*. Ramsbury: Crowood.

Matsumoto, B. (2010) *Practical Digital Photomicrography*. Santa Barbara, CA: Rockynook.

Matsumoto, B. and Roullard, C. (2017) *An Introduction to Digital Photomicrography*. Ramsbury: Crowood.

HISTORY

Bowler, C., Ibarbalz, F.M. and Karlusich, J.J.P. (2020) Exploration of marine phytoplankton: from their historical appreciation to the omics era, *Journal of Plankton Research* 42(6): 595–612. https://doi.org/10.1093/plankt/fbaa049

Damkaer, D. (2016) John Vaughan Thompson (1779–1847), pioneer planktonologist: a life renewed. *Journal of Crustacean Biology* 36(2): 256–262. https://doi.org/10.1163/1937240X-00002409

Franke, H.D., Buchholz, F. and Wiltshire, K. (2004) Ecological long-term research at Helgoland (German Bight, North Sea): retrospect and prospect – an introduction. *Helgoland Marine Research* 58: 223–229. https://doi.org/10.1007/s10152-004-0197-z

Gosse, P.H. (1853) *A Naturalist's Rambles on the Devonshire Coast*. London: John Van Voorst.

Gosse, P.H. (1855) *A Manual of Marine Zoology of the British Isles*. London: John Van Voorst. https://doi.org/10.5962/bhl.title.3987

Gosse, P.H. (1856) *Tenby*. London: John Van Voorst.

Thompson, J.V. (1828) *Natural History of Nondescript or Imperfectly Known Animals* Vol. 1, Pt 1. Cork: King and Ridings.

Thwaite, A. (2002) *Glimpses of the wonderful the life of Philip Henry Gosse*. London: Faber and Faber.

HUMAN INFLUENCES

Behera, P.R., Raju, S.S., Jishnudev, M.A., Ghosh, S. and Saravanan, R. (2020) Emerging jellyfish fisheries along Central South East coast of India, *Ocean and Coastal Management* 191: 105183. https://doi.org/10.1016/j.ocecoaman.2020.105183

Benedetti, F., Vogt, M., Hofmann Elizondo, U., Righetti, D., Zimmermann, N.E. and Gruber, N. (2021) Major restructuring of marine plankton assemblages under global warming, *Nature Communications* 12. https://doi.org/10.1038/s41467-021-25385-x

Botterell, Z.L.R., Lindeque, P.K., Thompson, R.C. and Beaumont, N.J. (2023) An assessment of the ecosystem services of marine zooplankton and the key threats to their provision. *Ecosystems Services* 63: 101542. https://doi.org/10.1016/j.ecoser.2023.101542

Ziska, R. (ed.) (2025) A global estimate of multi-ecosystem photosynthesis losses under microplastic pollution. *Proceedings of the National Academy of Sciences* 122. https://doi.org/10.1073/pnas.2423957122

Cardiff University (2023, 12 January) Recycling sludge could lead to large plastic pollution reservoirs, study finds. *Cardiff University News*. Accessed at: http://www.cardiff.ac.uk/news/view/2692907-recycling-sewage-sludge-plastic-pollution-reservoirs

Index